Science for Sustainable Societies

Scope of the Series

This series aims to provide timely coverage of results of research conducted in accordance with the principles of sustainability science to address impediments to achieving sustainable societies – that is, societies that are low carbon emitters, that live in harmony with nature, and that promote the recycling and re-use of natural resources. Books in the series also address innovative means of advancing sustainability science itself in the development of both research and education models.

The overall goal of the series is to contribute to the development of sustainability science and to its promotion at research institutions worldwide, with a view to furthering knowledge and overcoming the limitations of traditional discipline-based research to address complex problems that afflict humanity and now seem intractable.

Books published in this series will be solicited from scholars working across academic disciplines to address challenges to sustainable development in all areas of human endeavors.

This is an official book series of the Integrated Research System for Sustainability Science (IR3S) of the University of Tokyo and the Institute for Global Environmental Strategies (IGES).

More information about this series at http://www.springer.com/series/11884

Alexandros Gasparatos • Merle Naidoo •
Abubakari Ahmed • Alice Karanja •
Kensuke Fukushi • Osamu Saito •
Kazuhiko Takeuchi

Editors

Sustainability Challenges in Sub-Saharan Africa II

Insights from Eastern and Southern Africa

 Springer

Editors
Alexandros Gasparatos
Institute for Future Initiatives (IFI)
The University of Tokyo
Tokyo, Japan

Abubakari Ahmed
Department of Planning
University for Development Studies
Wa, Ghana

Kensuke Fukushi
Institute for Future Initiatives (IFI)
The University of Tokyo
Tokyo, Japan

Institute for the Advanced Study
of Sustainability (UNU-IAS)
United Nations University
Tokyo, Japan

Kazuhiko Takeuchi
Institute for Global Environmental
Studies (IGES)
Hayama, Japan

Institute for Future Initiatives (IFI)
The University of Tokyo
Tokyo, Japan

Merle Naidoo
Graduate Programme in Sustainability
Science – Global Leadership
Initiative (GPSS-GLI)
The University of Tokyo
Tokyo, Japan

Alice Karanja
Institute for Future Initiatives (IFI)
The University of Tokyo
Tokyo, Japan

Osamu Saito
Institute for Global Environmental
Studies (IGES)
Hayama, Japan

Institute for Future Initiatives (IFI)
The University of Tokyo
Tokyo, Japan

ISSN 2197-7348   ISSN 2197-7356   (electronic)
Science for Sustainable Societies
ISBN 978-981-15-5357-8   ISBN 978-981-15-5358-5   (eBook)
https://doi.org/10.1007/978-981-15-5358-5

This Springer imprint is published by the registered company Springer Nature Singapore Pte Ltd.
The registered company address is: 152 Beach Road, #21-01/04 Gateway East, Singapore 189721, Singapore

Acknowledgments

We wish to thank our Springer editorial team for the guidance, support, and patience throughout the development of the volume. We wish to thank particularly Sushil Kumar Sharma and Mei Hann Lee for their endless patience throughout the development of these two volumes. We also thank Marcin Jarzebski for designing many of the figures included in these two volumes.

The two volumes were conceptualized during the development and follow up to the 6th International Conference on Sustainability Science (ICSS 2016) in Stellenbosch, South Africa (March 2016). We sincerely thank the conference partners and participants for their suggestions and inspiration that kick-started the process of these books.

The two volumes would not have been possible without the generous support of the Japan Science and Technology Agency (JST) for project FICESSA, Japan International Cooperation Agency (JICA) for project USiA (Urban Sustainability in Africa based on inter-linkages of SDGs), and the Japan Society for the Promotion of Science (JSPS) for Grant-in-Aid for Young Scientists A (Project 17H05037) and the Core-to-core project "Establishment and Advancement of a Global Meta-Network on Sustainability Science" (Project 23001).

Merle Naidoo, Alice Karanja, and Abubakari Ahmed received Monbukagakusho PhD scholarships offered by the Japanese Ministry of Education, Culture, Sports, Science, and Technology (MEXT) through the Graduate Program in Sustainability Science—Global Leadership Initiative (GPSS-GLI).

Contents

Contributors

Abubakari Ahmed Department of Planning, University for Development Studies, Wa, Ghana.

Precious Akampumuza is a PhD student at the University of Tokyo.

Yukiko Amitani is a former graduate student at Ochanomizu University.

Pippin Anderson is a Senior Lecturer at the University of Cape Town.

Rivolala Andriamparany is a Biodiversity Calculation Supervisor at Ambatovy Joint Venture.

Yuko Caballero is a former graduate student at Ochanomizu University.

Tamara Chansa-Kabali is a Lecturer at the University of Zambia.

Moses Ngongo Chisola is a PhD Scholar at the University of Pretoria and a Lecturer at the University of Zambia.

Thomas Elmqvist is a Professor at Stockholm Resilience Centre, Stockholm University.

Kensuke Fukushi Institute for Future Initiatives (IFI), The University of Tokyo, Tokyo, Japan.
Institute for the Advanced Study of Sustainability (UNU-IAS), United Nations University, Tokyo, Japan.

Sara Gabrielsson is an Assistant Professor at Lund University.

Susan Gaskin is an Associate Professor at McGill University.

Alexandros Gasparatos Institute for Future Initiatives (IFI), The University of Tokyo, Tokyo, Japan.

Kasim Munyegera Ggombe is the Managing Director at the Mazima Research Consultancy.

Angela Huston is a PhD candidate at McGill University and a Programme Officer with the IRC International Water and Sanitation Centre, The Hague.

Hiroaki Imanishi is the Director of the Program/Operation Department at World Vision Japan.

Francis X. Johnson is a Senior Research Fellow at the Stockholm Environment Institute (SEI).

Alice Karanja Institute for Future Initiatives (IFI), The University of Tokyo, Tokyo, Japan.

Chibuye Florence Kunda-Wamuwi is a Lecturer at the University of Zambia.

Jakob Lundberg is the Head of policy at We effect.

James Manchisi is a Lecturer at the University of Zambia and a Post-Doctoral Research Fellow at the University of Witwatersrand.

Ntombini Marrengane is a Research Coordinator at the University of Cape Town.

Hirotaka Matsuda is an Associate Professor at the Tokyo University of Agriculture.

Takuya Matsuoka is a Senior Program Coordinator at the Program/Operation Department at World Vision Japan.

Musingo Tito Mbuvi is a Principal Research Scientist at the Kenya Forestry Research Institute (KEFRI).

Orleans Mfune is a Senior Lecturer at the University of Zambia.

Geetha Mohan is a Senior Research Fellow at the United Nations University Institute for the Advanced Study of Sustainability (UNU-IAS).

Shakespear Mudombi is a Sustainable Growth Economist at Trade and Industrial Policy Strategies (TIPS).

Merle Naidoo Graduate Programme in Sustainability Science – Global Leadership Initiative (GPSS-GLI), The University of Tokyo, Tokyo, Japan.

Leila Ndalilo is a Social Scientist at the Kenya Forestry Research Institute (KEFRI).

Anne Nyambane is a Research Associate at the Stockholm Environment Institute (SEI).

Caroline A. Ochieng is a Research Fellow at the National University of Ireland Galway and the Energy Sector Management Assistance Program (ESMAP) of the World Bank.

Zarina Patel is an Associate Professor at the University of Cape Town.

Markku Pyykonen is a Lecturer at the University of Gävle.

Carla Romeu-Dalmau is a Research Associate at the University of Oxford.

Osamu Saito Institute for Global Environmental Studies (IGES), Hayama, Japan.
Institute for Future Initiatives (IFI), The University of Tokyo, Tokyo, Japan.

Takayo Sasaki is a Manager at the Program/Operation Department at World Vision Japan.

Makiko Sekiyama is a Senior Researcher at the National Institute for Environmental Studies (NIES).

Warren Smit is a Senior Researcher at the University of Cape Town.

Noriko Sudo is an Associate Professor at Ochanomizu University.

Kazuhiko Takeuchi Institute for Global Environmental Studies (IGES), Hayama, Japan.
Institute for Future Initiatives (IFI), The University of Tokyo, Tokyo, Japan.

Graham Paul von Maltitz is a Research Group Leader at the South African Council for Scientific and Industrial Research (CSIR).

Chemuku Wekesa is a Landscape Ecologist at the Kenya Forestry Research Institute (KEFRI).

Sebastian Wurtz is a former graduate student at Stockholm University.

Ayumi Yanagisawa is a former graduate student at Ochanomizu University.

Chapter 1
Tackling Child Malnutrition by Strengthening the Linkage Between Agricultural Production, Food Security, and Nutrition in Rural Rwanda

Makiko Sekiyama, Hirotaka Matsuda, Geetha Mohan, Ayumi Yanagisawa, Noriko Sudo, Yukiko Amitani, Yuko Caballero, Takuya Matsuoka, Hiroaki Imanishi, and Takayo Sasaki

1.1 Introduction

Even though global food production is currently sufficient to feed the global population, in 2016, approximately 11% of the world's population (including children) was undernourished (FAO 2017). Nearly 3.1 million children died in 2011, with approximately 45% of those deaths being linked to undernutrition (Black et al. 2013). Undernutrition during childhood has many negative short-term and long-term consequences. Short-term consequences include disability, morbidity, and mortality during childhood (Black et al. 2008). Longer-term effects include lower body size, intellectual ability, economic productivity, reproductive performance, and higher cardiovascular disease risk during adulthood (Black et al. 2008).

M. Sekiyama (✉)
National Institute for Environmental Studies (NIES), Tsukuba, Japan
e-mail: sekiyama.makiko@nies.go.jp

H. Matsuda
Tokyo University of Agriculture, Atsugi, Japan
e-mail: hm206784@nodai.ac.jp

G. Mohan
United Nations University, Tokyo, Japan
e-mail: Geetha.Mohan@unu.edu

A. Yanagisawa · N. Sudo · Y. Amitani · Y. Caballero
Ochanomizu University, Tokyo, Japan
e-mail: sudo.noriko@ocha.ac.jp

T. Matsuoka · H. Imanishi · T. Sasaki
World Vision Japan, Tokyo, Japan
e-mail: takuya_matsuoka@worldvision.or.jp; hiroaki_imanishi@worldvision.or.jp; takayo_sasaki@worldvision.or.jp

© Springer Nature Singapore Pte Ltd. 2020
A. Gasparatos et al. (eds.), *Sustainability Challenges in Sub-Saharan Africa II*, Science for Sustainable Societies, https://doi.org/10.1007/978-981-15-5358-5_1

Sub-Saharan Africa (SSA) registers the highest rate of undernutrition in the world, affecting an alarming 22.7% of the population in 2016 (see also Chaps. 1 and 3 Vol. 1). The situation is especially troubling in Eastern Africa, where approximately one-third of the population is estimated to be undernourished, with the prevalence of undernutrition increasing from 31.1% in 2015 to 33.9% in 2016 (FAO 2017). Child undernutrition is also prominent across SSA, with the prevalence of child stunting standing at 34.1%, which is significantly higher than the global average of 22.2% (UNICEF et al. 2018) (Chap. 1 Vol. 1). Similarly, Eastern Africa has the highest rate of child stunting, with 35.6% of children estimated to be stunted (UNICEF et al. 2018).

Child undernutrition is one of the most important sustainability challenges that SSA countries currently face (UNICEF et al. 2018) (see also Chap. 1 Vol. 1). Due to its significant effect on young generations (see above) and their families, it is believed that child undernutrition can possibly take a toll on the economic growth and the overall development of the region for decades to come (Black et al. 2013). Furthermore, it can have synergistic effects with other major sustainability challenges such as good health, quality education, and gender equality.

As a result, tackling child undernutrition in SSA has been a major policy priority both at the national and international levels in the past decades. For example, tackling child undernutrition was a major target of the Millennium Development Goals (MDGs) and is currently featuring in two distinct targets of Sustainable Development Goal 2 (SDG2).[1] Furthermore, many SSA countries have put in place comprehensive food security plans and programs (see Chap. 3 Vol. 1), where the reduction of child undernutrition is a prominent goal (FAO 2017). However, there has been little to no change in many parts of SSA in the prevalence of undernutrition between 1990 and 2011 (Stevens et al. 2012), even though urbanization improves in some parts of the continent access to food (Garcia 2012).

Tackling child undernutrition can be a particularly difficult and multidimensional sustainability challenge to tackle given the lack of resources and capacity in most SSA countries (WHO 2017a) (Chap. 5 Vol. 1). For example, the UNICEF suggests that the nutritional status of children is determined by numerous immediate and underlying causes related to food security, health care, and adequate sanitation, among others (UNICEF 1990). Very different approaches targeting the different determinants of child undernutrition have been implemented across SSA in the past decades. Most of the intervention programs have taken nutrition-specific approaches (Bhutta et al. 2008) that address the immediate determinants of child undernutrition (i.e., dietary intake and disease) through micronutrient supplementation and

---

[1]SDG2 aims to end all forms of hunger by 2030. Two of its targets have a strong focus on child undernutrition. Target 1 stipulates that by 2030 countries should "end hunger and ensure access by all people, in particular the poor and people in vulnerable situations, including infants, to safe, nutritious and sufficient food all year round." Target 2.2 stipulates that by 2030 countries should "end all forms of malnutrition, including achieving, by 2025, the internationally agreed targets on stunting and wasting in children under 5 years of age, and address the nutritional needs of adolescent girls, pregnant and lactating women and older persons."

fortification, improvement of breastfeeding and complementary feeding, and strengthening of hygiene. However, such interventions tend to be highly project- or donor-dependent and thus difficult to sustain in local communities in the long term (Bhutta et al. 2008). On the other hand, nutrition-sensitive approaches tend to address the underlying determinants of undernutrition (i.e., food security, health access, healthy household environment, care practices), through improving practices in agriculture, water provision, sanitation, education, and social protection (Black et al. 2013).

The global interest on nutrition-sensitive approaches has been recently increasing. This is because in such interventions, apart from tackling child undernutrition, they might also offer sustainable solutions for local communities, thus having sustainability benefits. However, so far, there is weak evidence base on how to develop effectively nutrition-sensitive interventions that address such underlying determinants (Haddad et al. 2015). For example, a systematic review assessed the impact of agricultural interventions on child nutritional status and concluded that very little evidence supports positive outcomes on the prevalence of malnutrition, mostly due to methodological weaknesses (Masset et al. 2012). The 2014 and 2015 Global Nutrition Report provided some specific ideas for interventions on food systems and agriculture, but the evidence is limited to kitchen gardens and bio-fortification (IFPRI 2016).

Rwanda is one of the SSA countries that still experiences a significant level of malnutrition, despite its impressive economic and social progress in the past decades (Logie et al. 2008). Severe poverty is still prevalent in the country regardless of the strong and sustained economic growth following the 1994 genocide when 600,000–800,000 people were killed (World Bank 2018). A strong indication of chronic poverty and food insecurity is the high level of malnutrition, especially among poor and rural households (World Bank 2018). According to the Rwanda Demographic and Health Survey (RDHS), in 2014–2015, the prevalence of child stunting (below 5 years old) was 38% (NISR et al. 2015). Even though this signifies a substantial improvement from past level (i.e., 51% in 2005 and 44% in 2010), the overall child stunting rates are still high for global standards (UNICEF et al. 2018) (Chap. 1 Vol. 1).

However, as in other SSA countries, there is a distinct urban-rural disparity in the prevalence of child stunting, ranging from 24% in urban areas to 41% in rural areas in 2014–2015 (NISR et al. 2015). Furthermore, the prevalence of stunting varies between provinces, with the Western Province having the highest prevalence (45%), followed by the Southern (41%), Northern (39%), and Eastern Provinces (35%) and Kigali (23%) (NISR et al. 2015). Thus, stunting remains one of the major health and sustainability challenges in Rwanda, especially in rural areas. As a result, many interventions have been promoted across the country to tackle undernutrition (World Bank 2018).

The aim of this study is to examine the relationship between household crop production, diet diversity, and nutritional status of children in areas that have received related interventions from an NGO (World Vision) and areas that have not received such interventions. We focus on areas where the Gwiza Nutrition

Project (Sect. 1.2.1) was implemented and use data collected through several field survey waves that capture in a detailed manner the food intake, socioeconomic status, and height/weight of local community members. In this respect, the study aims to elicit and critically examine information that can inform subsequent studies that explore the effectiveness of nutrition interventions, which are similar to those promoted in many other parts of SSA (FAO 2013) but often with mixed results (World Vision 2017).

Section 1.2 outlines the study site, the characteristics of the nutritional intervention, and the methodological approach. Section 1.3 highlights the main results of the analysis, and Sect. 1.4 discusses the main findings, including the main policy implications.

1.2 Methodology

1.2.1 Study Area

The Republic of Rwanda (hereafter, Rwanda) is located in the central part of Easter Africa and borders Lake Kivu in the west. The landscape of Rwanda is quite hilly, and as a result, the country is often called the country of 1000 hills. Three landscape types dominate the country, namely, grasslands, gently sloping hills, and mountainous areas with volcanoes. The average elevation of the country is approximately 1600 m, and its climate is temperate. The country spans an area of 26,338 km^2, with a population density of 302 persons/km^2 in 2012, which is the highest in SSA.

As already mentioned in Sect. 1.1, Rwanda has been undergoing rapid economic growth but still exhibits a high prevalence of malnutrition, especially in rural areas. For the purpose of this chapter, we will focus on two rural areas, the Rukara sector and the Mwiri sector (Fig. 1.1) (see below for more details).

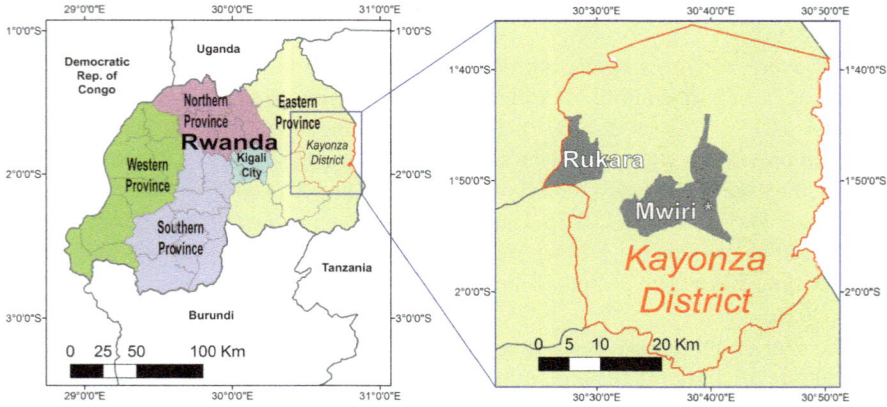

Fig. 1.1 Location of the Rukara and Mwiri sectors within the Kayonza District, Rwanda

The Rukara sector is located in a hilly area in the Kayonza District, with its altitude ranging between 1400 and 1600 m. The sector is characterized by hot and humid climate with an average temperatures of 18–26 °C and an average annual rainfall of 1000–1200 mm. The area of Rukara sector is approximately 64 km^2, and two of its three cells have access to Lake Muhazi and to an associated swampy area. Despite its high levels of impurities, Lake Muhazi is the largest source of water, supplying over 40% of Gwiza households, as other water sources in the area are quite limited.

As of 2013, the population of the Rukara sector is 28,422 people distributed in 6335 households. Their main livelihoods are related to agriculture, with the water from Lake Muhazi playing a rather important role, especially for horticultural and livestock production. The land and water shortages are particularly problematic given the high population density (444 people/km^2) and the prevalence of subsistence agriculture with traditional farming methods. These practices often lead to environmental degradation and hence low crop harvests, low incomes, and food shortages, especially among the highly vulnerable households that own very small piece of land. The Rukara sector suffers from considerable food shortages and high food prices rise, especially during the dry season.

It is in this context that the World Vision Rwanda (WVR), in collaboration with World Vision Japan (WVJ), a Christian development advocacy and relief organization, has implemented the Gwiza Area Development Program (ADP), which started in 2011 (Sect. 1.1). World Vision and its ADP office and staff based in the community for 10–15 years has a long-term commitment in improving the well-being of children and their families, especially those that are most vulnerable (e.g., orphans, widows, women-headed households, child-headed households, households with persons infected with HIV and AIDS). WVR and WVJ implement several projects in the area (e.g., food security, education, nutrition, healing, peace building, and reconciliation) in response to the local context characterized by low agricultural productivity, high population pressure, high school dropout rates, malnutrition, and mistrust within/among families following the 1994 genocide (see above and Sect. 1.1).

The Gwiza Nutrition Project was implemented between 2011 and 2015 as part of the Gwiza ADP activities. The aim of the program was to reduce malnutrition and child mortality in cooperation with the local leaders, health centers, and community health workers. Much like malnutrition, many different diseases are prevalent in the area including malaria, respiratory infections, diarrhea, HIV/AIDS, and sexually transmitted diseases. The Gwiza Nutrition Project conducted multiple interventions including (a) active feeding sessions for malnourished children under 5 years old and their caregivers through the Positive Deviance Hearth (PD Hearth) approach, (b) exclusive breast feeding and complementary feeding, and (c) knowledge sharing about community management of childhood illnesses including diarrhea, malaria, and acute respiratory infections. The Gwiza Nutrition Project also supported child growth monitoring on a monthly basis for all children under 5 years old. The goal was to identify moderately malnourished children for rehabilitation and refer severely malnourished children to the health center for proper treatment.

Further to Gwiza, we also conducted research in a reference sector with similar livelihood options that did not receive such interventions. In particular, we focused on the Mwiri sector that is also located in the Kayonza District. The Mwiri sector is much larger than the Rukara sector, covering an area of 540 km^2, although a large portion is occupied by Akagera National Park. The Mwiri sector is rather new and was opened in response to the rapidly increasing population density caused by the high population growth and the returnees from neighboring countries. However, with a population of 19,090 persons and 4162 households (in 2013), the sector's overall population and population density are much lower compared to the Rukara sector.

1.2.2 Data Collection

1.2.2.1 Preliminary Survey

Initially, we conducted several weighed food record (WFR) surveys in both the Rukara and Mwiri sectors to capture in detail the food intake patterns of local communities and to create a food frequency questionnaire (FFQ). Initially, the authors and staff of World Vision's Gwiza ADP selected 20 households in the Rukara sector and 12 households in the Mwiri sector (total of 32 households), with at least two adult members (i.e., aged 18 years or over). As there were many households that fulfilled the criterion, we therefore resorted to convenience sampling for the final selection. In particular, we selected households within a close distance, as the researchers had to visit at least two households in the same day, one to obtain a consent form and one to conduct the dietary survey (see below for the full WFR and FFQ procedure).

From these 32 households, we recruited a total of 162 participants of both sexes aged 1–69 years old (70 males and 92 females). We conducted the WFRs with all 162 participants in order to develop a food list and estimate the usual individual portion sizes for each food type, for both children and adults.

As the local communities in the study areas are highly dependent on locally cultivated food crops (Sect. 1.2.1), there are large seasonal variations of food availability both in terms of quantity and quality. Thus, in order to observe the seasonal variation in food intake, we conducted the WFRs twice in a year, in March and August. The first period coincides with the rainy season and the latter with the dry season. In order to assess intraindividual differences, we tried to visit the same households three times. Hence, the WFRs in the Rukara sector were conducted in March 2013, August 2013, and March 2014 and the WFRs in the Mwiri sector in August 2013, March 2014, and August 2014.

The WFRs were conducted on weekdays, as the local WVR staff was not permitted to work on weekends for religious reasons. One research team consisting of one Japanese researcher and a WVR staff member stayed in each household from early morning to evening in order to observe and record the quantity of ingredients, portion amounts, and, if any, leftovers for the three main meals (i.e., breakfast, lunch, dinner) and snacks.

The WFR that measured individual daily food intake followed a four-stage procedure. Firstly, we weighed all of the raw ingredients and seasonings that were used for cooking. After cooking, we weighed the cooked food and calculated the percentage of each of the raw ingredient included in the dish (proportion coefficient). Secondly, to determine individual portion sizes (i.e., consumption amount per meal), we weighed not only the served food but also, if any, the amount of extra servings and leftovers. Thirdly, we calculated the intake of each raw ingredient for each participant by multiplying the portion of the dish (in grams), with the proportional coefficient for each ingredient. We also asked absent family members whether they had consumed any food while away from home. Fourthly, we calculated the nutrient intake for each participant using the food composition tables for Uganda (a neighboring country of Rwanda) (Hotz et al. 2012), as this information was unavailable for Rwanda.

Any food item that was consumed by more than two participants (out of the 162 participants) was included in the FFQ. However, eventually, we excluded two food items whose median portion size contained only small amounts of energy, namely, biscuits and sugarcane. On the other hand, we added in the FFQ two food items, (eggs and commercial beverages), which were not observed in the WFRs but that are commonly sold at stores in the study areas and possibly contribute significantly to the energy/nutrient intake and differentiation of individuals. We examined the food list by percent contribution to energy and nutrient intake (Date et al. 2009; Jayawardena et al. 2012). More detailed information about the process followed to determine the portion size for the FFQ, validate the FFQ, and calculate the energy and nutrient intake from the FFQ can be found elsewhere (Yanagisawa et al. 2016).

1.2.2.2 Main Household Survey

The full field survey aimed at collecting information about the demographic, socio-economic, and nutritional status of households. Nutrition variables included the FFQ and anthropometric measurements that are a reliable long-term indicator of child undernutrition (Bogin 1999) (see below). The main household survey was conducted in four rounds, from August 2014 to March 2016, twice in the rainy season and twice in the dry season. The sample consisted of 120 households in the Rukara sector and 120 households in the Mwiri sector (Sect. 1.2.1). The targeted households were selected through stratified random sampling. In particular, each sector was divided into around 20 cells (i.e., the smallest community unit in Rwanda), with the households selected evenly from all cells through a resident list obtained from the local government. We visited each of the 240 households identified with the above process during each round of the household survey.

The demographic and socioeconomic variables in the household survey included (a) age, gender, education, and occupation of the household members, (b) household income and expenditure, (c) agricultural production practices, and (d) land tenure.

Specific expenditures include clothing and footwear for each household member, including the household head, father, wife, spouse, boys, and girls. These questions were answered basically by the household heads.

The survey also included the 18-food item FFQ developed during the preliminary surveys (Sect. 1.2.2.1). From the total of 26 food items (including single food items and mixed dishes) that were observed during the WFRs (see Sect. 1.3.1), 18 major items were selected for the food list of the FFQ (Yanagisawa et al. 2016). The 18 food items were maize flour porridge, mixed flour porridge, agatogo with animal food (e.g., meat, eggs) or beans, agatogo without animal food and beans, soup/sauce with animal food, soup/sauce with beans, soup/sauce without animal food and beans, boiled banana/cassava/potato, rice, ubugari (cassava), umutsima (maize), egg, milk/ tea with milk and sugar, mandazi, commercial beverages, sorghum alcohol, and sweet banana.

The consumption frequency for each of these 18 food items was elicited through nine frequency options (see below) and converted into weighing factors for statistical analysis. The frequency options and weighing factors include (1) almost never, 0; (2) one to three times per month, 0.07; (3) once per week, 0.1; (4) two to four times per week, 0.4; (5) five to six times per week, 0.8; (6) once per day, 1; (7) two to three times per day, 2.5; (8) four to six times per day, 5; and (9) more than six times per day, 6.

Finally, the Japanese researchers and local assistants conducted anthropometric measurements in August–September 2014 (R-I), March 2015 (R-II), and August– September 2015 (R-III). The researchers visited each of the 240 households during R-I, R-II, and R-III and measured the height and weight of all available household members at the time of the visit indoors and outdoors. However, only data for children under 12 years old are used for the analysis of this chapter (Sect. 1.3.2.4). The height was measured to the nearest 1 mm using a Martin anthropometer. Weight with minimal clothing was measured to the nearest 1 kg using an analog weighing scale, as uneven surfaces and scale instability rendered the use of a digital scale impossible in some households. For children who weighed less than 10 kg and were unable to stand, we measured lying weight and length using a digital baby scale (SECA 336). Mid-upper arm circumference was measured to the nearest 1 mm using a plastic tape measure (SECA 201). We used the height and weight of children collected during R-I, R-II, and R-III as cross-sectional data.

1.2.3   Data Analysis

We conducted a factor analysis for the 18 food items contained in the FFQ in order to assess the major dietary patterns in the study sample. When determining the number of factors to retain, we considered the results of the Scree test, eigenvalues greater than 1, and interpretability of the factors (Zazpe et al. 2014). The labeling of dietary patterns was based on the interpretation of the food items with high factor loadings for each dietary pattern (Newby and Tucker 2004), with only food types with a factor loading $\geq |0.25|$ included in this study.

For the analysis of the anthropometric measurements, we estimated two z-scores for children under 12 years of age: height-for-age z-score (HAZ) and weight-for-age z-score (WAZ). These indicators are calculated using each child's height and weight at R-I, R-II, and R-III based on the WHO Child Growth Standard (WHO CGS) using EPI-INFO (Version 7, Centers for Disease Control and Prevention, Atlanta). To determine the 6-month growth velocity of children under 2 years old, we use the increments from R-I to R-II and from R-II to R-III based on a mixed-longitudinal study design, which can provide information about growth velocity using repeated measurements from subjects of different starting ages (Bogin 1999). The 6-month growth velocity was compared with the WHO Child Growth Standard (WHO CGS).

The overall effects of dietary intake and other socioeconomic indicators (particularly agriculture) on nutritional status are estimated using Eq. 1.1:

$$
\begin{aligned}
HAZ = {} & constant + \beta_1 \times sex + \beta_2 \times age + \beta_3 \times factor1 + \beta_4 \times factor2 \\
& + \beta_5 \times land + \beta_6 \times number\_variety + \beta_7 \times total\_income + \beta_8 \\
& \times ex\_share + \beta_9 \times ex\_share2 + \beta_{10} \times edu\_year\_mother
\end{aligned} \tag{1.1}
$$

In Eq. 1.1, the variable *HAZ* denotes the z-score for children under 12 years old. *Age* is considered in months (for children below 5 years old) and in years (for children above 6 years old). The variables, *factor1* and *factor 2*, are estimated through the factor analysis of the FFQ (see above). The variable *land* denotes the total area of owned land (in ha), *number_variety* the number of crop varieties cultivated by the household, *total_income* the total household income (in RWF), and *ex_share* the expenditure share of private assignable goods for children (e.g., clothing. footwear) (in %). It is assumed that *ex_share* and *HAZ* have u-shaped relationships; hence, we include the quadratic terms of *ex_share*, *ex_share2*. The variable *edu_year_mother* denotes the education level of the mother in the household (in years of schooling). The variables *factor1*, *factor 2*, *land*, *number_variety*, *total_incomeex_share2*, and *edu_year_mother* are expected to have a positive effect to *HAZ* while *ex_share* to have a negative effect. Eq. 1.1 is estimated through an *ordinary least squares* (OLS) using Stata 14 (StataCorp LP, College Station, Texas, USA).

1.2.4 Ethical Approval and Permissions

We obtained written informed consent from each household before proceeding with participation in the study after explaining the purpose and procedures of the study. The study protocol was approved by the Ethics Committees of the Graduate School of Frontier Sciences (University of Tokyo), the Institutional Review Board of

Ochanomizu University (approval number 2013-77), and the Medical Ethical Committee of the Ministry of Health, Rwanda.

1.3 Results

1.3.1 Preliminary Survey

1.3.1.1 Dietary Patterns Through Weighed Food Record Surveys

We conducted 1-day WFR three times for 10 households, twice for 1 household, and once for 21 households. This means that overall we conducted 53 household/day WRF. Three of the 32 households were an extended family. The average number of household members and children were 5.3 and 3.5, respectively. The WFRs were collected from all of household members and amounted to a total of 260 WFRs from 162 participants. Almost all households depended on farming for their livelihoods, but three households (9.4%) owned small grocery stores in front of (or near to) their houses.

Respondents overwhelmingly ate from an individual dish, rather than from the same shared dish. A total of 47 out of the 53 household/day WFRs (83.0%) represented households that cooked and consumed two or three meals per day. Three households (5.7%) cooked large quantities once per day for lunch and dinner to save firewood, while six households (11.3%) ate only lunch.

Twenty-three children aged 1–15 years ate snacks between meals, mainly cassava, biscuits, sugarcane, porridge, boiled sweet potato, mandazi (fried bread), agatogo (a typical Rwandan dish, see below), boiled beans, carrot, tomato, sweet banana, papaya, and avocado.

The WFR results suggest that the diet in the study area is highly dependent on carbohydrates. For our sample, approximately 77.7% of the dietary energy was obtained from carbohydrates and only 9.4% from protein. Thirty-six households (67.3%) drank a cup of porridge for breakfast, whose ingredients were only flour and hot water. About half of the 32 households used mixed flour for making porridge, while the rest used only maize flour. All households consumed agatogo at least once per day (i.e., a simmered dish with green banana or potato, with sometimes added beans and dodo, a green leafy vegetable). They also consumed soup/sauce about five to six times per week, commonly served with starchy staple foods such as umutsima (made from maize flour), ubugari (made from cassava flour), boiled potatoes, green bananas, or rice.

The WFR results also reveal that diet diversity in the study sites is quite low. A total of 26 food items (including single food items and mixed dishes) were observed during the WFRs, of which 16 amounted to almost 100% of the total energy and protein intake (Table 1.1). "Agatogo with beans" accounted for a considerable percentage of the energy and nutrient intake of the surveyed households (Table 1.1).

Table 1.1 Contribution of the 16 main food items to the total energy and protein intake

	Energy			Protein	
	Food items	(%)		Food items	(%)
1	Agatogo with beans	44.1	1	Agatogo with beans	42.7
2	Agatogo without beans	8.8	2	Soup with beans	11.7
3	Boiled banana, cassava, and sweet potato	8.7	3	Porridge with mixed flour	8.3
4	Porridge with mixed flour	7.7	4	Agatogo without beans	7.7
5	Ubugari	5.8	5	Boiled banana, cassava, and sweet potato	6.4
6	Soup with beans	5.5	6	Soup with meat	4.6
7	Rice	5.3	7	Porridge with maize flour	4.3
8	Porridge with maize flour	4.9	8	Rice	4.2
9	Umutsima	2.4	9	Soup without meat and beans	2.7
10	Soup with meat	2.1	10	Ubugari	2.5
11	Soup without meat and beans	1.7	11	Umutsima	1.9
12	Milk/tea with milk and sugar	1.0	12	Milk/tea with milk and sugar	1.5
13	Mandazi	0.4	13	Sorghum alcohol	0.4
14	Avocado	0.3	14	Avocado	0.2
15	Sweet banana	0.3	15	Mandazi	0.2
16	Sorghum alcohol	0.3	16	Sweet banana	0.2
	SUM	**99.3**		**SUM**	**99.5**

Source: Adapted from (Yanagisawa et al. 2016)

1.3.1.2 Dietary Quality and Participation in World Vision Activities

We compare the dietary data obtained through the WFR in terms of energy and nutrient intake between households that have joined the World Vision's activity (at least once even if they discontinued) and household that have never been targeted by a World Vision intervention in the study areas. As shown in Fig. 1.2, for both mothers and children under 5 years old, energy and vitamin A intakes were significantly higher in households that have joined a World Vision activity in the past.

1.3.2 Main Household Survey

1.3.2.1 Household Characteristics and Livelihoods

In Rukara, the average household size is 5.57 persons, with a maximum household size of 11 and a minimum household size of 2. In Mwiri, the average household size is 5.46, with a maximum household size of 15 and a minimum household size of 1. The above suggests the relatively large family size and variation in the study sites. Figure 1.3 indicates the main crops cultivated in the surveyed households in the two areas. Overall, households in Rukara plant a larger diversity of crop varieties

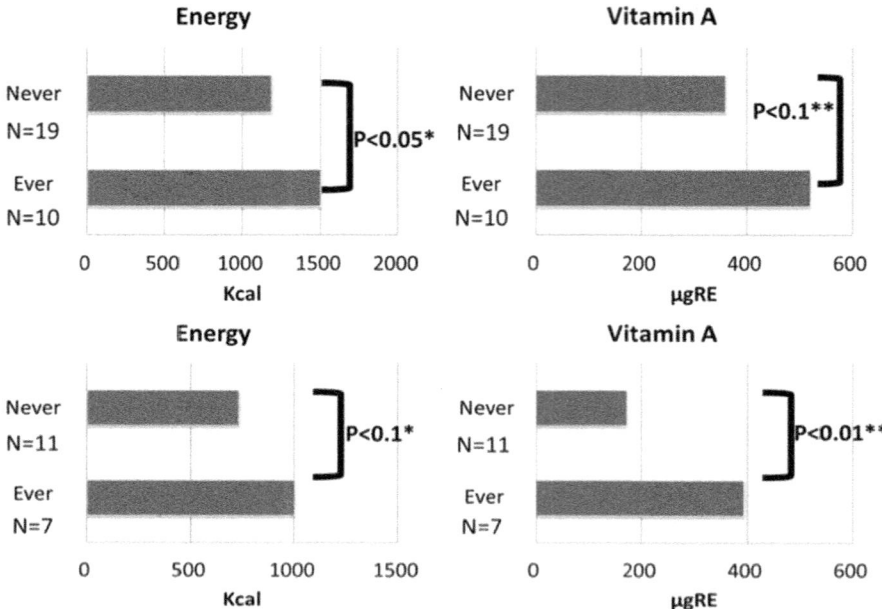

Fig. 1.2 Energy and vitamin A intake for mothers and children for the households participating at World Vision activities

Note: Upper figure refers to mothers and bottom figure to children below 5 years old. The statistical differences for energy intake were compared through Mann-Whitney's U-test while for vitamin A through t-test. $**p < 0.05$; $*p < 0.1$

compared to Mwiri, possibly due to the better market access (both inside and outside the sector), and the many relevant projects are funded by external organizations. Figure 1.3 also includes the average market selling prices for the main crops as collected through the survey questionnaire.

Table 1.2 includes the average total household income estimated from crops sales and the revenue obtained through nonagricultural sources (e.g., other livelihood activities, land rent, support from outside the household). Figure 1.4 highlights the proportion of the value of crops that were self-consumed within the household compared to the different income sources. Results suggest that incomes in Rukara are higher than Mwiri, but in both areas, income is derived overwhelmingly from agricultural activities, with the value of household crops being many times higher to the actual income obtained through different livelihood activities. This suggests the important role of subsistence agriculture in the study areas (Sect. 1.2.1).

Figure 1.5 indicates the relationships between farm productivity (RWF/m^2) and the number of crop varieties in each farm. Increases in the number of planted crop varieties seems to improve farm productivity to some degree but then decreases after some threshold level in the number of planted varieties.

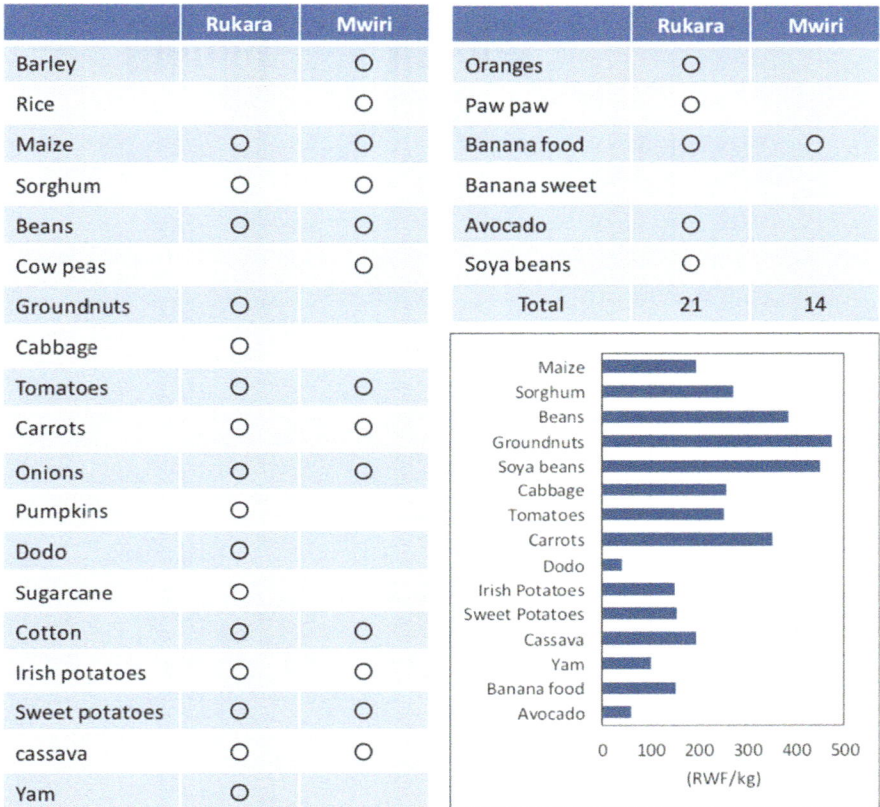

Fig. 1.3 Crop varieties and selling prices (in RWF/kg)

Table 1.2 Average total household income and share by income source

	Rukara	Mwiri
Non-agricultural income (RWF)	2941.0	2579.2
% of total income	13.2	16.9
Agricultural income (RWF)	18,986.1	12,570.6
% of total income	85.2	82.5
Other income (RWF)	356.0	100.7
% of total income	1.6	0.6
Total	22,283.1	15,250.5

Figure 1.6 indicates the share of expenditures (dotted lines) on private assignable goods such as clothing and footwear for children under 12 years old, categorized by family size and the number of family members. The expenditure share for clothing and footwear for children under 12 years old is decreasing with family size to some degree. This is because larger families, which are expected to have more land, tend to allocate higher budgets for these goods for children. Overall, we find a u-shaped

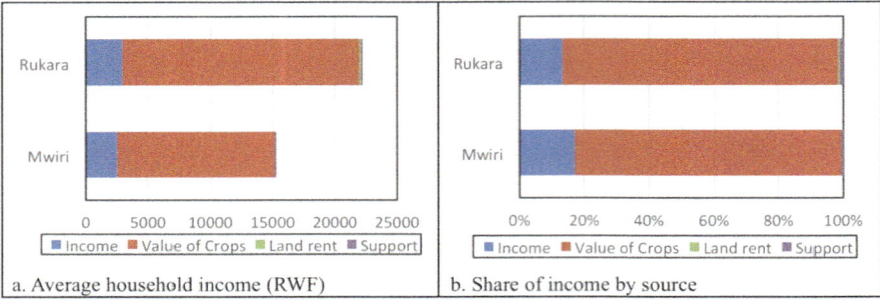

Fig. 1.4 Average household income and share by income source

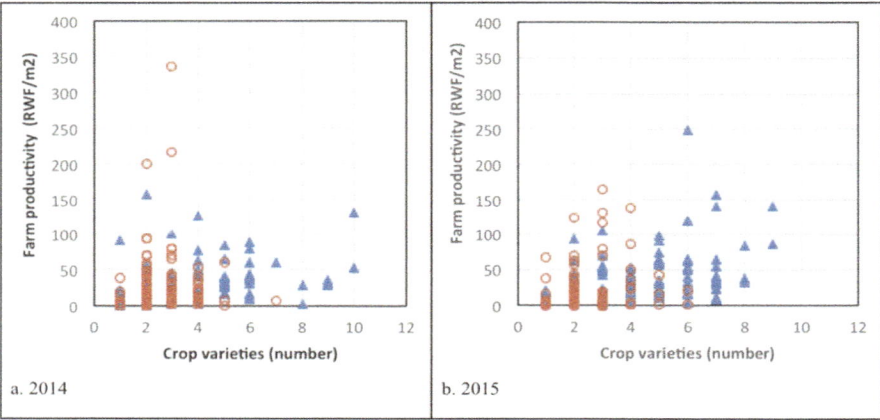

Fig. 1.5 Farm productivity and number of crop cultivated varieties in 2014 and 2015
Note: Crops value is estimated from Fig. 1.3 and Table 1.2

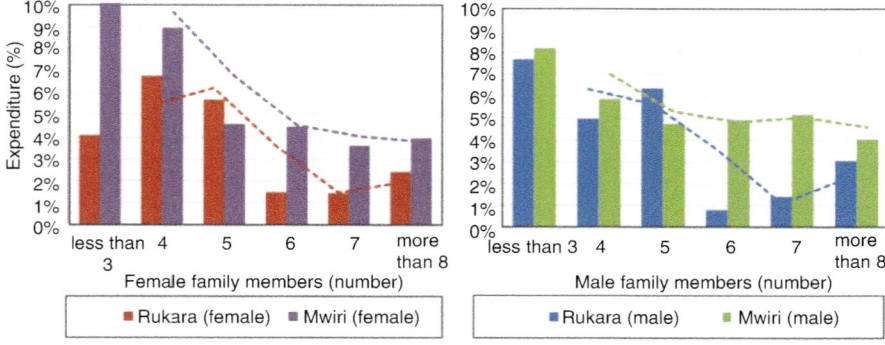

Fig. 1.6 Share of expenditure for private assignable goods for children by family size

relationship between the expenditure share of private assignable goods for children under 12 and family size.

1.3.2.2   Food Diversity

The FFQ survey indicates the types of food that the surveyed households consume (and the frequency of this consumption) across 18 major food types. Factor analysis, one of the types of major multivariable analysis, is employed to capture unobserved common correlated variables (called factors) among observed variables. Through the factor analysis, we identify two main factors (Fig. 1.7). Factor 1 has a larger value and contains rather basic types of food consumed by almost all households. However, factor 2 contains types of food that exhibit higher food diversity. Factors 1 and 2 are used as indicators of basic food and diverse food in Sect. 1.3.2.4.

1.3.2.3   Anthropometric Measurements

Through the anthropometric measurements conducted over the three time intervals (Sect. 1.2.2.2), we obtain (a) 682 data points for height and 683 data points for weight among boys under 12 years old and (b) 624 data points for height and 626 for weight for girls under 12 years old.

For children under 5 years old, we calculate the prevalence of stunting and underweight to compare with the RDHS datasets. The prevalence of stunting among this age group is 37.7%, which is quite similar to the national average

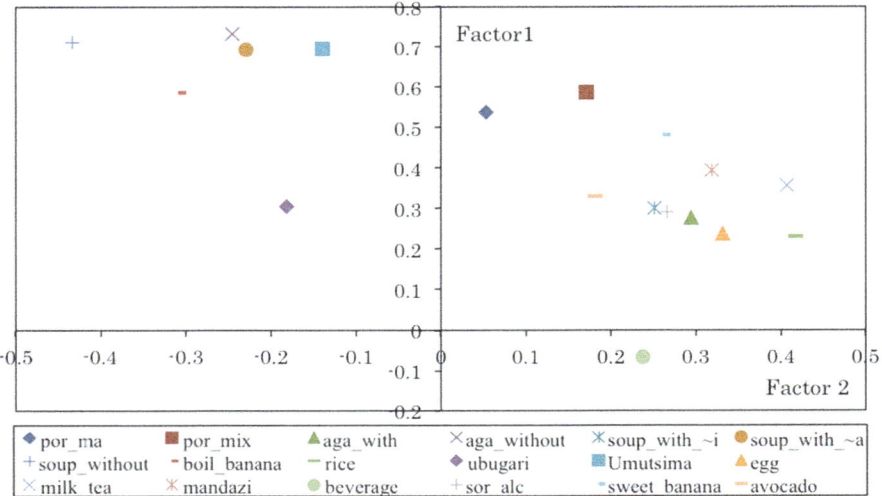

Fig. 1.7 Factor analysis indicating basic and diverse types of food

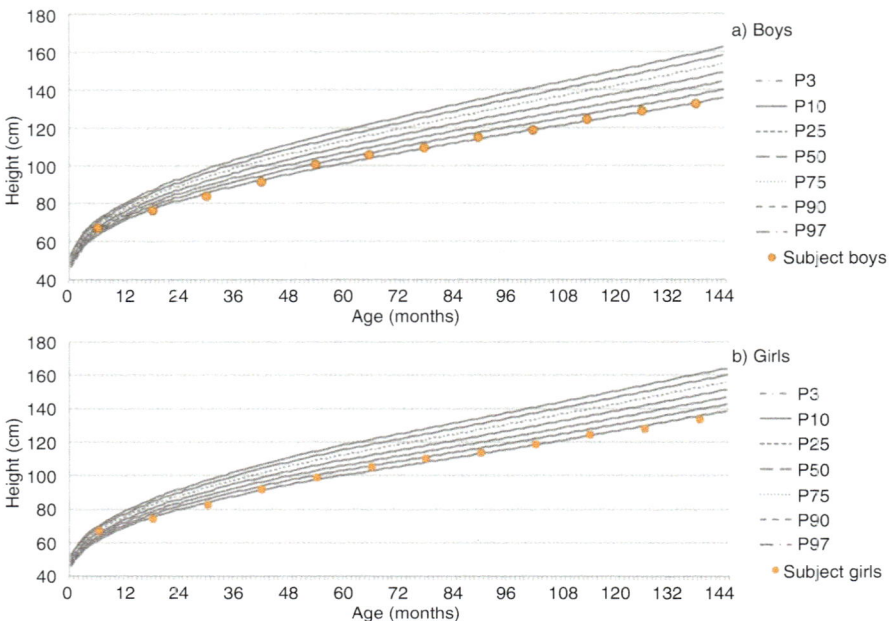

Fig. 1.8 Height growth curves of subject boys (**a**) and girls (**b**) across the WHO growth standard percentiles

(38%) and slightly higher than observed stunting rates in the Eastern Province (35%). Similarly, approximately 16.2% of the surveyed sample was underweight.

Figure 1.8 plots the height growth of the sampled boys and girls relative to the WHO growth standard. Results suggest that the sampled children have a far lower than average height considering the WHO growth standard. In more detail, for sampled boys, the mean height is between the 25th and the 50th percentile of the WHO standard at age 0 years. However, it drops below the 3rd percentile for ages 1–3 years. It slightly improves by age 4 years, ranging between the 10th and 25th percentiles, but then drops again and remains between the 3rd and 10th percentiles between ages 5 and 7 years. From ages 8–12 years, the height remains below the 3rd percentile. Conversely, for sampled girls, the mean height is between the 75th and 90th percentiles of the WHO standard at age 0 years. However, height drops below the 3rd percentile between ages 1 and 2 years. Height improves slightly at age 3 years, ranging between the 3rd and 10th percentiles, and remains in that range until age 9 years. At the age of 10 years, however, the height again drops below the 3rd percentile due to the later timing of the secondary growth spurt.

Figure 1.9 plots the weight growth of the sampled boys and girls relative to the WHO growth standard but only up to the age of 10 years, as the WHO weight standard percentile is not available for higher ages. Again, the results suggest that the sampled children had a far lower weight considering the WHO standard. For boys aged 0–1 years, the mean weight is between the 25th and 50th percentiles of the

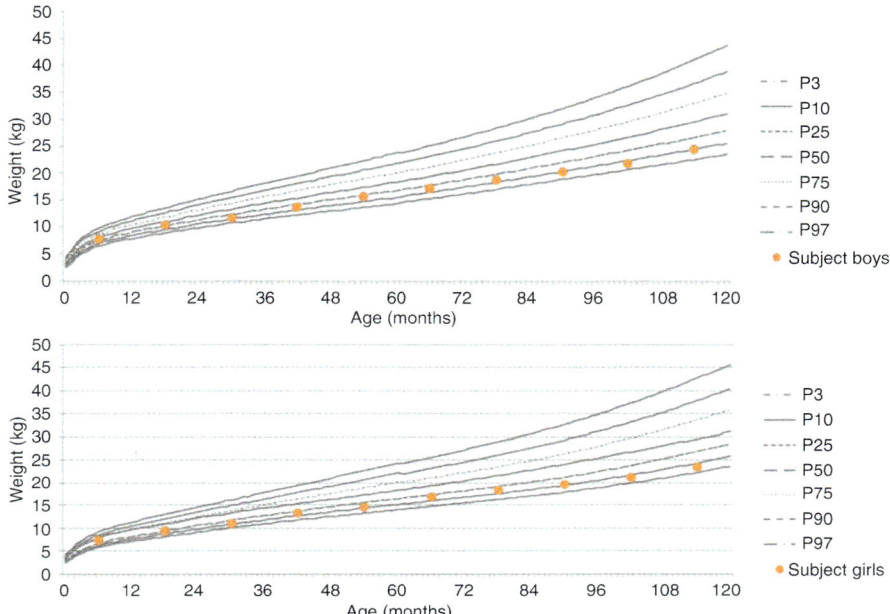

Fig. 1.9 Weight growth curves of subject boys (**a**) and girls (**b**) across the WHO growth standard percentiles

WHO growth standard. However, it then drops slightly and remains between the 10th and 25th percentiles from ages 3–9 years (except at age 8 years, when it drops to between the 3rd and 10th percentiles). For the sampled girls, the mean weight is between the 50th and 75th percentiles of the WHO growth standard at the age of 0 years. However, it drops to between the 25th and 50th percentiles at the age of 1 years and then again to between the 10th and 25th percentiles from the age of 3 to 7 years. Weight drops further again to between the 3rd and 10th percentiles for girls between 8 and 9 years old.

For children under 2 years old, the 6-month growth increment was compared with the WHO Child Growth Velocity Standard, as shown in Fig. 1.10. More than 40% of the surveyed children fell below -3 standard deviation (SD) of the WHO growth standard. The overall growth velocity of most surveyed children (71%) was below the mean value of the WHO standard.

The findings from the anthropometric measurement reveal that the growth of the subject children can be summarized as follows: a good start, followed by a sharp drop during infancy, and then a slight catchup growth during childhood. The sharp drop observed during infancy is critical to improving the growth of subject children.

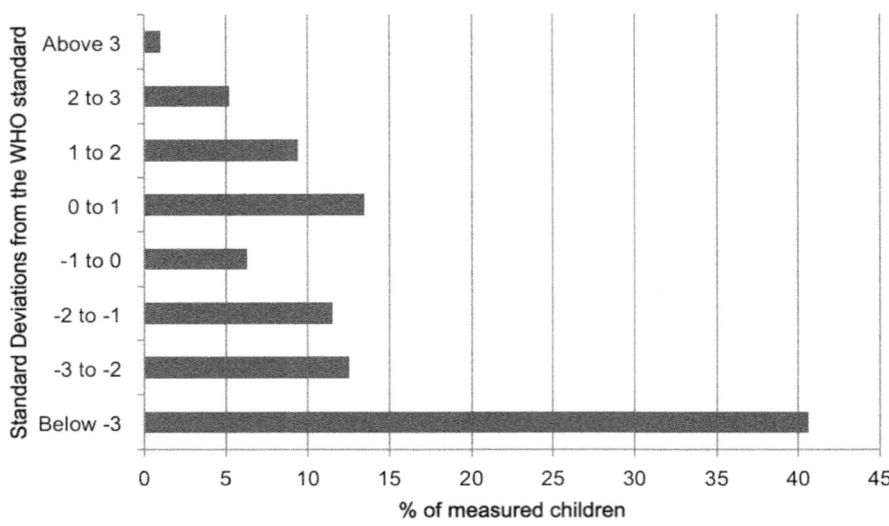

Fig. 1.10 Growth velocity relative to the WHO standard

Table 1.3 Estimation of nutritional status with OLS and logistic regression

Variables	OLS	Logistic regression
Cons	−0.884 (−1.570)	0.092 (0.110)
Sex	0.038 (0.130)	0.389 (0.432)
Age	−0.016∗∗∗ (−3.060)	−0.015∗ (−1.750)
Factor 1	−0.124 (−1.450)	−0.245∗∗ (−1.990)
Factor 2	0.655∗∗ (2.160)	0.844∗∗ (1.880)
Land	0.000 (0.300)	0.000 (1.810)
Number_variety	0.153∗∗ (2.140)	0.215∗ (1.920)
Total_income	0.000∗∗∗ (2.810)	0.000 (1.090)
Ex_share	−8.986∗ (−1.830)	−16.485∗∗∗ (−1.830)
Ex_share_2	38.299∗∗ (2.370)	49.804∗∗ (2.070)
Edu_year_mother	−0.822 (−1.720)	−0.044 (−0.410)
Adj. R^2	0.103	0.152
F-value	2.720∗∗∗	25.140∗∗∗
Obs.	121	121

Note: t-values and z-values are shown in parentheses for OLS and logistic regressions, respectively:
∗∗∗$p < 0.01$; ∗∗$p < 0.05$; ∗$p < 0.1$

1.3.2.4 Relationship Between Crop Production, Food Security, and the Nutritional Status of Children

Table 1.3 indicates the results of the analysis for Eq. 1.1 (Sect. 1.2.3). In the first equation, the variables "*factor 2*" and "*number_variety*" are positive and significantly different from zero, while "*total_income*" is positive and significantly different from zero. Those results indicate that having access to a variety of food types is an important factor in improving height-for-age z-score (HAZ). The results of "*ex_share*" and "*ex_share2*" indicate that the intra-household resource allocation has an impact on the nutritional status, following a u-shaped relation. However, the variables related to land ownership "*land*" and female education "*edu_year_mother*" are not significant. This suggests that planting many crop varieties is more important than the overall amount of land owned by a household. As indicated in Sect. 1.3.2.1, larger land ownership is associated with lower farm economic productivity; thus, the efficient management and use of cropland are possibly more important in terms of nutritional outcomes.

1.4 Discussion

1.4.1 Main Patterns

Similar to other parts of SSA (Sect. 1.1), child malnutrition is still a big sustainability challenge in the surveyed area. The prevalence of stunting among children under 5 years old is 37.7%, which is quite similar to the national average (38%) and slightly higher compared to the Eastern Province (35%) (NISR et al. 2015). The results in Sect. 1.3.1 suggest that involvement in World Vision activities seem to indeed be associated with better dietary quality for some indicators, but deeper research is needed to establish the actual effects and the mechanisms through which this happens.

Our growth analysis reveals that the prevalence of malnutrition highly relates to the developmental stage of the children, with their growth patterns summarized as a good start, followed by a sharp drop during infancy, and then a slight catchup growth during childhood (Sect. 1.3.2.3). The sharp drop in growth rates observed during infancy is critical to improving child growth and could be due to (a) the poor nutritional quality/quantity of both breast milk and complementary foods and (b) a high disease burden based on life history theory (McDade 2003).

The above results can be put into perspective using the life history theory, which provides a comparative evolutionary framework for understanding developmental and reproductive strategies. This theory assumes that resources are limited and that energy is allocated to three primary life functions, namely, growth, reproduction, and maintenance (McDade 2003). The immune system is an essential component of the maintenance function, and from birth to the juvenile period, many resources are allocated to either growth or maintenance considering that the reproductive effort is negligible (McDade 2003). Children growing in nutritionally and epidemiologically privileged environments have sufficient fuel for growth and effective immune responses to infections, whereas children growing in adverse environments exhibit higher resource allocation competition, resulting in poor growth and impaired immunity (McDade et al. 2008). Based on this theory, the sharp height deficit (Sect. 1.3.2.3) can be due to the (a) energy deficit from the poor nutritional quality/quantity of breast milk and complementary foods and (b) considerable resources allocated to maintenance rather than growth due to a high disease burden. However, from the results, it is not possible to judge which factor is more important in the study area.

The high disease burden can be possibly interpreted through the rapid decline in infant mortality, with children surviving by sacrificing growth. Even so, it is important to decrease the disease burden as much as possible in the study area by further improving the health-care system and household sanitation (Null et al. 2018) (see Chap. 4 Vol. 2). However, as we did not obtain any data on sanitation and disease, it is not possible to discuss the possible effect of the lack/underperformance of health and sanitation on tackling child malnutrition. Future studies in the area should aim to explore these linkages further.

When it comes to energy deficit, it is important to discuss the duration of breastfeeding and the quality/quantity of weaning food supplies (Victor et al. 2014). In terms of the duration of breastfeeding, traditional pastoralist groups in SSA are well known for the prolonged duration of breastfeeding (approximately 4 years) during which mothers are infertile due to lactational amenorrhea. This minimizes the natural fertility of pastoralists to approximately four to five children, which has been interpreted as a survival strategy of these groups (Ohtsuka et al. 2012). Based on our observations, our study sample seems to also undergo prolonged breastfeeding periods, even though the WHO guidelines recommend initiating complementary feeding at 6 months (WHO 2017b). As a result, it would be important to educate mothers to start complementary feeding no later than the age of 6 months in order to improve the child growth during infancy.

In terms of the quality/quantity of weaning food, the results of the WFR reveal that the diets of our sample are characterized by limited food variety (Sect. 1.3.1). The total number of food items observed during the WFRs is as few as 26 with a high dependence on starchy food (Sect. 1.3.1). This is similar to other rural communities across SSA, where the main staple crops are starchy crops such as maize and cassava (Sheeh et al. 2014) (see also Chap. 6, 10 Vol. 1; Chap. 2 Vol. 2).

As discussed in Sect. 1.3.1, a common breakfast among the respondents was porridge that contained only flour and hot water. In fact, about half of the surveyed households only use maize flour to make porridge, which includes little amounts of protein. The porridge is also commonly consumed as a weaning food, as it is easy to swallow even for the small children. During the field survey, the authors often observed that children drink porridge poured in a big plastic cup. In order to improve the nutritional quality of porridge, it can be recommended to add milk or change maize flour to mixed flour, as far as it is available and affordable in the targeted area. The results of the WFR reveal that porridge with mixed flour includes higher amounts of protein and calcium compared to that of maize flour (Table 1.1), especially considering that mixed flour includes soy powder. During the site visits and interviews, we also observed that the price of mixed flour and maize flour was almost the same. Hence, this kind of recommendation should be properly conveyed to the targeted population to improve the nutritional quality of porridge.

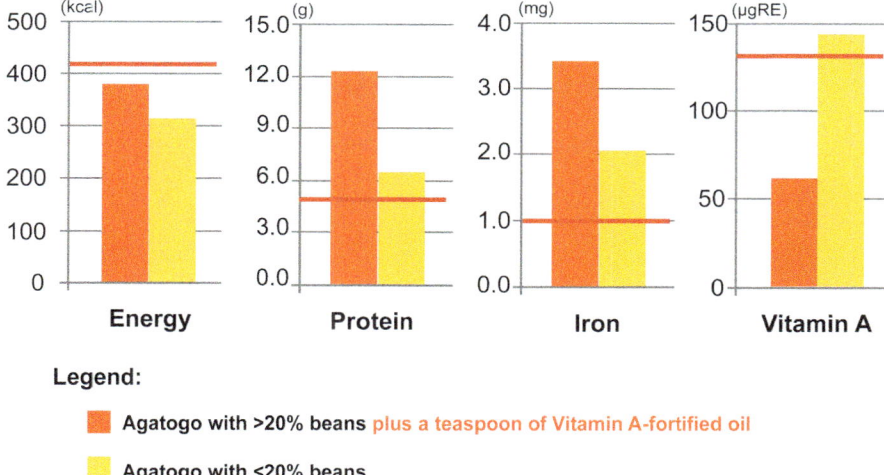

Fig. 1.11 Energy and nutrient content of an average agatogo portion for 1–4-year-old boys
Note: The red lines denote 1/3 of the daily nutritional requirement for 3-year-old boys

Agatogo is another important type of food in the study area, with the WFR revealing that every household consumes agatogo at least once per day (Sect. 1.3.1.1). As the nutritional quality of porridge is not high (see above), people in the targeted area need to receive enough energy and nutrients from agatogo. Agatogo is also commonly consumed by small children and thus offers an important opportunity and avenue for improving nutritional quality.

Figures 1.11 and 1.12 and Table 1.4 provide three recommendations toward improving the nutritional quality of agatogo. The first recommendation is to include large portions of boiled beans, i.e., >20% of the total dish amount. Figure 1.11 illustrates the possible effects for 3-year-old boys. In particular, according to the WRF (Sect. 1.3.1.1), the average size of an agatogo portion is 343 g for 1–4 year old boys, and by adding boiled beans at about 20% of the total portion, the energy, protein, and iron content will increase, almost reaching or exceeding one-third of the daily nutritional requirement for 3-year-old boys (Fig. 1.11). However, the amount of vitamin A will decrease, as beans include lower amounts of vitamin A. Thus, adding oil fortified with vitamin A could overcome this deficit (Fig. 1.12). Adding dodo (i.e., a green leafy vegetable) to agatogo can further enhance the calcium and vitamin A content (Table 1.3).

As shown in Fig. 1.2, dodo (alongside other crops such as groundnuts, cabbage, pumpkins, soybeans) is currently planted only in the Rukara sector, probably because many projects funded by organizations such as World Vision were implemented in the sector. Thus, it would be easier for local communities in the Rukara sector to improve the nutritional quality of agatogo, as the recommended ingredients are easily accessible in the area. Furthermore, as shown in Table 1.3, our OLS model suggests that such food varieties play an important role in relation to the

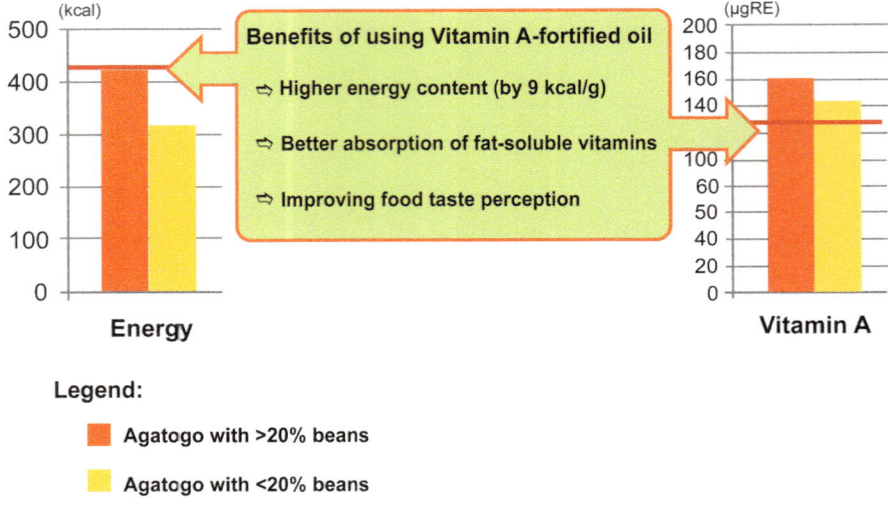

Fig. 1.12 Energy and vitamin A content of an average agatogo portion to 1–4-year-old boys
Note: The red lines denote 1/3 of daily nutritional requirements for 3-year-old boys

Table 1.4 Energy and nutrient content per 100 g of agatogo

	Energy (kcal)	Protein (g)	Ca (mg)	Fe (mg)	Vitamin A (µgRE)
Agatogo without dodo	109	3.7	16.5	1.1	73
Agatogo with dodo	109	3.8	24.8	1.1	84

HAZ of children below 12 years old. Thus, the crop varieties cultivated in local households could improve the nutritional status of children, through improving the quality of their diet. This has been observed in many parts of SSA and has been highlighted as an important avenue for tackling undernourishment, child undernourishment in particular (Gelli et al. 2018; Huang et al. 2018; Solomon et al. 2017).

The above results highlight the important linkages between household agricultural production, food security, and nutritional status of children in rural Rwanda (Agho et al. 2019). The diversification of cultivated plants seems to be a key strategy for improving the nutritional status of children. Thus, combining appropriate crop selection and farming management approaches could further help improve land productivity, which may alter the living standards of households. In the short and long term, this could be applicable not only in rural Rwanda but also in other SSA countries where local communities still rely heavily on subsistence agriculture (e.g., Chap. 10 Vol. 1, Chaps. 2–3 Vol. 2). In this sense, nutrition-specific approaches (whether project- or donor-dependent) may not offer a sustainable solution toward alleviating undernutrition in rural SSA (Nsabuwera et al. 2016). Finally, as intra-household resource allocation seems to also be a significant factor affecting the nutritional status of children in families with limited resources, much more attention should be paid to improve intra-household resource allocation, especially by empowering mothers (Alaofè et al. 2017).

1.4.2 Policy and Practice Recommendation

Tackling child undernutrition was major target in the Millennium Development Goals (MDGs) and is currently featuring in two distinct SDG targets, namely, to end hunger and ensure access by all people to safe and nutritious food (Target 2.1) and to end all forms of malnutrition (Target 2.2) (Sect. 1.1). These targets have very strong links with multiple other SDGs such as SDG1 (No Poverty) and SDG3 (Good Health and Well-being). In several SSA contexts with low availability of natural resources, social services, and overall capacity, tackling child undernutrition can be particularly challenging. Very different approaches addressing different undernutrition determinants have been implemented to reduce child undernutrition in SSA in the past decades, with most of the relevant interventions following nutrition-specific approaches that are very project- and/or donor-dependent and cannot be sustained easily by local communities when the funding is over.

The findings of this chapter are at the interface of SDG1, SDG2, and SDG3 and have highlighted the importance of the linkages between household agricultural production, food security, and nutritional status of children in rural Rwanda. We identify three major implications and recommendations at this interface, namely, the need to (a) diversify crop production, (b) empower women, and (c) support more strongly nutrition-sensitive approaches.

In particular the diversification of cultivated plants and the improvement of intra-household resource allocation seem to be two key factors for improving the nutritional status of children in families with limited resources. In this respect, there should be stronger efforts to promote these linkages locally, for example, through building capacity and promoting the cultivation of various types of plants and utilizing kitchen gardens for these purposes at each household. The empowerment of mothers would be indispensable for improving intra-household resource allocation, especially for purchasing high-quality and nutritious food. This could be achieved by training women on breastfeeding and weaning foods and especially targeting mothers who actually prepare and cook meals. Finally, the global community would need to support more strongly nutrition-sensitive approaches that address local agricultural production and food security, as a means of offering sustainable solutions for combating. This would require viewing undernutrition in a broader lens that considers multiple relevant aspects such as agricultural production and sanitation among others (Sect. 1.1).

1.5   Conclusions

Similar to most SSA contexts, child malnutrition still remains a big sustainability challenge in rural Rwanda. Our study found that the prevalence of stunting among children below 5 years is as high as 37.7% in the study areas. Our growth analysis reveals that the prevalence of stunting highly relates to the developmental stage of the children. Reversing the sharp growth drop observed during infancy would be critical for improving the long-term growth of children. A major avenue to achieve this is to elucidate the need for appropriate education for mothers regarding breastfeeding and weaning foods.

Results from the dietary survey reveal that diets in rural Rwanda (including weaning foods) are characterized by a limited variety and high dependency on starchy foods. In order to improve diet quality, it is critical to increase the variety of consumed food that contains various types of nutrients. Furthermore, as revealed by the OLS results, the plant varieties cultivated in rural households can enhance the nutritional status of children, through improving the quality of their diet.

This result highlights the important linkage between household agricultural production, food security, and nutritional status of the children in rural Rwanda. This finding would be applicable not only in rural Rwanda but also in other SSA contexts where local communities still rely heavily on subsistence agriculture. Nutrition-specific interventions, which tend to be project- and/or donor-dependent, may not offer a sustainable solution for combating undernutrition in rural areas of SSA. More effort should be invested to improve household agricultural production and intra-household resource allocation to combat child malnutrition across sub-Saharan Africa.

Acknowledgments Marcin Jarzebski created Fig. 1.1. The authors thank the participants for their cooperation. They are grateful to World Vision Rwanda for making arrangements for this study.

References

Agho KE, Mukabutera C, Mukazi M, Ntambara M, Mbugua I, Dowling M, Kamara JK (2019) Moderate and severe household food insecurity predicts stunting and severe stunting among Rwanda children aged 6–59 months residing in Gicumbi district. Matern Child Nutr 15(3): e12767. https://doi.org/10.1111/mcn.12767. Epub 2019 Jan 13

Alaofè H, Zhu M, Burney J, Naylor R, Douglas T (2017) Association between women's empowerment and maternal and child nutrition in Kalalé District of Northern Benin. Food Nutr Bull 38 (3):302–318. https://doi.org/10.1177/0379572117704318. Epub 2017 Apr 26

Bhutta ZA, Ahmed T, Black RE, Cousens S, Dewey K, Giugliani E, Haider BA, Kirkwood B, Morris SS, Sachdev HP, Shekar M, Maternal and Child Undernutrition Study Group (2008) What works? interventions for maternal and child undernutrition and survival. Lancet 371 (9610):417–440. https://doi.org/10.1016/S0140-6736(07)61693-6

Black RE, Allen LH, Bhutta ZA, Caulfield LE, De Onis M, Ezzati M, Mathers C, Rivera J, Maternal and Child Undernutrition Study Group (2008) Maternal and child undernutrition: global and regional exposures and health consequences. Lancet 371(9608):243–260

Black RE, Victora CG, Walker SP, Bhutta ZA, Christian P, de Onis M, Ezzati M, Grantham-McGregor S, Katz J, Martorell R, Uauy R, Maternal and Child Nutrition Study Group (2013) Maternal and child undernutrition and overweight in low-income and middle-income countries. Lancet 382(9890):427–451

Bogin B (1999) Patterns of human growth. Academic Press, London

Date C, Tokudome Y, Yoshiike N (2009) The manual for diet survey, 2nd edn. NANZANDO Co., Ltd, Tokyo. (in Japanese)

FAO (2013) Synthesis of guiding principles on agriculture programming for nutrition. Food and Agriculture Organisation (FAO), Rome

FAO (2017) The state of food security in the world 2017. Food and Agriculture Organisation (FAO), Rome

Garcia V (2012) Children malnutrition and horizontal inequalities in sub-Saharan Africa: a focus on contrasting domestic trajectories (UNDP Working Paper 2012-091)

Gelli A, Margolies A, Santacroce M, Roschnik N, Twalibu A, Katundu M, Moestue H, Alderman H, Ruel M (2018) Using a community-based early childhood development center as a platform to promote production and consumption diversity increases children's dietary intake and reduces stunting in Malawi: a cluster-randomized trial. J Nutr 148(10):1587–1597. https://doi.org/10.1093/jn/nxy148

Haddad L, Achadi E, Bendech MA, Ahuja A, Bhatia K, Bhutta Z et al (2015) The global nutrition report 2014: actions and accountability to accelerate the world's progress on nutrition. J Nutr 145(4):663–671

Hotz C, Lubowa A, Sison C, Moursi M, Loechl C (2012) A food composition table for central and eastern Uganda, HarvestPlus technical monograph series 9. HarvestPlus, Washington DC. Available at: http://www.harvestplus.org/content/food-composition-table-central-and-east ern-uganda. Accessed 10 Jan 2015

Huang M, Sudfeld C, Ismail A, Vuai S, Ntwenya J, Mwanyika-Sando M, Fawzi W (2018) Maternal dietary diversity and growth of children under 24 months of age in rural Dodoma, Tanzania. Food Nutr Bull 39(2):219–230

IFPRI (2016) Global nutrition report 2016: actions and accountability to advance nutrition and sustainable development. International Food Policy Research Institute (IFPRI), Washington, DC

Jayawardena R, Swaminathan S, Byrne NM, Soares MJ, Katulanda P, Hills AP (2012) Development of a food frequency questionnaire for Sri Lankan adults. Nutr J 11(1):63

Logie DE, Rowson M, Ndagije F (2008) Innovations in Rwanda's health system: looking to the future. Lancet 372:256–261

Masset E, Haddad L, Cornelius A, Isaza-Castro J (2012) Effectiveness of agricultural interventions that aim to improve nutritional status of children: systematic review. BMJ 344:d8222

McDade TW (2003) Life history theory and the immune system: steps toward a human ecological immunology. Am J Phys Anthropol Suppl 37:100–125

McDade TW, Reyes-Garcia V, Tanner S, Huanca T, Leonard WR (2008) Maintenance versus growth: investigating the costs of immune activation among children in lowland Bolivia. Am J Phys Anthropol 136:478–484

Newby PK, Tucker KL (2004) Empirically derived eating patterns using factor or cluster analysis: a review. Nutr Rev 62:177–203

NISR, MOH, ICF International (2015) Demographic and health survey 2014–15. National Institute of Statistics of Rwanda (NISR) [Rwanda], Ministry of Health (MOH) [Rwanda], and ICF International, Rwanda, Rockville

Nsabuwera V, Hedt-Gauthier B, Khogali M, Edginton M, Hinderaker SG, Nisingizwe MP, Tihabyona Jde D, Sikubwabo B, Sembagare S, Habinshuti A, Drobac P (2016) Making progress towards food security: evidence from an intervention in three rural districts of Rwanda. Public Health Nutr 19(7):1296–1304

Null C, Stewart CP, Pickering AJ, Dentz HN, Arnold BF, Arnold CD, Benjamin-Chung J, Clasen T, Dewey KG, Fernald LCH, Hubbard AE, Kariger P, Lin A, Luby SP, Mertens A, Njenga SM,

Nyambane G, Ram PK, Colford JM Jr (2018) Effects of water quality, sanitation, handwashing, and nutritional interventions on diarrhoea and child growth in rural Kenya: a cluster-randomised controlled trial. Lancet Glob Health 6(3):e316–e329

Ohtsuka R, Kawabe T, Takasaka K, Watanabe C, Abe T (2012) Human ecology, 2nd edn. University of Tokyo Press, Tokyo. (In Japanese)

Sheeh T, Kolahdooz F, Mtshali TL, Khamis T, Sharma S (2014) Development of a quantitative food frequency questionnaire for use among rural South Africans in KwaZulu-Natal. J Hum Nutr Diet 27:443–449

Solomon D, Aderaw Z, Tegegne TK (2017) Minimum dietary diversity and associated factors among children aged 6-23 months in Addis Ababa, Ethiopia. Int J Equity Health 16(1):181

Stevens GA, Finucane MM, Paciorek CJ, Flaxman SR, White RA, Donner AJ, Ezzati M, Nutrition Impact Model Study Group (Child Growth) (2012) Trends in mild, moderate, and severe stunting and underweight, and progress towards MDG 1 in 141 developing countries: a systematic analysis of population representative data. Lancet 380(9844):824–834

UNICEF (1990) Strategy for improved nutrition of children and women in developing countries. Policy Review Paper E/ICEF/1990/1.6. UNICEF, New York; JC 27/UNICEF-WHO/89.4. New York

UNICEF, WHO, World Bank (2018) Levels and trends in child malnutrition: key findings of the 2018 edition of the joint child malnutrition estimates. United Nations Children's Fund (UNICEF), World Health Organization (WHO), World Bank, Geneva

Victor R, Baines SK, Agho KE, Dibley MJ (2014) Factors associated with inappropriate complementary feeding practices among children aged 6-23 months in Tanzania. Matern Child Nutr 10(4):545–561

WHO (2017a) Nutrition in the WHO African region. World Health Organization, Brazzaville

WHO (2017b) Guideline: protecting, promoting and supporting breastfeeding in facilities providing maternity and newborn services. World Health Organization, Geneva

World Bank (2018) World Bank Rwanda Economic Update. https://www.worldbank.org/en/news/press-release/2018/06/21/world-bank-rwanda-economic-update

World Vision (2017) Annual Review 2017. https://www.wvi.org/sites/default/files/ISH_25030_WV_AR_2017_FA_Digital.pdf. Accessed 29 Nov 2018

Yanagisawa A, Sudo N, Amitani Y, Caballero Y, Sekiyama M, Mukamugema C, Matsuoka T, Imanishi H, Sasaki T, Matsuda H (2016 Jul 6) Development and validation of a data-based food frequency questionnaire for adults in eastern rural area of Rwanda. Nutr Metab Insights 9:31–42

Zazpe I, Sánchez-Tainta A, Toledo E, Sánchez-Villegas A, Martínez-González MÁ (2014) Dietary patterns and total mortality in a Mediterranean cohort: the SUN project. J Acad Nutr Diet 114:37–47

Chapter 2
Weather Shocks, Gender, and Household Consumption: Evidence from Urban Households in the Teso Sub-region, Uganda

Precious Akampumuza, Kasim Ggombe Munyegera, and Hirotaka Matsuda

2.1 Introduction

Weather shocks such as intense or prolonged droughts and floods are becoming more prevalent in sub-Saharan Africa (SSA) due to climate change (IPCC 2018; Niang et al. 2014). At the same time agricultural, the seasons are becoming increasingly unpredictable in the continent (Patricola and Cook 2011), often affecting negatively crop yields and food consumption (Akampumuza and Matsuda 2017) and food security (Wheeler and Von Braun 2013), especially among rural farming communities (Fafchamps and Lund 2003; Dercon et al. 2005; Kurosaki 2006; Lobell and Burke 2009; Lema and Majule 2009; Ringler et al. 2010; Jack and Suri 2014) (Chap. 1 Vol. 1; Chaps. 1, 3 Vol. 2). Such climatic effects are also expected to have negative effects on food production – and especially cereals – (Fraser et al. 2013; McMichael et al. 2007; Parry et al. 1999; Xiong et al. 2010; Rosenzweig and Parry 1994; Parry et al. 2004) and exert upward pressures on food prices (Vermeulen et al. 2012). This could eventually reduce food affordability and calorie availability and increase childhood malnutrition in the region (Jankowska et al. 2012). It is also projected that food security and livelihoods could be affected in SSA due to the loss of access to drinking water (Wheeler and Von Braun 2013). In urban areas, weather shocks often disrupt food flows from rural areas, further affecting food security in the continent (Gasper et al. 2011).

P. Akampumuza (✉)
University of Tokyo, Tokyo, Japan

K. G. Munyegera
Mazima Research Consultancy, Kigali, Rwanda

H. Matsuda
Tokyo University of Agriculture, Tokyo, Japan
e-mail: hm206784@nodai.ac.jp

The ability of households to insure against such weather shocks is limited given the market imperfections in African insurance markets (Alderman and Haque 2007; Townsend 1995). In such contexts, households strive to maintain food security and their livelihoods through different coping strategies such as increasing their partic- ipation in the labor market (Beegle et al. 2006; Ito and Kurosaki 2009), selling livestock (Hoddinott 2006), adjusting grain stocks (Kazianga and Udry 2006), receiving remittances from family members and friends (Jack and Suri 2014; Munyegera and Matsumoto 2016), and diversifying their income sources (Kochar 1999; Porter 2012) (see Chap. 3 Vol. 2).

However, the negative outcomes of weather shocks on household livelihoods and food security are sometimes far too strong to be fully offset through common coping strategies (Akampumuza and Matsuda 2017; Fafchamps et al. 1998; Dercon 2002). The effectiveness of coping strategies is often limited by the technological, environ- mental, and economic constraints faced by the affected households (Fafchamps 1999). It is also noteworthy that climatic and environmental hazards affect urban residents differently, depending on their assets and coping capabilities, which in turn depend on multiple factors such as income (Mendelsohn et al. 2006; Bohle et al. 1994), age (Striessnig et al. 2013), level of education (Muttarak and Lutz 2014), and gender (Akampumuza and Matsuda 2017; Asfaw and Maggio 2018). Thus when designing pro-poor adaptation strategies aimed at protecting individual, household, and community assets and capabilities, it is necessary to understand the factors that give rise to such differentiated vulnerabilities (Muttarak et al. 2016).

The above suggest that securing urban livelihoods and ensuring food security in the context of climate change are major sustainability challenges in SSA. This is because ensuring food security and livelihood resilience, two major sustainability challenges in their own right (Chap. 1 Vol. 1), is further compounded in the context of urbanization and climate change, two of the major changes facing the region (Chap. 1 Vol. 1). This interface spans multiple sustainable development goals (SDGs) such as SDG1 (No poverty), SDG2 (Zero hunger), SDG11 (Sustainable cities and communities), and SDG13 (Climate action), to mention some. This constitutes a multifaceted sustainability challenge for which most SSA countries lack capacity and resources to prepare against (Chap. 5 Vol. 1).

Although many studies have analyzed the effect of (and coping strategies against) weather shocks in SSA (Akampumuza and Matsuda 2017; Jack and Suri 2014; Ito and Kurosaki 2009; Kazianga and Udry 2006; Beegle et al. 2006; Hoddinott 2006), there is little evidence on whether common coping strategies are effective enough to restore the pre-shock levels of livelihoods and/or food security. Besides, most existing studies in SSA have analyzed the impacts of weather shocks in rural contexts (Kazianga and Udry 2006; Dercon et al. 2005), while impacts to urban households remain less studied. Achieving the comprehensive understanding of such phenomena is quite important because rural-urban disparities imply that weather shocks could pose different impacts and available coping strategies to rural and urban households.

Another major knowledge gap is whether female-headed households have differ- ent vulnerability to (and ability to cope with) weather shocks compared to male- headed households (Klasen et al. 2014). Female-headed households in SSA tend to have lower access to productive assets such as land (Deere and León 2003) and education (World Bank 2012), as well as face more restrictive entry requirements into the formal labor market due to prevailing economic and socio-cultural inequality (Contreras and Plaza 2010) (Chap. 1 Vol. 1). Such differences might reduce the ability of urban female-headed households to cope with changes in food consumption due to weather shocks.

Uganda is one of the SSA countries characterized by high gender inequality and urban vulnerability to extreme weather events, which create certain preconditions for the disruption of livelihoods and food security. The climate is generally bimodal, with two rainfall seasons (March–May and October–November) and two dry seasons (June–August and December–February) (Egeru 2012; Mubiru et al. 2012; Nimusiima et al. 2013). However, climate change over the past three decades has affected the onset, offset, and duration of the rainy seasons, making it increasingly unpredictable (Mcsweeney et al. 2010; Funk et al. 2012; Lipper et al. 2014). There is also a notable increase in surface temperature (+1.5 °C between 1960 and 2030), with the number of extremely hot days expected to further increase by 15–43% by 2060. Extreme weather events such as droughts and floods are equally changing in both frequency and severity (Irish Aid 2017). For example, between 2001 and 2011, Uganda experienced five major droughts in 2001, 2002, 2005, 2008, and 2010 (Masih et al. 2014). It has been argued that the increasing occurrence and intensity of droughts and floods also increased socioeconomic risks in a country where 33.2% of the households (representing 43% of the population) were below the international poverty line of USD 1.9 per day in 2015 (World Bank 2018). It has been estimated that between 2006 and 2013, two thirds of Ugandan households that had escaped poverty fell back into it, partly due to weather shocks (World Bank 2018). At the same time, women constitute most of the workforce in agriculture (which is a highly climate-sensitive sector), but do not have the same access to resources compared to males (Hill and Vigneri 2014). Urban livelihoods are also becoming increasingly vulnerable to weather shocks, especially considering the adverse consequences on urban infrastructure (Mcsweeney et al. 2010). In fact, damages in urban infrastructure often contribute to the disruption of food flows from rural areas to urban areas, threatening urban food security (Akampumuza and Matsuda 2017).

The aim of this study is to assess the impact of weather shocks on household consumption (including food-related consumption) and identify the coping strategies employed by urban residents against such shocks. We focus on the Kumi district of Uganda, as it is located in the Teso sub-region, which is one of the most vulnerable sub-regions in the country. We explore four interrelated objectives as follows: (a) the effect of exposure to weather shocks on household welfare (in terms of consumption expenditure); (b) the gender-differentiated impacts of (and response strategies against) weather shocks; (c) the types of coping strategies that the affected households adopt to mitigate potential consumption loss due to weather shocks; and (d) the extent to which the coping strategies effectively safeguard the affected households from consumption declines due to exposure to one or more weather shocks. Section 2.2 explains the key methodological aspects of this study including the study site (Sect. 2.2.1) and the data collection and analysis methods (Sects. 2.2.2

and 2.2.3). Section 2.3 outlines the main results across the objectives outlined above. Section 2.4 identifies the main patterns and outlines the policy and practice implications and recommendations of this study.

2.2 Methodology

2.2.1 Study Site

The Teso sub-region is located in the Eastern Region of Uganda and comprises of eight districts: Kumi, Ngora, Soroti, Serere, Amuria, Bukedea, Kaberamaido, and Katakwi (Fig. 2.1). According to the 2014 population census, the sub-region has a total land area of 13,027 km^2 and is inhabited by 1,819,790 people, implying an average population density of 140 residents/km^2 (Uganda Bureau of Statistics 2014). Relative to other sub-regions, Teso has a high incidence of poverty, with approximately 28% of the inhabitants being categorized as "poor" in 2018, compared to a national poverty rate of 19.7% for the same year (World Bank 2018). Furthermore, the Teso sub-region is one of the most vulnerable sub-regions in Uganda, characterized by frequent floods and prolonged droughts (Akampumuza and Matsuda 2017; Kisauzi et al. 2012; Majaliwa et al. 2015).

Farming is the main economic activity in the sub-region, with the main cultivated crops being cassava, sorghum, millet, sweet potatoes, and groundnuts. Like in many other parts of Uganda, the vast majority of the residents are smallholder farmers relying directly on rain-fed agriculture for livelihoods and subsistence. This has raised concern due to the associated high vulnerability to extreme weather events such as prolonged droughts, floods, and changing and/or unpredictable seasons (see below) (see Chap. 3 Vol. 2).

The main focus of this study is Kumi Town (the administrative capital of the Kumi district) and its surrounding areas.[1] Similar to the Teso sub-region (see above), the Kumi district is one of the most vulnerable districts to extreme weather events in Uganda and especially droughts (UNDP 2014). Kumi district is characterized by a bimodal rainfall pattern, with peaks in April–May and July–August. The annual mean temperature is 24 °C, and the total rainfall is 800–1000 mm. However, the rainfall seasons have become less predictable and less stable over recent years (Sect. 2.1). The risk of droughts, floods, and food insecurity varies substantially within the Kumi district. For example, the northern sub-counties such as Ongino, are at particularly high risk of flooding mainly due to the siltation of Lake Bisina from human activities such as farming and logging. However, there is very low capacity in

---

[1] Kumi Town Council consists of eight local councils 1 (LC1s). LC1s are the lowest administrative units in Uganda's administrative structure and are headed by chairmen. These chairmen are tasked to address village-level issues before they are escalated to higher administrative levels (i.e., parish, sub-county, county, and district in ascending order) if critical decisions cannot be taken at the LC1 level.

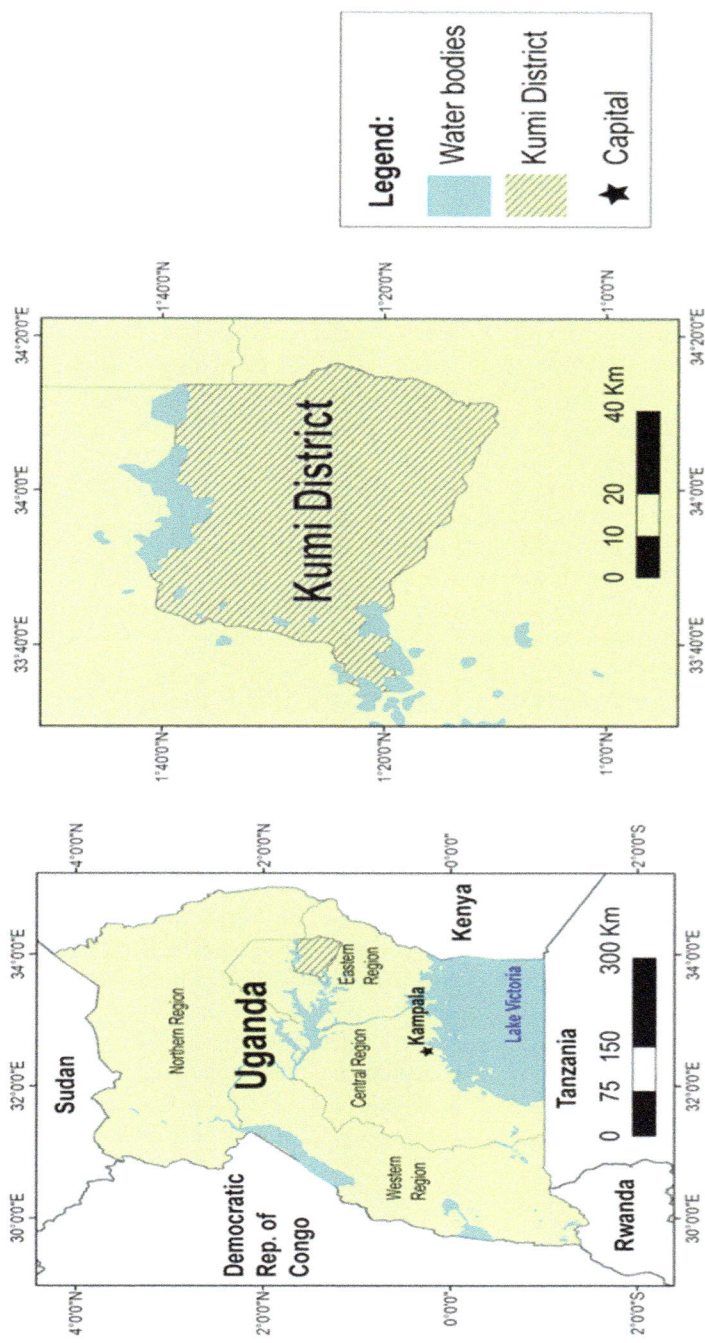

Fig. 2.1 Location of the study site

the district to maintain, operate, analyze, interpret, and predict weather data. In particular, there is no resident weather expert, with the closest expert assisting with weather information in the region being based at the Serere Agricultural and Animal Research Institute (SAARI), located 45.9 km away.

Frequent droughts and floods (due to erratic rainfall with heavy storms) have become common in Kumi district and often result into crop loss and the destruction of infrastructure. In particular the sub-region has experienced five major droughts between 1990 and 2010, specifically in 1998, 1999, 2002, 2005, and 2008. In one of the biggest floods that swept through Eastern Africa and the Horn of Africa in 2007, about 30 people died, and an estimated 8500 acres (3440 ha) of cropland was affected in the sub-region. On the aftermath of the flood, many smallholder farmers experienced bad harvests, which clearly contributed to the outbreak of acute famine in 2008 and the further deterioration of food security in many villages across the sub-region.[2] Even as late as mid-2008, approximately 135,987 people in 4 highly affected districts (i.e., Amuria, Katakwi, Bukedea, Soroti) needed food assistance due to crop failure, poor harvest, and surging food prices (Rukandema et al. 2008). Shortly after, the Teso region experienced a prolonged drought in 2009, which was followed by a short drought and a flood in 2010.

Reports reveal the tremendous decline in food production in the Teso region during (and following) these flood and drought events, as well as the reduction of food flows to Kumi Town from nearby villages following these weather shocks. This reduction of food flows to Kumi Town is often associated with the combined effect of crop loss in surrounding villages (i.e., reduced food production) and the disruption of communication channels due to road destruction from flooding. Outbreaks of crop and animal diseases have also been linked to weather shocks and have posed major challenges to smallholder farmers in the district. Common animal epidemics include swine fever, foot and mouth disease, nagana, and bird flu, while common crop diseases include coffee wilt, banana bacterial wilt, cassava mosaic, and cassava brown streak disease.

Gender issues permeate smallholder farming in Kumi district. Women provide more than 70% of agricultural labor in Kumi, yet only 30% have control over means of production, and only 7% own land (KUMI District Hazard, Risk and Vulnerability Profile Report 2014). Given their limited access to agricultural resources and low decision-making power, women are possibly more vulnerable to the numerous weather shocks discussed above. However, despite their relatively higher vulnerability, women still spend more on education and health than their male counterparts. For example, according to the 2014 KUMI District Hazard, Risk and Vulnerability Profile (UNDP 2014), women and children tend to seek health services more often

---

[2]Famine is defined as the sudden and sharp reduction in food supply resulting in widespread hunger (Buringh 1977). Other scholars define famine as a sudden collapse in the level of food consumption of a large population (Scrimshaw 1987) or a set of conditions that occurs when a large population in a region cannot obtain sufficient food, resulting in widespread, acute malnutrition (Cuny and Hill 1999).

than men, with 67% of the total outpatient department attendance for persons above 5 years of age in 2013 in the district being female.

2.2.2   Data Collection

To tackle the four research objectives outlined in Sect. 2.1, we collected between February and March 2015 (a) structured household surveys, (b) focus group discussions (FGDs), and (c) secondary datasets from relevant organizations.

First, we randomly selected 25 households through transect walks in the urban and peri-urban areas of each of the eight LC1s (N = 200). These households were surveyed through structured questionnaires that targeted the main decision-maker (i.e., the household head). These household surveys aimed at capturing household welfare dynamics and how they evolve amidst weather shocks. For most of the key variables, retrospective data was also collected for 2009, when a major drought affected the Teso sub-region (and most parts of the Horn of Africa) (Sect. 2.2.1). Thus, the main analysis presented in this chapter is based on a random sample of 200 households, constituting a quasi-panel of 400 household-year observations (i.e., for the year 2009 and 2015).

The main variables included in the household questionnaire were expenditure, exposure to each of the four weather-related shocks (i.e., drought, flood, pests, and diseases), household assets, land endowment, distance to market, and demographic characteristics (e.g., household size, gender, level of schooling, the age of the household head). We broadly categorized consumption expenditure into three main categories: (a) food and food-related items, (b) semi-durable household items referred to as basics (e.g., education, health, transportation, clothing, cooking/lighting items, fuel), and (c) contributions to socio-cultural and religious activities (e.g., funeral, wedding, churches, mosques) (hereafter called contributions).

Due to concerns over data reliability for questions using recall periods, we allowed respondents to use different recall periods of each of the key household variables. For example, for food consumption expenditure, the surveyed households were asked to estimate the monetary value of the consumed food items in the past 7 days prior to the survey (for 2015) and in a typical week (for 2009). For food-related expenditures such as sugar, salt, beverages, and tobacco, we established a monthly recall period. To capture properly the food and food-related expenditures, respondents were asked to estimate the market value of any self-produced items such as food crops from their family farms. Basic expenditures, contributions, as well as savings [e.g., to Savings and Credit Cooperative Organizations (SACCOs)] were asked for the year before the survey.

Additionally we conducted three FGDs as follows: (a) one FGD with eight LC1 chairpersons; (b) one FGD with food transporters and suppliers, and (c) one FDD with market vendors. Each FGD consisted of five participants and provided information about broader phenomena in the study areas such as the dynamics of weather shocks and food security at LC1 and district levels. Lastly, we gathered secondary data that mainly relates to food production in the study area. This information was collected through visits to relevant district offices and agencies, such as those related to agriculture and the environment.

2.2.3   Data Analysis

2.2.3.1   Weather Shocks and Household Consumption Expenditures

We presume a linear association between consumption expenditure and household-level covariates following the specification below (Eq. 2.1):

$$Cons_{ijt} = \alpha + f_j + f_t + \beta_1 H_{ijt} + \beta_2 Shock_{ijt} + \varepsilon_{ijt} \qquad (2.1)$$

...where $Cons_{ijt}$ is the aggregate real expenditure on all consumption categories (i.e., food-related and basic expenditures and social contributions, Sect. 2.2.2). Subscripts i, j, and t indicate household, LC1, and year, respectively; H_{ijt} is a vector of household characteristics (i.e., age, gender, years of schooling of the household head, land and household asset endowments, and household size). f_j and f_t are dummy variables that, respectively, capture LC1 and bimodal effects., $Shock_{ijt}$ is a binary variable that takes the value one if a household was affected by at least one of the weather shocks and zero otherwise.

In order to investigate potential reallocation among expenditure categories (especially when a household is affected by a weather shock), we also estimate separate regressions for expenditures on each of these consumption components. By expressing household consumption expenditure in adult equivalent units (rather than per capita units), this allows us to adjust for differences in expenditure needs due to the demographic composition of households. Otherwise, this would account for part of the observed consumption difference between affected and unaffected households. We then deflate all consumption values using the consumer price indices (CPIs) obtained from the Uganda Bureau of Statistics (UBOS) for the 2 years used in this analysis (i.e., 2009 and 2015). For both 2009 and 2015, the UBO uses a common financial year (2005/2006) as the base year for the CPIs, which facilitates the comparability of our results across survey years.

The parameter β_2 is expected to have a negative sign to reflect the expected negative impact of exposure to shocks on household consumption. It is also expected that the coefficient for the female headship dummy will be negative, as the literature already points to high female vulnerability to poverty (Klasen et al. 2014). Household size is presumed to have a negative sign because a larger number of household members are expected to place a marginal burden on the household budget (Akampumuza and Matsuda 2017; Munyegera and Matsumoto 2016). On the contrary, we expect a positive sign for education of the household head because education generally increases the chance of paid employment, which, in turn, increases consumption and ability to cope with shocks (Muttarak and Lutz 2014).

Similarly, asset and land endowments ought to boost household consumption and are expected to have positive signs (Munyegera and Matsumoto 2016).

To enhance the reliability of the results and ameliorate heteroscedasticity concerns that would affect the results, we report robust, heteroscedasticity-free standard errors in all regression specifications. Given the quasi-panel nature of the data used in this study (Sect. 2.2.2), various analytical methods can be used, including household fixed effects and random effects. However, for brevity, we present only the fixed effects results. Fixed effects are a better method based on a formal Hausman test for model selection. The two models differ only in the assumption made about the association between household-specific characteristics and covariates. The random effects approach assumes no such association, while the fixed effects approach assumes that individual households have unobserved constant attributes that could be correlated with covariates and in turn affect the relationship between covariates and the dependent variable.

The null hypothesis of the Hausman test is that there is no systematic difference between fixed effects and random effects estimates. A p-value lower than 0.05 points to the rejection of the null hypothesis, implying that the estimates of the two models are significantly different. This difference could be attributed to the effect of unobserved fixed characteristics, which could affect the results. In that case, as a rule of thumb, fixed effects estimation should be adopted to smooth out their potentially confounding effect.

2.2.3.2 Heterogeneous Shock Impacts and Coping Strategies by Gender of the Household Head

One critical factor mediating the potentially heterogeneous impacts of weather shock is the gender of the household head. This is reflected by Eq. 2.2, which is an extension of Eq. 2.1 outlined in Sect. 2.2.3.1:

$$Cons_{ijt} = \alpha + f_j + f_t + \beta_1 H_{ijt} + \beta_2 Shock_{ijt} + \lambda_1 Fehead_{ijt}$$
$$+ \lambda_2 Fehead_{ijt} \text{X} Shock_{ijt} + \varepsilon_{ijt} \tag{2.2}$$

...where $Fehead_{ijt}$ is a dummy variable that takes the value one if the household head is female and zero otherwise. The coefficient λ_2 on the interaction term between the female head dummy and the shock dummy captures the potentially heterogeneous impact of shocks by gender of the household head. It is expected to have a negative sign, which reflects the relatively higher vulnerability of female-headed households, and is indicative of the higher consumption poverty of female-headed households (see Klasen et al. 2014). In other words, the food consumption decline due to shock exposure is expected to be larger among female-headed households than among their male-headed counterparts. All other explanatory variables are as explained in Eq. 2.1.

2.2.3.3 Coping Strategies Against Weather Shocks

We presume that households respond to exposure to weather shocks by adopting one (or more) coping strategy. We adopt a probit model (Eq. 2.3) to estimate the likelihood of the household adopting a particular coping strategy, conditional on shock exposure, and other covariates.

$$Strategy_{ijt}^k = \theta_0 + f_j + f_t + \theta_1 H_{ijt} + \theta_2 Shock_{ijt} + v_{ijt} \tag{2.3}$$

…where $Strategy_{ijt}^k$ is a binary indicator taking one if the *i-th* household in LC1 j adopts a coping strategy k in year t. Coefficient θ_2 captures the extent to which shock exposure induces the household to adopt in a certain coping strategy. It is expected to have a positive sign, indicating that the occurrence of a weather shock ideally induces the affected household to adopt the respective coping strategies.

We investigated coping strategies that are commonly cited in the literature. These include the receipt of remittances from family members and friends (Jack and Suri 2014), off-farm employment (Ito and Kurosaki 2009), livestock sales (Kazianga and Udry 2006), and household assets (Akampumuza and Matsuda 2017). All explanatory variables are as explained in Eq. 2.1.

2.2.3.4 Effectiveness of Coping Strategies

Section 2.2.3.3 assumes that households respond to weather shocks by adopting a particular coping strategy and that this would be a positive signal of its ability to offset the effects of the shock on consumption expenditure. However, it is quite possible that even after adopting and employing coping strategies, the affected households may still experience a significant decline in household consumption (Fafchamps 1996). We therefore slightly modify Eq. 2.1 to assess the effectiveness of each of the studied coping strategies (Eq. 2.4):

$$Cons_{ijt} = \alpha + f_j + f_t + \beta_1 H_{ijt} + \beta_2 Shock_{ijt} + \mu Strategy_{ijt}^k$$
$$+ \psi Strategy_{ijt}^k \, XShock_{ijt} + \varepsilon_{ijt} \tag{2.4}$$

…where the additional term is an interaction between the shock dummy and dummies for each coping strategy k, with its coefficient ψ capturing the effectiveness of the coping strategies. A positive and statistically significant coefficient would imply that the particular coping strategy is effective in offsetting the negative impact of the shock on consumption expenditure. Similarly, a significant negative or insignificant coefficient would suggest that the coping strategy either exacerbates consumption decline or does not offer any significant protection to consumption against weather shocks.

2.2.3.5   Robustness Check and Propensity Score Matching

The observed differences in consumption expenditure and coping strategies against weather shocks between households affected and unaffected by weather could be due to severe differences between household characteristics across the two categories. The household fixed effects results (Sect. 2.3.3) ameliorate time-invariant unobserved heterogeneity, but do not rigorously smooth out differences in time-variant observed heterogeneity, which, if present, could confound our results. To overcome this challenge and appropriately attribute consumption differences between households affected and unaffected by weather shocks, we perform propensity score matching (PSM) to identify comparable counterpart households between the two household categories along observed characteristics (Rosenbaum and Rubin 1985). In non-randomized observational studies where covariate balance is often challenging, PSM provides a remedy to the treatment (shock exposure in this case) that is potentially prone to selection bias (Morgan 2017).

2.3   Results

2.3.1   Key Regional Patterns

According to the FGDs, food production declines whenever the Kumi district is affected by weather shocks due to crop loss and crop failure. Furthermore, crucial infrastructure is destroyed, which cuts off food supply routes from the surrounding villages to Kumi Town. During these occasions, food prices immediately skyrocket, especially in towns, as the reduced food production in nearby villages necessitates costly food imports from distant districts. Additionally the infrastructure breakdown exacerbates transportation costs.

   The food production data obtained from the district agriculture office corroborated the above results from the FGDs. Figure 2.2 shows the food production trends for the Kumi district between 2005 and 2013 for the main food crops such as rice, finger millet, sorghum, sweet potatoes, groundnuts, and cassava. It is important to note the sharp decline in food production for each of the seven crops between 2010 and 2013, with the combined production decreasing substantially from well above 120,000 t to below 30,000 t for that period. As mentioned in Sect. 2.2.1, the region experienced a major drought in 2009 and a severe flood in 2011. FGD participants suggested that these extreme weather events could be partially responsible for the observed decline in food production.

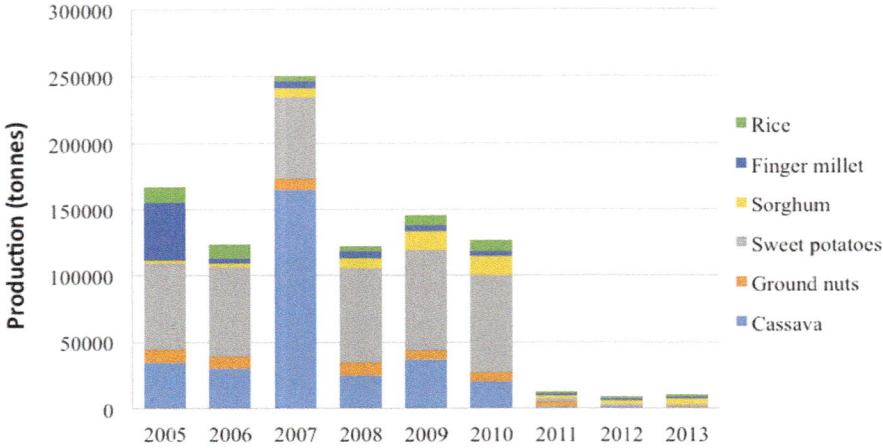

Fig. 2.2 Production of major food crops in the Kumi District (2005–2013)
Note: Based on food production data from Kumi District Office

2.3.2 Main Household Characteristics

Table 2.1 outlines household characteristics based on whether households were reported of having been affected by any of the four weather shocks studied in this chapter. Households that were affected by any (or a combination) of these weather shocks tend to have significantly larger household sizes ($p < 0.05$). Furthermore, a higher proportion of female-headed households have been affected by weather shocks, although this difference is not statistically significant at conventional levels. There are also differences in the age and education level of the household heads in relation to exposure to weather shocks. Household heads that self-reported exposure to weather shocks are more than 3 years older and completed 1 year less of schooling relative to the household heads of unaffected households.

The monthly per capita consumption expenditure of unaffected households by weather shocks is UGX 179,902 (USD 62), which is approximately 1.5 times larger than that of self-reported weather shock victims (UGX 117,069, USD 41) ($p < 0.05$).[3] Also remittance flows are significantly different between households affected and unaffected by weather shocks ($p < 0.05$), with the proportion of remittance recipients being twice as high among affected households. This could partly reflect the receipt of remittances from family members and friends as an ex-post strategy to cope with weather shocks.

Table 2.2 provides summary statistics stratified by the gender of the household head. In this respect, it crudely illustrates differential vulnerability to weather shocks and heterogeneity in key household characteristics between households headed by

---

[3]The official exchange rate used in this study was obtained from the Bank of Uganda. The exchange rate around the survey month (February 2015) was around USD 1 = UGX 2890.

Table 2.1 Summary statistics by exposure to weather shocks

| Variables | Affected | | Unaffected | | |
	Mean	SD	Mean	SD	Difference
Distance to market (km)	0.91	0.41	0.82	0.46	0.09
Household size (people)	6.24	3.20	5.24	3.48	0.99***
Savings (1 = belongs to SACCO)	0.36	0.48	0.27	0.45	0.08*
Age of household head (years)	39.05	12.12	35.73	12.18	3.32**
Gender of household head (1 = female head)	0.26	0.44	0.21	0.41	0.05
Education of household head (years of schooling)	9.97	5.14	11.34	5.72	−1.36**
Education of household head (1 = head attended secondary school or higher)	0.58	0.03	0.58	0.04	0.00
Food-related expenditure (UGX/month/person)	39,020	44,336	37,337	55,531	1682
Basic expenditure (UGX/month/person)	37,947	52,394	66,631	88,068	−28,684***
Social contributions (UGX/month/person)	13,395	19,130	26,716	80,373	−13,320**
Total expenditure (UGX/month/person)	117,069	103,236	179,902	297,538	−62,833***
Asset value ('000 UGX)	4029	11,325	3324	8774	705
Asset ownership (1 = owns mobile phone)	0.76	0.43	0.74	0.44	0.02
Remittances (1 = received remittance)	0.42	0.49	0.24	0.43	0.17***
Number of observations	262	–	138	–	–

Note: Student t-test is used to establish the significance of difference in the means of key variables between households affected and unaffected by shocks ***$p < 0.01$; **$p < 0.05$; *$p < 0.1$

males and females. Approximately 65% and 70% of male-headed and female-headed households, respectively, reported having been exposed to weather shocks in the past 5 years.

There is a significant difference in the years of schooling, with female household heads having 3 years of schooling less compared to their male counterparts. Additionally, while 65% of male household heads attained at least secondary school, only 39% of female household heads achieved so. This reflects common patterns of gender gap in education that are observed elsewhere in SSA (World Bank 2012) (see Chap. 1 Vol. 1). Nonetheless, the overall literacy level in our sample (11.35 years for male heads, 7.74 years for female heads) is notably higher than the average years of schooling reported nationally in 2012 (4.7 years) (UNDP 2013). Female household heads are also significantly older and spend more on household basics, although this difference in expenditure is marginally significant.

Table 2.2 Summary statistics by gender of the household head

Variables	Male-headed		Female-headed		
	Mean	SD	Mean	SD	Difference
Distance to market (km)	0.86	0.44	0.94	0.38	−0.08
Household size (number of members)	5.90	3.41	5.93	3.06	0.02
Savings (1 = belongs to SACCO)	0.33	0.47	0.35	0.48	−0.03
Age of household head (years)	36.87	11.28	41.17	14.20	−4.30**
Education of household head (years of schooling)	11.35	4.99	7.74	5.51	3.61***
Education of household head (1 = head attended secondary school or higher)	0.65	0.02	0.39	0.05	0.26***
Exposure to weather shock (1 = experienced drought)	0.55	0.50	0.60	0.45	−0.04
Exposure to weather shock (1 = experienced flood)	0.15	0.36	0.14	0.35	0.007
Exposure to weather shock (1 = experienced pests/diseases)	0.42	0.49	0.44	0.50	−0.02
Food-related expenditure (UGX/month/person)	39,872	51,466	34,518	35,653	−5354
Basic expenditure (UGX/month/person)	43,215	67,552	56,466	62,302	−13,250*
Social contributions (UGX/month/person)	19,578	53,921	11,157	14,961	8420
Total expenditure (UGX/month/person)	138,281	207,048	129,958	111,078	8323
Asset value ('000 UGX)	4292	11,530	2276	6447	2016
Asset ownership (1 = owns mobile phone)	0.77	0.42	0.72	0.45	0.04
Remittances (1 = received remittances)	0.35	0.48	0.38	0.49	−0.02
Exposure to weather shock (1 = experienced weather shock)	0.65	0.48	0.70	0.46	−0.05
Livestock ownership (number)	8.12	1.49	5.61	0.76	2.51
Livelihood sources (1 = household member involved in off-farm employment)	0.43	0.04	0.17	0.05	0.27***
Income (UGX/year/person)	830,442	147,230	248,560	42,928	581,882**
Number of observations	302	–	98	–	–

Note: The student t-tests were used to establish the significance of difference in the means of key variables between male-headed and female-headed households ***$p < 0.01$; **$p < 0.05$; *$p < 0.1$

There are also some notable differences in asset ownership and monthly per capita consumption. Female-headed households report a lower value of household assets and expenditures on food and social contributions, albeit these differences are statistically insignificant. Female-headed households are less likely to engage in the off-farm employment, and especially waged employment, and thus report significantly lower per capita income than male-headed households. Additionally, female-headed households own significantly fewer livestock, which could further curtail their ability to offset the adverse effects of weather shocks through livestock sales. This implies that fewer coping options are available to female-headed households, possibly influencing them to adopt more severe coping strategies such as missing meals and reallocating from other expenditure components such as education.

Table 2.3 categorizes households by educational attainment of the household head (secondary education or higher). The main economic activities carried out in the study area are broadly classified as off-farm paid labor, although some households are engaged in smallholder farming in nearby villages. However, further disaggregation reveals striking heterogeneity in access to opportunities that could potentially augment the ability to cope with weather shocks. First, households whose heads have secondary education (and above) are more likely to have at least one member working in the off-farm sector, particularly regular waged jobs. They also earn significantly higher per capita income than households headed by members with primary education or lower. In an urban setting where off-farm employment is the most usual and important source of livelihoods, it is thus not surprising that households headed by less educated members earn significantly lower per capita income and are more likely to report missing meals after experiencing a weather shock.

2.3.3 Impact of Shocks on Household Consumption Per Adult Equivalent

We find a negative and significant association between self-reported exposure to weather shocks and real household consumption per adult equivalent. We first present the OLS estimates of this association in Table 2.4. Column 1 reveals exposure to at least one of the weather shocks is associated with a 15.1% decline in real household consumption per adult equivalent. Disaggregating consumption into food-related expenditures, basic expenditures, and social contributions indicates that weather shocks have no significant impact on food-related consumption. Rather households seem to reallocate basic expenditures and social contribution to supplement food consumption. In fact, exposure to weather shocks reduces expenditure on basics and social contributions by 38% and 40%, respectively.

The education level of the household head augments household consumption, which perhaps indicates their relatively better access to productive resources and income opportunities including paid employment. Likewise, wealthy households in terms of asset ownership have significantly higher expenditures on non-food basics and social contributions. For every 1% increase in the value of household assets, expenditure on the two consumption categories increases by 8.2% and 14.1%, respectively. On the contrary, household size reduces consumption expenditure, which indicates the financial burden of maintaining larger households.

Table 2.3 Summary statistics by education level of the household head

Variables	Primary or lower		Secondary or higher		Difference
	Mean	SD	Mean	SD	
Distance to market (km)	0.79	0.35	0.91	0.46	0.12
Exposure to weather shock (1 = experienced weather shock)	0.73	0.44	0.66	0.48	0.07
Livestock ownership (number)	4.72	20.02	3.44	5.17	1.28
Household size (number of members)	6.25	0.29	5.67	0.20	0 0.58*
Savings (1 = belongs to SACCO)	0 0.30	0.03	0.35	0.03	−0.06
Age of household head (years)	39.67	1.13	36.87	0.71	2.80**
Food-related expenditure (UGX/month/person)	29,760	3074	44,476	3614	−14,716***
Basic expenditure (UGX/month/person)	42,390	5725	49,490	4350	−7100
Social contributions (UGX/month/person)	12,290	2050	20,782	3897	−8492*
Total expenditure (UGX/month/person)	114,700	11,561	150,186	14,487	−35,486*
Asset value ('000 UGX)	3947	1153	3682	398	264
Asset ownership (1 = owns mobile phone)	0.62	0.03	0.85	0.02	−0.23***
Remittances (1 = received remittance)	0.21	0.03	0.46	0.03	−0.25***
Livelihood sources (1 = household member involved in off-farm employment)	0.44	0.03	0.62	0.03	−0.18***
Income (1000UGX/year/person)	730	131	2518	286	−1787***
Weather shock impact (1 = missed meals during weather shock)	0 0.89	0.07	0.40	0.11	0.49***
Number of observations	169	–	231	–	–

Note: The student t-tests were used to establish the significance of difference in the means of key variables between households whose heads attained at least secondary education and households headed by members that attained primary education and below ***$p < 0.01$; **$p < 0.05$; *$p < 0.1$

Table 2.4 Relation between weather shocks and household consumption using ordinary least squares (OLS)

Variables	(1) Log (cons)	(2) Log (food)	(3) Log (basics)	(4) Log (contribution)
Exposure to weather shock (1 = experienced weather shock)	−0.151** (0.0699)	0.0205 (0.136)	−0.385*** (0.128)	−0.408** (0.207)
Dependency ratio	0.0390 (0.0341)	0.148** (0.0629)	0.0222 (0.0646)	0.0157 (0.0851)
Age of household head (years)	0.00223 (0.0122)	−0.0269 (0.0312)	0.0196 (0.0278)	0.0826** (0.0381)
Age squared of household head (years^2)	5.88e-05 (0.000136)	0.000318 (0.000343)	4.36e-05 (0.000319)	−0.000725* (0.000428)
Gender of household head (1 = female)	0.111* (0.0639)	−0.124 (0.139)	0.506*** (0.125)	−0.110 (0.182)
Education of household head (years of schooling)	0.0216*** (0.00622)	0.0241** (0.0111)	0.0402*** (0.0120)	0.0476*** (0.0156)
Asset value (log)	0.0367 (0.0328)	0.0339 (0.0440)	0.0829** (0.0363)	0.141** (0.0696)
Asset ownership (1 = owns mobile phone)	−0.102 (0.104)	−0.184 (0.179)	−0.0851 (0.188)	0.199 (0.289)
Household size (number of members)	−0.0345*** (0.0100)	−0.0517*** (0.0189)	−0.0210 (0.0220)	−0.0390 (0.0277)
Livelihood sources (1 = household member involved in off-farm employment)	0.0597 (0.0625)	−0.0405 (0.129)	0.0314 (0.118)	0.521*** (0.190)
Year (2015)	0.511*** (0.0629)	2.744*** (0.132)	0.0333 (0.115)	0.371** (0.187)
Constant	10.52*** (0.441)	8.194*** (0.805)	7.342*** (0.709)	3.439*** (1.092)
Observations	376	376	376	376
R-squared	0.359	0.676	0.272	0.286

Note: Robust standard errors are reported in parentheses. Dummy variables for community and year are controlled for in all specifications for location- and time-specific effects ***$p < 0.01$; **$p < 0.05$; *$p < 0.1$

The OLS results are corroborated by the fixed effects estimates indicated in Table 2.5. The smaller coefficients are suggestive of a positive bias due to confounding time-invariant household characteristics in the OLS estimates which was smoothed out by fixed effects estimation. Although we do not find systematically lower aggregate consumption among female-headed, their food-related expenditures and social contributions are significantly lower. This could indicate their relative higher vulnerability to poverty and shocks.

Table 2.5 Relation between weather shocks and household consumption using fixed effects

Variables	(1) Log (consumption)	(2) Log (food)	(3) Log (basics)	(4) Log (contribution)
Exposure to weather shock (1 = experienced weather shock)	−0.132* (0.0743)	0.00362 (0.211)	−0.185** (0.0882)	−0.206 (0.205)
Gender of household head (1 = female)	−0.163 (0.139)	−1.005*** (0.336)	−0.264 (0.191)	−0.432* (0.261)
Age of household head (years)	0.0312 (0.0230)	−0.0218 (0.0600)	0.0398 (0.0255)	0.159*** (0.0467)
Age squared of household head (years^2)	−0.000352 (0.000415)	0.000766 (0.00110)	−0.000588 (0.000495)	−0.00248*** (0.000893)
Education of household head (years of schooling)	−0.00575 (0.0391)	−0.00328 (0.101)	−0.0351 (0.0337)	0.00286 (0.0793)
Asset value (log)	−0.0112 (0.0421)	0.00738 (0.0704)	−0.0241 (0.0270)	−0.0339 (0.0617)
Asset ownership (1 = owns mobile phone)	−0.0431 (0.114)	−0.257 (0.319)	−0.00737 (0.165)	0.194 (0.265)
Household size (number of members)	−0.00333 (0.0159)	−0.00715 (0.0435)	0.00705 (0.0216)	−0.0286 (0.0395)
Livelihood sources (1 = household member involved in off-farm employment)	−0.0562 (0.0841)	−0.0808 (0.201)	−0.0550 (0.102)	0.180 (0.221)
Constant	11.11*** (0.822)	8.540*** (1.805)	9.886*** (0.654)	6.181*** (1.410)
Observations	376	376	376	376
R-squared	0.487	0.798	0.097	0.256
Number of households	188	188	188	188

Note: Robust standard errors are reported in parentheses. Dummy variables for community and year are controlled for in all specifications for location- and time-specific effects ***$p < 0.01$; **$p < 0.05$; *$p < 0.1$

2.3.4 Heterogeneity of the Impacts of Shocks by Gender of the Household Head

We explore potential heterogeneities in the impact of weather shocks by gender of the household head. The results suggest that female-headed households are more severely affected by exposure to weather shocks, with the difference being significant ($p < 0.05$). Table 2.6 contains fixed effects estimates of Eq. 2.2 (Sect. 2.2.3.3) for both total and disaggregated consumption. The negative and statistically significant coefficient on the interaction term between the female headship dummy and the shock exposure dummy indicates that (conditional on shock exposure) the consumption decline is larger for female-headed households than male-headed counterparts. This reflects studies, which found that the drought-induced consumption decline in Ethiopian villages was significantly more pronounced among female-headed households (Dercon et al. 2005).

Table 2.6 Heterogeneous weather shock impacts by gender of the household head

Variables	(1) Log (consumption)	(2) Log (food)	(3) Log (basics)	(4) Log (contribution)
Exposure to weather shock (1 = experienced weather shock)	−0.0590** (0.0300)	−0.0737** (0.036)	−0.316** (0.146)	−0.285 (0.244)
Gender of household head (1 = female)	0.276* (0.142)	−0.486* (0.293)	0.751*** (0.279)	0.0740 (0.341)
Interaction of female head dummy and shock dummy	−0.186** (0.087)	0.586* (0.319)	−0.263** (0.125)	−0.180* (0.411)
Education of household head (years of schooling)	0.0236*** (0.00648)	0.0247** (0.0118)	0.0429*** (0.0119)	0.0562*** (0.0165)
Household size (number of members)	−0.0351*** (0.0109)	−0.0493** (0.0216)	−0.0178 (0.0225)	−0.0472* (0.0286)
Age of household head (years)	0.00370 (0.0122)	−0.0261 (0.0327)	0.0186 (0.0295)	0.0870* (0.0451)
Savings (1 = belongs to SACCO)	0.177*** (0.0673)	0.123 (0.131)	0.270** (0.117)	0.970*** (0.180)
Age squared of household head (years^2)	3.55e-05 (0.000135)	0.000280 (0.000359)	4.03e-05 (0.000339)	−0.000795 (0.000504)
Asset value (log)	0.0304 (0.0312)	0.0313 (0.0433)	0.0761** (0.0341)	0.135** (0.0618)
Asset ownership (1 = owns mobile phone)	−0.139 (0.104)	−0.219 (0.182)	−0.160 (0.197)	0.0608 (0.277)
Constant	10.28*** (0.388)	8.136*** (0.895)	7.238*** (0.700)	3.151*** (1.073)
Observations	376	376	376	376
R-squared	0.384	0.678	0.300	0.348

Note: Robust standard errors are reported in parentheses. Dummy variables for community and year are controlled for in all specifications for location- and time-specific effects ***$p < 0.01$; **$p < 0.05$; *$p < 0.1$

As noted above, the likelihood of exposure to shocks does not significantly differ with the gender of the household head. This implies that the observed differences in the impact of weather shocks would be indicative of the relative inability of female-headed households to cope with the shocks. A partial confirmation of this premise is that female-headed households are less likely to engage in the off-farm sector (especially waged employment) and that they own significantly fewer livestock compared to male-headed households. This implies that fewer coping strategies are available to female-headed household, possibly leading female-headed households to adopt more severe coping strategies such as missing meals (Sect. 2.3.1). In fact, female-headed households report significantly fewer years of schooling and are less likely to have household members engaged in off-farm employment. Column 3 in Table 2.6 further reveals that female-headed households spend significantly more money on basic expenditure including education and health.

2.3.5   Exposure to Shocks and Livelihood Coping Strategies

In this section, we investigate whether exposure to weather shocks influences the affected households to adopt more in ex-post coping strategies. Results reveal a positive and significant relationship between exposure to weather shocks, on one hand, the likelihood of engagement on waged employment, and receipt of credit/ remittances on the other hand (Table 2.7). Since the outcome variables are binary

Table 2.7 Exposure to shocks and likelihood of adopting individual coping strategies

Variables	(1) Sell livestock	(2) Sell assets	(3) Work off-farm	(4) Receive credit	(5) Receive remittance
Exposure to weather shock (1 = experienced weather shock)	0.0786 (0.0547)	0.00552 (0.0522)	0.213*** (0.0637)	0.142** (0.0603)	0.143** (0.0579)
Dependency ratio	0.0274 (0.0284)	0.0521* (0.0282)	0.0119 (0.0348)	−0.0522 (0.0326)	0.0196 (0.0333)
Household size (number of members)	0.0225** (0.00915)	−0.00385 (0.00780)	−0.00286 (0.0102)	0.0102 (0.0108)	−0.00380 (0.00967)
Savings (1 = belongs to SACCO)	0.0792 (0.0558)	0.000329 (0.0499)	0.0777 (0.0616)	0.200*** (0.0615)	0.0455 (0.0603)
Age of household head (years)	0.000130 (0.0126)	0.0140 (0.0115)	0.0172 (0.0140)	0.0499*** (0.0166)	−0.00795 (0.0125)
Age squared of household head (years²)	−8.12e-05 (0.000153)	−0.000125 (0.000132)	−0.000216 (0.000159)	−0.000575*** (0.000193)	0.000145 (0.000145)
Gender of household head (1 = female)	−0.0496 (0.0592)	0.112* (0.0648)	0.108 (0.0700)	0.0588 (0.0714)	0.0645 (0.0688)
Education of household head (years of schooling)	−0.00112 (0.00523)	0.00533 (0.00397)	0.0138** (0.00617)	0.00418 (0.00572)	0.0188*** (0.00596)
Asset value (log)	0.0215* (0.0122)	0.00297 (0.00755)	0.0395*** (0.0125)	0.0503*** (0.0163)	0.0308*** (0.0106)
Asset ownership (1 = owns mobile phone)	0.0557 (0.0725)	−0.188** (0.0818)	0.116 (0.0866)	0.00622 (0.0804)	0.0745 (0.0774)
Log-likelihood	−162.23	−108.62	191.87	199.78	206.63
Pseudo R-squared	0.1250	0.1382	0.1853	0.1739	0.1236
Observations	345	345	345	345	345

Note: Robust standard errors are reported in parentheses. Dummy variables for community and year are controlled for in all specifications for location- and time-specific effects ***$p < 0.01$; **$p < 0.05$; *$p < 0.1$

indicators (i.e., whether or not a household adopts a particular coping strategy), we estimate probit regressions and present their marginal effects in Table 2.7.

Columns 1 and 2 in Table 2.7 indicate that exposure to weather shocks is associated with an increased, albeit statistically insignificant, likelihood of selling livestock. This is contrary to previous studies that have found a positive and statistically significant effect of weather shock exposure on the likelihood of engaging in livestock sales (Hoddinott 2006). This difference is possibly due to the relatively higher representation of urban residents in our sample, whose main source of livelihood is off-farm employment (combined with minimal livestock ownership). In fact, results in Column 3 support this conjecture, as exposure to weather shocks increases by 21% the probability that at least one of the household members is engaged in paid off-farm employment (Table 2.7).

Households with a high dependency ratio are more likely to sell household assets when experiencing weather shocks, perhaps due to their relative lack of alternative coping strategies. Columns 4 and 5 reveal that the probability of borrowing (from formal and/or informal sources) and receiving remittances (from family members/ friends) increases by 14% following exposure to weather shocks. Membership to a savings and credit association (SACCO) increases borrowing probability, something that is consistent with the main functions of such groups (i.e., rotational saving and borrowing).

Wealth, in terms of asset ownership, augments the likelihood of adopting all types of coping strategies, with the exception of asset sales. This further confirms that relatively poorer households with no access to alternative coping strategies resort to selling household assets to cope with the impacts of weather shocks. Column 3 of Table 2.7 further reveals that for each additional year of schooling of the household head, the probability of at least one household member engaging in off-farm waged employment and receiving remittances increases by 1.4% and 1.9%, respectively. This implies that these households tend to invest more in education, making it in turn easier to find off-farm employment (both within Kumi Town and other urban centers). This ultimately enhances the ability to assist financially the household in the form of remittances.

2.3.6 Effectiveness of Coping Strategies Against Weather Shocks

Often the coping strategies discussed in Sect. 2.3.4 may not enable the effective offsetting of the devastating impacts of weather shocks on consumption. Below, we assess the effectiveness of each coping strategy using interaction terms between the dummy variable for weather shocks and dummy variables for each of the respective coping strategies.

Table 2.8 indicates that households receiving remittances can offset shock impacts. This is indicated by the positive and significant interaction term between

Table 2.8 Effectiveness of remittances as a coping strategy

Variables	(1) Log (consumption)	(2) Log (food)	(3) Log (basics)	(4) Log (contribution)
Exposure to weather shock (1 = experienced weather shock)	−0.166* (0.0889)	−0.00745 (0.186)	−0.439*** (0.159)	−0.279 (0.242)
Remittances (1 = received remittances)	−0.0319 (0.155)	−0.0688 (0.290)	0.0635 (0.0435)	0.188 (0.404)
Interaction of shock dummy and remittance receipt dummy	0.150** (0.073)	0.216** (0.110)	0.122 (0.292)	−0.162 (0.449)
Gender of household head (1 = female)	0.150** (0.0693)	−0.0780 (0.146)	0.566*** (0.128)	−0.0601 (0.184)
Education of household head (years of schooling)	0.0227*** (0.00674)	0.0232* (0.0127)	0.0406*** (0.0121)	0.0549*** (0.0165)
Household size (number of members)	−0.0342*** (0.0110)	−0.0513** (0.0211)	−0.0157 (0.0226)	−0.0473* (0.0283)
Age of household head (years)	0.00559 (0.0121)	−0.0272 (0.0316)	0.0218 (0.0298)	0.0887** (0.0448)
Savings (1 = belongs to SACCO)	0.165** (0.0659)	0.142 (0.131)	0.255** (0.117)	0.964*** (0.179)
Age squared of household head (years^2)	6.97e-06 (0.000135)	0.000296 (0.000350)	−7.39e-06 (0.000342)	−0.000817 (0.000500)
Asset value (log)	0.0306 (0.0306)	0.0257 (0.0431)	0.0756** (0.0339)	0.135** (0.0620)
Asset ownership (1 = owns mobile phone)	−0.144 (0.101)	−0.259 (0.181)	−0.160 (0.196)	0.0769 (0.276)
Constant	10.33*** (0.394)	8.154*** (0.889)	7.280*** (0.712)	3.119*** (1.086)
Observations	345	345	345	345
R-squared	0.387	0.674	0.303	0.348

Note: Robust standard errors are reported in parentheses. Dummy variables for community and year are controlled for in all specifications for location- and time-specific effects ***$p < 0.01$; **$p < 0.05$; *$p < 0.1$

the dummy variables for weather shock and remittance receipt for total and food-based consumption expenditure (see Columns 1 and 2 in Table 2.8, respectively). However, although the coefficient is positive for basic expenditures, it is statistically indistinguishable from zero. This implies that remittances are perhaps allocated to food consumption during crises, rather than non-food consumption categories like durable household items and social contributions.

Table 2.9 presents a similar analysis for asset sales. The negative coefficient on the dummy variable for asset sales points to the possibility that poor households (in terms of assets) generally have lower consumption per adult equivalent. However, the positive and significant coefficient on the interaction term between the dummy variables for exposure to weather shocks and asset sales shows that (conditional to shock exposure) selling household assets provides temporary insurance against consumption decline. The coefficient is higher for food-related consumption, possibly indicating that asset sales are made to temporarily supplement food intake.

Table 2.9 Effectiveness of asset sales as a coping strategy

Variables	(1) Log (consumption)	(2) Log (food)	(3) Log (basics)	(4) Log (contribution)
Exposure to weather shock (1 = experienced weather shock)	−0.154** (0.0764)	−0.0773 (0.149)	−0.460*** (0.139)	−0.403* (0.224)
Coping strategy (1 = sold assets)	−0.325* (0.185)	−0.973*** (0.343)	−0.375 (0.371)	−0.563 (0.465)
Interaction of shock dummy and asset sale dummy	0.380* (0.197)	1.226*** (0.388)	0.603 (0.395)	0.579 (0.497)
Gender of household head (1 = female)	0.154** (0.0688)	−0.0652 (0.141)	0.571*** (0.127)	−0.0363 (0.183)
Education of household head (years of schooling)	0.0249*** (0.00647)	0.0279** (0.0120)	0.0444*** (0.0118)	0.0584*** (0.0165)
Household size (number of members)	−0.0346*** (0.0110)	−0.0517** (0.0209)	−0.0160 (0.0227)	−0.0471* (0.0284)
Age of household head (years)	0.00312 (0.0123)	−0.0334 (0.0334)	0.0166 (0.0285)	0.0863* (0.0441)
Savings (1 = belongs to SACCO)	0.164** (0.0649)	0.130 (0.128)	0.252** (0.114)	0.954*** (0.178)
Age squared of household head (years^2)	3.95e-05 (0.000135)	0.000371 (0.000368)	5.64e-05 (0.000327)	−0.000787 (0.000491)
Asset value (log)	0.0321 (0.0303)	0.0270 (0.0425)	0.0776** (0.0333)	0.137** (0.0603)
Asset ownership (1 = owns mobile phone)	−0.135 (0.102)	−0.240 (0.181)	−0.135 (0.200)	0.0551 (0.279)
Constant	10.33*** (0.386)	8.222*** (0.903)	7.303*** (0.683)	3.227*** (1.051)
Observations	326	332	333	328
R-squared	0.390	0.685	0.305	0.351

Note: Robust standard errors are reported in parentheses. Dummy variables for community and year are controlled for in all specifications for location- and time-specific effects ***$p < 0.01$; **$p < 0.05$; *$p < 0.1$

However, this strategy may not necessarily be used to safeguard other components of consumption.

Finally, Table 2.10 presents the results for the remaining coping strategies, i.e., credit access, livestock sales, and participation in both farm and off-farm waged employment. There is no evidence to suggest that the households that adopt each of these coping strategies experience lower consumption decline following exposure to weather shocks. This finding is not surprising for livestock sales due to the predominately urban sample used in this study. However, when it comes to farm and off-farm employment, the results point that wages might be too low to offset the strong impact of weather shocks on consumption. Moreover, demand for farm labor in nearby villages reduces due to the declining productivity during periods of weather shocks, thus likely causing the wages to decline further. We also find similar results for the disaggregated consumption measures, but these results are not reported for reasons of brevity.

Table 2.10 Effectiveness of access to credit, off-farm employment, and on-farm employment as coping strategies

Variables	(1) Credit	(2) Off-farm labor	(3) Farm labor	(4) Livestock sale
Exposure to weather shock (1 = experienced weather shock)	−0.117 (0.103)	−0.203* (0.105)	−0.108 (0.0821)	−0.182** (0.0826)
Coping strategy (1 = received credit)	0.0829 (0.120)			
Interaction of shock dummy and credit dummy	−0.0144 (0.141)			
Livelihood sources (1 = household member involved in off-farm employment)		−0.0274 (0.128)		
Interaction of shock dummy and off-farm employment dummy		0.141 (0.141)		
Livelihood sources (1 = household member involved in on-farm employment)			−0.0797 (0.197)	
Coping strategy (1 = sold livestock)				−0.105 (0.144)
Interaction of shock dummy and livestock sale dummy				0.217 (0.157)
Education of household head (years of schooling)	0.0226*** (0.00636)	0.0223*** (0.00635)	0.0215*** (0.00616)	0.0224*** (0.00628)
Household size (number of members)	−0.0344*** (0.0102)	−0.0339*** (0.0101)	−0.0337*** (0.0102)	−0.0357*** (0.0102)
Age of household head (years)	8.02e-05 (0.0125)	0.00160 (0.0124)	0.00230 (0.0125)	0.00234 (0.0122)
Gender of household head (1 = female)	0.126* (0.0657)	0.126* (0.0655)	0.120* (0.0665)	0.140** (0.0664)
Asset value (log)	0.0347 (0.0317)	0.0363 (0.0309)	0.0342 (0.0334)	0.0372 (0.0312)
Asset ownership (1 = owns mobile phone)	−0.102 (0.105)	−0.111 (0.106)	−0.126 (0.104)	−0.104 (0.105)
Observations	(0.432)	(0.421)	(0.434)	(0.402)
Observations	335	335	335	335
R-squared	0.374	0.376	0.376	0.377

Note: Robust standard errors are reported in parentheses. Dummy variables for community and year are controlled for in all specifications for location- and time-specific effects ***$p < 0.01$; **$p < 0.05$; *$p < 0.1$

2.3.7 Robustness Checks and Propensity Score Matching

The OLS and fixed effects results presented above assume no systematic differences in household characteristics between households affected and unaffected by weather shocks. Although the fixed effects estimates control for unobserved time-invariant

Table 2.11 Propensity score matching (PSM) of weather shocks and household consumption

Variables	(1) Log (consumption)	(2) Log (food)	(3) Log (basics)	(4) Log (contribution)
ATE (weather shock)	−0.173∗∗∗ (0.0612)	0.150 (0.125)	−0.227∗∗ (0.103)	−0.148 (0.225)
ATE (drought)	−0.0499 (0.0786)	0.140 (0.151)	−0.221∗ (0.128)	−0.417∗ (0.220)
ATE (flood)	−0.00744 (0.0493)	0.0805 (0.150)	−0.308∗ (0.165)	−0.240∗∗ (0.0970)
ATE (pests/ diseases)	−0.0532 (0.0642)	−0.111 (0.134)	−0.0886 (0.126)	0.0736 (0.203)
Observations	342	342	342	342

Note: Robust standard errors are reported in parentheses ∗∗∗$p < 0.01$; ∗∗$p < 0.05$; ∗$p < 0.1$

Table 2.12 Propensity score matching (PSM) of weather shocks and coping strategies

Variables	(1) Sell livestock	(3) Sell assets	(5) Work off-farm	(6) Receive credit	(7) Receive remittances
ATE (weather shock)	0.0261 (0.0610)	0.00290 (0.0402)	0.125∗ (0.0753)	0.116∗ (0.0621)	0.171∗∗ (0.0665)
ATE (drought)	0.0551 (0.0641)	0.0696 (0.0467)	0.183∗∗∗ (0.0595)	0.183∗∗ (0.0725)	0.0377 (0.0625)
ATE (flood)	0.110 (0.0693)	0.0580 (0.0712)	0.270∗∗∗ (0.0757)	0.110 (0.0921)	0.322∗∗∗ (0.0960)
ATE (pests/ diseases)	0.0145 (0.0596)	−0.0232 (0.0436)	0.0580 (0.0573)	0.0551 (0.0604)	0.0116 (0.0603)
Observations	345	345	345	345	345

Note: Robust standard errors are reported in parentheses ∗∗∗$p < 0.01$; ∗∗$p < 0.05$; ∗$p < 0.1$

household heterogeneity, observed and time-variant unobserved heterogeneity could confound our results. In fact, Table 2.1 indicates that households affected and unaffected by weather shocks are systematically different along key characteristics such as household size and age and education of the household head. These differences could be responsible for the observed consumption difference between these groups of households. Similarly, such differences could also imply variations in the ability of households to adopt coping strategies, even in the absence of weather shocks.

The average treatment effect (ATE) presented in Tables 2.11 and 2.12 is used to compare consumption expenditures and the likelihood of adopting a certain coping strategy between comparable households that have been affected and unaffected by weather shocks. The negative impact of weather shocks identified through the OLS and fixed effects analyses is also confirmed through the PSM analysis (Table 2.11). Exposure to weather shocks reduces household consumption expenditure per adult equivalent by 17%. When disaggregating consumption expenditures, there is a 22% decline in expenditure for household basics and a negative but statistically insignificant effect for social contributions. When analyzing individual weather shocks, droughts and floods significantly reduce expenditure on basics and social contributions, while shocks related to animal/crop pests and diseases have no significant impact on any of the consumption categories.

Table 2.12 confirms the findings of Table 2.7 and particularly the positive and significant impact of exposure to weather shock on the likelihood to be involved in off-farm waged employment, borrow money, and receive remittance, as well as a positive but insignificant impact on the sale of livestock and household assets.

Finally, Table 2.13 reports the results of covariate balance tests to assess the comparability of covariates before and after matching. P-values for the equality of means covariates such as distance to the market, SACCO membership dependency ratio, age, and years of schooling of the household head are lower than 0.05 before matching but higher than 0.1 after matching. This indicates that covariates were unbalanced before matching but became balanced after matching. Failure to reject the hypothesis of joint equality of means after matching (as indicated by a p-value higher than 0.05) shows that covariates for households affected and unaffected by weather shocks are drawn from comparable distributions (Caliendo and Kopeinig 2008). Additionally, the mean absolute bias of 3.3% is lower than the recommended value of 5% to yield reliable estimates (Rosenbaum and Rubin 1985). This implies that the propensity score matching technique reliably compared shock-affected households with unaffected households sharing similar observable characteristics, hence ameliorating the issue of observed heterogeneity that would confound the results.

2.4 Discussion

2.4.1 Synthesis of Findings

As already discussed in Sect. 2.1, the interface of climate change, food security, and livelihoods is a very important sustainability challenge in SSA (see also Chaps. 1–3 Vol. 1). Climate variability can increase the frequency and intensity of droughts and floods, which can have multiple negative socioeconomic effects (Sect. 2.1). Agriculture is one of the most vulnerable sectors (Howden et al. 2007), with many studies having identified or predicted possibly severe yield declines in SSA due to climate change (Ringler et al. 2010; Dinar et al. 2012; Schlenker and Lobell 2010; Kurukulasuriya and Mendelsohn 2007; Roudier et al. 2011).

Our analysis suggests that exposure to weather shocks significantly reduces the aggregate household consumption expenditure per adult equivalent by approximately 17% (Sect. 2.3.7). By disaggregating consumption to its components, we can deduce that food-related consumption is not affected by weather shocks but non-food expenditures are severely reduced (Sect. 2.3.3). This indicates a potential reallocation of household resources upon exposure to shocks away from other consumption components to supplement food purchases. In fact, empirical literature has revealed that households tend to reallocate internal resources when faced with

Table 2.13 Covariate balance check before and after propensity score matching (PSM)

Variables	Mean before			Mean after			%|Bias| Reduction
	Shock = 1	Shock = 0	P-value	Shock = 1	Shock = 0	P-value	
Dependency ratio	1.0326	1.3944	0.062	1.0326	0.82387	0.168	73.3
Education of household head (years of schooling)	9.97	11.34	0.041	11.347	12.405	0.137	35.3
Age of household head (years)	39.05	35.73	0.037	39.438	39.884	0.801	28.4
Household size (number of members)	6.24	5.24	0.009	5.7273	5.5868	0.709	26.1
Gender of household head (1 = female)	0.26	0.21	0.366	0.2562	0 0.22314	0.549	980.0
Asset value (log)	14.283	13.989	0.309	14.283	14.456	0.357	41.1
Distance to market (km)	0.91694	1.043	0.044	0.91694	0.81481	0.138	23.4
Savings (1 = belongs to SACCO)	0.36	0.27	0.067	0.34711	0.38889	0.597	176.9
Asset ownership (1 = owns mobile phone)	0.76	0.74	0.812	0.75	0.75	1.000	100.0
Pseudo R^2	–	–	0.088	–	–	0.0043	–
Mean bias	–	–	11.8	–	–	3.3	–
P-value (joint mean equality)	–	–	0.040	–	–	0.154	–

Note: Balance check before and after PSM for observations for which $0.1 < e(X) < 0.9$. Pseudo R^2 indicates how well covariates explain treatment probability. A small value after matching indicates goodness of the matching technique (Sianesi 2004). A standardized absolute mean bias <5 after matching indicates effective matching (Rosenbaum and Rubin 1985). A non-significant p-value for the joint mean equality test after matching shows significant similarity between treatment and control groups after matching (Caliendo and Kopeinig 2008) negative income shocks from weather events (Akampumuza and Matsuda 2017; Sawada and Shimizutani 2011). Even in the case of positive income shocks, changes in the relative importance of consumption components may necessitate intra-household resource reallocation. For example, Prina (2015) finds that households obtaining access to a banking account for the first time tend to increase their spending on education, meat, and fish and reduce spending on health and dowry.

We also find significant gender differences in poverty and vulnerability to weather shocks. Although the gendered impacts of climate change and weather shocks have attracted increasing policy and academic attention in SSA (Akampumuza and Matsuda 2017; Asfwa and Maggio 2018), empirical evidence based on micro-data is still scanty. Generally, our study finds that irrespective of the exposure status to weather shocks, female-headed households are, on average, poorer in terms of consumption (including food-related consumption) relative to male-headed households (Sect. 2.3.3). Following their exposure to weather shocks, female-headed households decrease consumption expenditure per adult equivalent more sharply relative to male-headed households (Sect. 2.3.4). This finding is consistent with some of the existing evidence attributing such gender-differentiated vulnerability to the relative lack of access to coping resources and opportunities (Akampumuza and Matsuda 2017; Klasen et al. 2014). Although our chapter focuses on the short-term effects of weather shocks, it is likely that the food consumption among households affected by weather shocks will divert severely from its long-term trajectory. For example, panel evidence from Malawi indicates that the overall consumption, food consumption, and calorie intake divert significantly from their long-term trajectories after periods of abnormally high temperatures (Asfaw and Maggio 2018).

With regard to the coping strategies, we find that affected households are more likely to engage in off-farm waged employment (Sect. 2.3.5). This reflects existing literature that has stressed the importance of participation in labor market as an ex-post coping strategy (Mathenge and Tschirley 2015; Ito and Kurosaki 2009). For example, studies have found that off-farm employment has increased for adults and children during shocks in Tanzania (Beegle et al. 2006). However, studies often fail to examine the effectiveness of such strategies in safeguarding consumption and income against shocks. Some of the other coping strategies identified in this study include remittances and the sale of household assets (Sect. 2.3.5). However, contrary to studies that have identified livestock selling as a crucial coping strategy (Hoddinott 2006), this was not observed in our case. This is possibly due to the relatively larger representation of urban residents in our sample, which mainly engage in off-farm employment and had low level of livestock ownership.

Finally, we find that most of the adopted coping strategies do not effectively safeguard the consumption of the affected households (Sect. 2.3.6). The only exception is remittances, which seem to fully offset the potential adverse effect of weather shocks on consumption (Sect. 2.3.6). This stresses the importance of both domestic and international remittances for poverty alleviation and vulnerability reduction, irrespective of the rural or urban contexts. This finding is consistent with a study that found that predominately rural Kenyan households that use

M-PESA (i.e., Safaricom's mobile money platform) are able to offset the negative effect of shocks from weather and illness, by receiving remittances from family and friends (Jack and Suri 2014).

2.4.2 Policy Implications and Recommendations

As mentioned in Sect. 2.1, the results of this study are relevant to various SDGs such as SDG1 (No poverty), SDG2 (Zero hunger), SDG5 (Gender and equality), SDG11 (Sustainable cities and communities), and SDG13 (Climate action) to mention some. Thus the findings outlined above carry important policy and practice implications for designing strategies and interventions to reduce the negative outcomes of weather shocks in urban contexts of SSA.

Firstly, they point to the need for designing and implementing comprehensive strategies to increase the resilience of food systems against weather shocks. To achieve this it would be necessary to achieve a much broader understanding of food systems spanning from the production to the distribution and consumption of food, especially in the context of urbanization and climate change (Chap. 1 Vol. 1). Considering that smallholder-based farming is a crucial source of livelihoods and food for a large portion of the Ugandan population (and SSA more generally), increasing its resilience to weather shocks would not only enhance food security but also safeguard their income against weather shocks. Strategies that simultaneously support climate-smart food crop production through the use of drought-tolerant crop varieties, soil management practices, and seasonal weather forecast would be particularly important (Chaps. 6, 10 Vol. 1; Chap. 3 Vol. 2). Beyond food crop production, it would be important to develop an efficient food supply and distribution system, especially for urban households that are often not engaged in agriculture and rely on food crops produced in surrounding rural areas. This would necessitate the strengthening of critical infrastructure to maintain steady rural-urban food flows and prevent any disruptions to the food supply chain during weather shocks and especially floods. Ultimately, such resilient food production systems and efficient and reliable food distribution channels could enhance food availability, avoid unnecessary food price hikes (especially during times of weather shocks), and sustain food consumption especially among poorer urban households. Such policy and practice interventions could contribute to SDG 2 (Zero hunger) and SDG 13 (Climate action) but also catalyze poverty alleviation and vulnerability reduction, thereby directly contributing to the achievement of SDG1 (No poverty).

Secondly, effective livelihood support strategies need to specifically target female-headed households, as these are disproportionately affected by weather shocks. For example, improving female education and skills, easing labor market restrictions, and providing relevant and timely weather forecasts to female farmers could potentially increase their resilience to weather shocks. Such strategies could increase women's access to alternative coping options and opportunities, which could in turn reduce both their vulnerability to weather shocks and the severity of impacts. Additionally, by increasing coping capacity, this could help female-headed households (and women in general) to recover faster from the adverse effects of weather shocks. This could contribute to wider national efforts to achieve gender equality (and overall equality of opportunities) and achieve more inclusive growth as stipulated in SDG5 and SDG10, respectively.

Thirdly, as most coping strategies do not seem to adequately safeguard household consumption against weather shocks, there is need to reinvigorate credit and insurance markets. In principle promoting agricultural insurance could safeguard smallholder farmers from severe income loss due to extreme weather events. These, coupled with credit guarantee schemes, reduced interest rates and strategies to increase access to credit could increase the ex-post coping ability of households affected by weather shocks. This is particularly important to households that lack the necessary collaterals to enter formally the credit market. Indeed, improved and inclusive agricultural insurance and credit markets could go a long way reducing poverty and vulnerability to climate change (SDG1) and reduce food insecurity and hunger (SDG2).

2.5   Conclusions

Urban households in many parts of SSA are potentially susceptible to food insecurity and the disruption of livelihoods due to extreme weather events. Weather-related shocks such as droughts, floods, and severe incidence of animal and crop pests and diseases have direct adverse impacts on food production in surrounding rural areas. Furthermore, some weather shocks such as floods can cause the breakdown of critical road infrastructure and disrupt rural-urban food flows. The ultimate outcome is the significant compromise of livelihoods and the decline of food availability, which, coupled with increased food prices, can curtail the ability of many poor urban households to access sufficient food following weather shocks. Sometimes this situation is even worse among female-headed households, as they generally have access to fewer resources and a lower ability to adopt appropriate coping strategies.

However, there is a lack of robust empirical analysis of these dynamics in the existing literature, as most available studies either focus on rural areas where the impact of weather shocks is direct (i.e., through crop loss and crop failure) or aggregate households and ignore gender heterogeneity for household heads. This chapter therefore contributes to the existing literature by (a) analyzing the impact of weather shocks on consumption expenditure among urban households, (b) investigating gender dynamics related to vulnerability to weather shocks, (c) assessing the effectiveness of different coping strategies against weather shocks, and (d) assessing potential reallocation of household expenditures among its different sub-components during weather shocks.

The results suggest that self-reported exposure to weather shocks reduces household consumption expenditure per adult equivalent by 17%. The consumption decline is particularly severe among female-headed households, partly owing to their relative lack of access to productive resources and opportunities including land, credit, and off-farm employment. By disaggregating consumption elements, we highlight how households reallocate expenditures from health, education, semi-durables, and contributions toward social, cultural, and religious functions, to augment food intake following weather shocks. We also find that shock exposure is associated with a higher likelihood of engaging in off-farm employment, borrowing and receiving remittances as coping strategies against weather shocks. However, we also find that most of the coping strategies adopted by affected households do not effectively safeguard against household consumption decline with the exception of remittance receipt.

These findings suggest the critical need to strengthen food production, distribution, and supply systems and increase the resilience of household consumption against extreme weather events. This, in return, could safeguard urban livelihoods and food security following weather shocks. Policy interventions should also target highly vulnerable households, for example, those headed by females, considering their lower access to resources (e.g., land), opportunities (e.g., access to credit, formal employment), and ability to adopt many of the appropriate coping strategies.

Acknowledgments   We extend our gratitude toward the University of Tokyo, Graduate Program in Sustainability Science-Global Leadership Initiative (GPSS-GLI) for funding the field survey upon which the analysis in this chapter was based. We commend the integral input by members of the field research team who worked hard to ensure the success of the data collection exercise.

References

Akampumuza P, Matsuda H (2017) Weather shocks and urban livelihood strategies: the gender dimension of household vulnerability in the Kumi District of Uganda. J Dev Stud 53(6):953–970

Alderman H, Haque T (2007) Insurance against covariate shocks: the role of index-based insurance in social protection in low-income countries of Africa. The World Bank, Washington, DC

Asfaw S, Maggio G (2018) Gender, weather shocks and welfare: evidence from Malawi. J Dev Stud 54(2):271–291

Beegle K, Dehejia RH, Gatti R (2006) Child labor and agricultural shocks. J Dev Econ 81(1):80–96

Bohle HG, Downing TE, Watts MJ (1994) Climate change and social vulnerability: toward a sociology and geography of food security. Glob Environ Change 4(1):37–48

Buringh PETER (1977) Food production potential of the world. World Dev 5(5–7):477–485

Caliendo M, Kopeinig S (2008) Some practical guidance for the implementation of propensity score matching. J Econ Surv 22(1):31–72

Contreras D, Plaza G (2010) Cultural factors in women's labor force participation in Chile. Fem Econ 16(2):27–46

Cuny FC, Hill RB (1999) Famine, conflict and response. Kumarian Press, West Hartford

Deere CD. León M (2003) The gender asset gap: land in Latin America. World Dev 31(6):925–947

Dercon S (2002) Income risk, coping strategies, and safety nets. World Bank Res Obs 17 (2):141–166

Dercon S, Hoddinott J, Woldehanna T (2005) Shocks and consumption in 15 Ethiopian villages, 1999–2004. J Afr Econ 14(4):559–585

Dinar A, Hassan R, Mendelsohn R, Benhin J (2012) Climate change and agriculture in Africa: impact assessment and adaptation strategies. Routledge, London

Egeru A (2012) Role of indigenous knowledge in climate change adaptation: a case study of the Teso sub-region, Eastern Uganda

Fafchamps M (1996) Risk sharing, quasi-credit, and the enforcement of informal contracts. Department of Economics, Stanford University

Fafchamps M (1999) Risk sharing and quasi-credit. J Int Trade Econ Dev 8(3):257–278

Fafchamps M, Lund S (2003) Risk-sharing networks in rural Philippines. J Dev Econ 71(2):261–287

Fafchamps M, Udry C, Czukas K (1998) Drought and saving in West Africa: are livestock a buffer stock? J Dev Econ 55(2):273–305

Fraser ED, Simelton E, Termansen M, Gosling SN, South A (2013) "Vulnerability hotspots": Integrating socio-economic and hydrological models to identify where cereal production may decline in the future due to climate change induced drought. Agric For Meteorol 170:195–205

Funk C, Rowland J, Eilerts G, White L, Martin TE, Maron JL (2012) A climate trend analysis of Uganda. US Geol Surv Fact Sheet 3062:4

Gasper R, Blohm A, Ruth M (2011) Social and economic impacts of climate change on the urban environment. Curr Opin Environ Sustain 3(3):150–157

Hill RV, Vigneri M (2014) Mainstreaming gender sensitivity in cash crop market supply chains. In: Gender in agriculture. Springer, Dordrecht, pp 315–341

Hoddinott J (2006) Shocks and their consequences across and within households in rural Zimbabwe. J Dev Stud 42(2):301–321

Howden SM, Soussana JF, Tubiello FN, Chhetri N, Dunlop M, Meinke H (2007) Adapting agriculture to climate change. Proc Natl Acad Sci 104(50):19691–19696

Intergovernmental Panel on Climate Change (2018) Global warming of 1.5°C: an IPCC special report on the impacts of global warming of 1.5°C above pre-industrial levels and related global greenhouse gas emission pathways, in the context of strengthening the global response to the threat of climate change, sustainable development, and efforts to eradicate poverty. Intergovernmental Panel on Climate Change, Geneva

Irish Aid (2017) Uganda climate action report for 2016. [online] Available at: https://www.irishaid.ie/media/irishaid/allwebsitemedia/30whatwedo/climatechange/Uganda-Country-Climate-Action-Reports-2016.pdf?fbclid=IwAR0Tc8ll3JZ0aEyxWomS3SyCsjWNoRObFoX2KzO3BbkGmQAXMjYv6w-5tM. Accessed 17 Feb 2019

Ito T, Kurosaki T (2009) Weather risk, wages in kind, and the off-farm labor supply of agricultural households in a developing country. Am J Agric Econ 91(3):697–710

Jack W, Suri T (2014) Risk sharing and transactions costs: evidence from Kenya's mobile money revolution. Am Econ Rev 104(1):183–223

Jankowska MM, Lopez-Carr D, Funk C, Husak GJ, Chafe ZA (2012) Climate change and human health: spatial modeling of water availability, malnutrition, and livelihoods in Mali, Africa. Appl Geogr 33:4–15

Kazianga H, Udry C (2006) Consumption smoothing? livestock, insurance and drought in rural Burkina Faso. J Dev Econ 79(2):413–446

Kisauzi T, Mangheni MN, Sseguya H, Bashaasha B (2012) Gender dimensions of farmers' perceptions and knowledge on climate change in Teso sub-region, eastern Uganda. Afr Crop Sci J 20(2):275–286

Klasen S, Lechtenfeld T, Povel F (2014) A feminization of vulnerability? female headship, poverty, and vulnerability in Thailand and Vietnam. World Dev 71:36–53

Kochar A (1999) Smoothing consumption by smoothing income: hours-of-work responses to idiosyncratic agricultural shocks in rural India. Rev Econ Stat 81(1):50–61

Kurosaki T (2006) Consumption vulnerability to risk in rural Pakistan. J Dev Stud 42(1):70–89

Kurukulasuriya P, Mendelsohn R (2007) A Ricardian analysis of the impact of climate change on African cropland. The World Bank, Washington, DC

Lema MA, Majule AE (2009) Impacts of climate change, variability and adaptation strategies on agriculture in semi arid areas of Tanzania: the case of Manyoni District in Singida Region, Tanzania. Afr J Environ Sci Technol 3(8):206–218

Lipper L, Thornton P, Campbell BM, Baedeker T, Braimoh A, Bwalya M et al (2014) Climate-smart agriculture for food security. Nat Clim Change 4(12):1068

Lobell DB, Burke M (eds) (2009) Climate change and food security: adapting agriculture to a warmer world, vol 37. Springer Science & Business Media, New York

Majaliwa JGM, Tenywa MM, Bamanya D, Majugu W, Isabirye P, Nandozi C et al (2015) Characterization of historical seasonal and annual rainfall and temperature trends in selected climatological homogenous rainfall zones of Uganda. Glob J Sci Res 15(4):21–40

Masih I, Maskey S, Mussá FEF, Trambauer P (2014) A review of droughts on the African continent: a geospatial and long-term perspective. Hydrol Earth Syst Sci 18(9):3635

Mathenge MK, Tschirley DL (2015) Off-farm labor market decisions and agricultural shocks among rural households in Kenya. Agric Econ 46(5):603–616

McMichael AJ, Powles JW, Butler CD, Uauy R (2007) Food, livestock production, energy, climate change, and health. Lancet 370(9594):1253–1263

Mcsweeney C, New M, Lizcano G, Lu X (2010) The UNDP climate change country profiles: improving the accessibility of observed and projected climate information for studies of climate change in developing countries. Bull Am Meteorol Soc 91(2):157–166

Mendelsohn R, Dinar A, Williams L (2006) The distributional impact of climate change on rich and poor countries. Environ Dev Econ 11(2):159–178

Morgan CJ (2017) Reducing bias using propensity score matching. J Nucl Cardiol 25:404–406

Mubiru DN, Komutunga E, Agona A, Apok A, Ngara T (2012) Characterising agrometeorological climate risks and uncertainties: crop production in Uganda. S Afr J Sci 108(3–4):108–118

Munyegera GK, Matsumoto T (2016) Mobile money, remittances, and household welfare: panel evidence from rural Uganda. World Dev 79:127–137

Muttarak R, Lutz W (2014) Is education a key to reducing vulnerability to natural disasters and hence unavoidable climate change? Ecol Soc 19(1):42

Muttarak R, Lutz W, Jiang L (2016) What can demographers contribute to the study of vulnerability? Vienna Yearb Popul Res 2015(13):1–13

Niang I, Ruppel OC, Abdrabo MA, Essel A, Lennard C, Padgham J, Urquhart P (2014) Africa. In: Barros VR, Field CB, Dokken DJ, Mastrandrea MD, Mach KJ, Bilir TE, Chatterjee M, Ebi KL, Estrada YO, Genova RC, Girma B, Kissel ES, Levy AN, MacCracken S, Mastrandrea PR, White LL (eds) Climate change 2014: impacts, adaptation, and vulnerability. Part B: regional aspects. Contribution of working group II to the fifth assessment report of the Intergovernmental Panel on Climate Change. Cambridge University Press, Cambridge, pp 1199–1265

Nimusiima A, Basalirwa CPK, Majaliwa JGM, Otim-Nape W, Okello-Onen J, Rubaire-Akiiki C, Konde-Lule J, Ogwal-Byenek S (2013) Nature and dynamics of climate variability in the Uganda cattle corridor. Afr J Environ Sci Technol 7(8):770–782

Parry M, Rosenzweig C, Iglesias A, Fischer G, Livermore M (1999) Climate change and world food security: a new assessment. Glob Environ Change 9:S51–S67

Parry ML, Rosenzweig C, Iglesias A, Livermore M, Fischer G (2004) Effects of climate change on global food production under SRES emissions and socio-economic scenarios. Glob Environ Change 14(1):53–67

Patricola CM, Cook KH (2011) Sub-Saharan Northern African climate at the end of the twenty-first century: forcing factors and climate change processes. Clim Dyn 37(5–6):1165–1188

Porter C (2012) Shocks, consumption and income diversification in rural Ethiopia. J Dev Stud 48 (9):1209–1222

Prina S (2015) Banking the poor via savings accounts: evidence from a field experiment. J Dev Econ 115:16–31

Ringler C, Zhu T, Cai X, Koo J, Wang D (2010) Climate change impacts on food security in sub-Saharan Africa. Insights from Comprehensive Climate Change Scenarios

Ringler C, Zhu T, Cai X, Koo J, Wang D (2010) Climate change impacts on food security in sub-Saharan Africa. International Food Policy Research Institute (IFPRI), Washington, DC

Rosenbaum PR, Rubin DB (1985) Constructing a control group using multivariate matched sampling methods that incorporate the propensity score. Am Stat 39(1):33–38

Rosenzweig C, Parry ML (1994) Potential impact of climate change on world food supply. Nature 367(6459):133

Roudier P, Sultan B, Quirion P, Berg A (2011) The impact of future climate change on West African crop yields: What does the recent literature say? Glob Environ Change 21 (3):1073–1083

Rukandema M, Ameziane T, Bhattacharyya D (2008) FAO/WFP assessment of the impact of the floods on food and agriculture in eastern and northern Uganda. Food and Agriculture Organisation (FAO), Rome

Sawada Y, Shimizutani S (2011) Changes in durable stocks, portfolio allocation, and consumption expenditure in the aftermath of the Kobe earthquake. Rev Econ Househ 9(4):429

Schlenker W, Lobell DB (2010) Robust negative impacts of climate change on African agriculture. Environ Res Lett 5(1):014010

Scrimshaw NS (1987) The phenomenon of famine. Annu Rev Nutr 7(1):1–22

Sianesi B (2004) An evaluation of the Swedish system of active labor market programs in the 1990s. Rev Econ Stat 86(1):133–155

Striessnig E, Lutz W, Patt AG (2013) Effects of educational attainment on climate risk vulnerability. Ecol Soc 18(1):16

Townsend RM (1995) Consumption insurance: an evaluation of risk-bearing systems in low-income economies. J Econ Perspect 9:83–102

Uganda Bureau of Statistics (2014) National population and housing census 2014

UNDP (2013) Uganda poverty maps 2012/13. United Nations Development Programme (UNDP), Kampala

UNDP (2014) Teso, Kumi District hazard, risk and vulnerability profile 2014. United Nations Development Programme (UNDP), Kampala

Vermeulen SJ, Campbell BM, Ingram JS (2012) Climate change and food systems. Annu Rev Environ Resour 37(1):195–222

Wheeler T, Von Braun J (2013) Climate change impacts on global food security. Science 341 (6145):508–513

World Bank (2012) World development indicators 2012. World Bank, Washington, DC

World Bank (2018) Uganda Economic Update, 11th edn. World Bank, Washington, DC

Xiong W, Holman I, Lin E, Conway D, Jiang J, Xu Y, Li Y (2010) Climate change, water availability and future cereal production in China. Agric Ecosyst Environ 135(1–2):58–69

Chapter 3
Indigenous and Local Knowledge Practices and Innovations for Enhancing Food Security Under Climate Change: Examples from Mijikenda Communities in Coastal Kenya

Leila Ndalilo, Chemuku Wekesa, and Musingo T. E. Mbuvi

3.1 Introduction

Agricultural production is increasingly becoming vulnerable to climate change. Studies have predicted that the average global temperature may increase by 1.4–5.8 °C by the end of the twenty-first century, leading to substantial declines in freshwater resources and agricultural yields of major crops such as maize, rice and wheat (FAO 2011; IPCC 2014a). At the same time, a growing consensus is forming that global population will grow rapidly, possibly reaching nearly ten billion people by 2050 (FAO 2009). Thus, providing adequate and nutritious food for this expanding population will pose a further significant challenge to the global agriculture system (IPCC 2014b).

Sub-Saharan Africa (SSA) is one of the world's regions with the lowest food security (FAO 2015) (Chap. 1 Vol. 1). Estimates from the Food and Agriculture Organization (FAO) suggest that one in four people in SSA lacks adequate food to sustain an active and healthy life (Bremner 2012; FAO 2015) (Chap. 1 Vol. 2). While this is an improvement on the 1990 levels (one in three), the total undernourished population has increased from 182 to 287 million, largely due to the rapid population growth (FAO 2015) (Chap. 1 Vol. 1). SSA continues to lag behind the rest of the world in terms of reducing chronic hunger (see Chap. 1 Vol. 1), while combating hunger and achieving food security remain a complex sustainability challenge, particularly in light of climate change (Hall et al. 2017) (see Chap. 2 Vol. 2).

L. Ndalilo (✉)
Kenya Forestry Research Institute, Malindi, Kenya

C. Wekesa
Kenya Forestry Research Institute, Wundanyi, Kenya

M. T. E. Mbuvi
Kenya Forestry Research Institute, Nairobi, Kenya

© Springer Nature Singapore Pte Ltd. 2020
A. Gasparatos et al. (eds.), *Sustainability Challenges in Sub-Saharan Africa II*,
Science for Sustainable Societies, https://doi.org/10.1007/978-981-15-5358-5_3

The impacts of climate change are likely to be very severe in SSA because of the high dependence on agriculture for livelihoods and subsistence (Chap. 1 Vol. 1) and the limited adaptive capacity (Collier et al. 2008) (see Chap. 2 Vol. 2). Furthermore, the high population growth is likely to exacerbate food insecurity in rural areas of SSA, whose populations mainly depend on agriculture and have limited capacity for alternative livelihoods (Rosenzweig and Hillel 2008). Already, there is an observed decline in crop yields in arid and semi-arid areas of SSA, which has caused food shortages and large food inflation due to the relatively large economic dependence on natural resource sectors such as forestry, agriculture, water and fisheries (Poya et al. 2002; Prasad et al. 2014). Projections show that by 2055, the yield of cereals in SSA could decrease by 10–20% relative to yields in the 1990s if appropriate adaptation mechanisms are not developed and implemented (Mutegi et al. 2018).

Traditionally, extreme climate events such as droughts and floods were relatively predictable, but in the last three decades, the variability and unpredictability of the patterns of rainfall, temperature, flooding and droughts have increased (Mutegi et al. 2018) (Chap. 6 Vol. 1; Chap. 2 Vol. 2). For instance, in Kenya, prior to the 1990s, droughts and famines occurred in a cyclic pattern once every 10 years, i.e. on the fourth year of every decade (i.e. 1964, 1974, 1984). However, in the last two decades, droughts and floods have become rather irregular and more frequent, disrupting the traditional systems of disaster prediction and preparedness (Mutegi et al. 2018). Such climatic extremes and unpredictable events have caused declines in agricultural productivity and created uncertainties for stakeholders involved in agricultural value chains, particularly farmers, policy-makers, extension workers and donors (Mutegi et al. 2018).

The nexus of agricultural production, livelihoods, food security and climate change is a key sustainability challenge in many SSA countries including Kenya. Actually it spans many Sustainable Development Goals (SDGs), such as SDG1 (No Poverty), SDG2 (Zero Hunger) and SDG13 (Climate Action), among others. For example, climate change affects agricultural systems in multiple ways and has a direct impact on food security and agricultural productivity (IPCC 2014b). Climate change is also likely to affect agrobiodiversity through drying up of streams/rivers, loss of crop storage quality, loss of pastureland and land degradation, among other mechanisms (Enete 2009; IPBES 2018). Agricultural systems, on the other hand, contribute to climate change both through anthropogenic greenhouse gases (GHG) emissions and the conversion of non-agricultural land uses, including forests (Chap. 3 Vol. 1; Chap. 5 Vol. 2). In fact, agriculture is directly responsible for as much as 14% of total GHG emissions, with agriculture-driven deforestation accounting for an additional 18% of emissions (IPCC 2014a).

Thus, agricultural production systems in SSA have to be designed and maintained to provide effectively sufficient and nutritious food for a growing population in an environmentally, socially and economically sustainable manner (Roué et al. 2016). Indigenous and Local Knowledge (ILK) practices and innovations can potentially offer solutions to some of these challenges posed by climate change to the agricultural sector, while it can be directly applied for weather forecasting, vulnerability assessment and climate change adaptation (Chaps. 6, 10 Vol. 1). Many different ILK practices and innovations[1] are relevant at the interface of climate change and food security in SSA, such as the good knowledge and ability to grow and use different types of food in times of crisis (e.g. Chap. 6 Vol. 2). For example, the IPBES Africa Assessment Report recognizes and respects the significant contribution of ILK for the conservation and sustainable use of biodiversity and its growing role at the interface of climatic change and food security in SSA (IPBES 2018).

Considering the above, the aim of this chapter is to identify and document ILK-based practices and innovations that enhance or maintain agricultural productivity in the face of climate change in rural Kenya. In the context of this chapter, ILK-based practices and innovations refer to the knowledge, technologies and practices of indigenous and local people, which often emanate from their customary laws, cultural values and spiritual beliefs. We focus on ILK-based practices and innovations used by the Mijikenda community in rural Kenya to enhance livelihoods and their food security, as well as agrobiodiversity conservation.

Section 3.2 provides a brief description of the study site, target communities and the data collection methods. Section 3.3 outlines the main livelihood/food security patterns, adaptation strategies for food security, agrobiodiversity conservation strategies and ILK-based practices and innovations. Section 3.4 discusses the main findings and highlights some relevant policy implications and recommendations.

3.2 Methodology

3.2.1 Study Site

The study was undertaken at the coast region of Kenya. The coast region stretches approximately 150 km inland, covering an area of about 67,500 km^2, and approximately accounts for 11.5% of the total area of Kenya (Republic of Kenya 2009; Wekesa et al. 2016). The region is endowed with vast natural resources that include coral reefs, mangroves, lowland and *Kaya* forests, Afromontane forests and historical sites, which collectively provide the foundation for the regional economy. Approximately 8.4% of the total land area of the coast region is under forest cover (KEFRI 2016). However, despite being rich in natural resources, the coast region is still characterized by high levels of poverty, with up to 70–80% of residents living below the poverty line (Republic of Kenya 2013a, b; Wekesa et al. 2015; Wekesa and Ndalilo 2018). The heavy dependence on natural resources, coupled with the high poverty rates, puts a significant pressure on natural resources, while the region is low lying and thus quite vulnerable to the impacts of climate change (Wekesa et al. 2016).

---

[1]According to the Intergovernmental Science-Policy Platform on Biodiversity and Ecosystem Services (IPBES), ILK-based skills, experiences and practices are generated from the direct and long-term interactions arising from local needs, situations, circumstances and specific environments (Roue et al. 2016).

   The Mijikenda ethnic groups mainly inhabit the coast region of Kenya, which is comprised of six counties, namely, Mombasa, Kilifi, Kwale, Taita Taveta, Lamu and Tana River. The study was undertaken in Kilifi and Kwale Counties (Fig. 3.1). The average annual rainfall in Kilifi County ranges from 300 mm in the hinterland to 1300 mm at the coastal belt, while the annual temperature ranges between 21 and 30 °C in the coastal belt and between 30 and 34 °C in the hinterland (Republic of Kenya 2013a). Kwale County has an average annual rainfall of about 400–1200 mm and an average temperature of 24.2 °C (Republic of Kenya 2013b).

   In particular the study targeted 31 villages in the communities of Giriama, Chonyi, Rabai (Kilifi County) and Digo and Duruma (Kwale County). These communities are spread along the Kenyan Coast in different agro-ecosystems, characterized by wet, semi-arid, and dryland conditions. The Giriama, Rabai and Digo communities are located in wetter areas near the coastline, while the Chonyi and Duruma communities are located inland in the semi-arid and dryland areas, respectively. These communities are also characterized by rich traditional knowledge and agrobiodiversity (indigenous vegetables and *Kaya* forests). Table 3.1 provides some of the key characteristics of the study communities.

3.2.2   Data Collection and Analysis

The survey used a combination of literature review, household surveys, focus group discussions (FGDs) and key informant interviews to elicit the local livelihoods, as well as prevailing patterns of climate variability, food security and ILK practices (including agrobiodiversity conservation). Purposive sampling was used to select 5 households with rich knowledge of traditional practices and agrobiodiversity conservation in each of the 31 villages in the five target communities (N = 155; Table 3.2). Furthermore, five FGDs were conducted (one in each community), and 50 key informant interviews were conducted across the five communities (Table 3.2). The study was conducted from January 2013 to February 2014.

   The study defined a village as the area under the smallest possible administration system, where village elders form the first decision-making body. Seven villages were selected in each target community with the exception of Digo and Duruma communities (6 and 4 villages, respectively). The criteria used for selecting the villages were (a) diverse socio-economic activities; (b) adherence to traditional culture; (c) development level and proximity to urban areas (i.e. villages with varying development levels were selected for comparison purposes); (d) geographical characteristics and unique landscapes; (e) linguistic/dialect differences; and (f) geographical positioning and distribution in the overall area. Villages with a rich tradition and history of employing diverse and established traditional practices in agriculture were prioritized for selection. Household selection was performed through purposive sampling targeting 5 households that had a rich knowledge of traditional practices and agrobiodiversity conservation in each of the 31 villages.

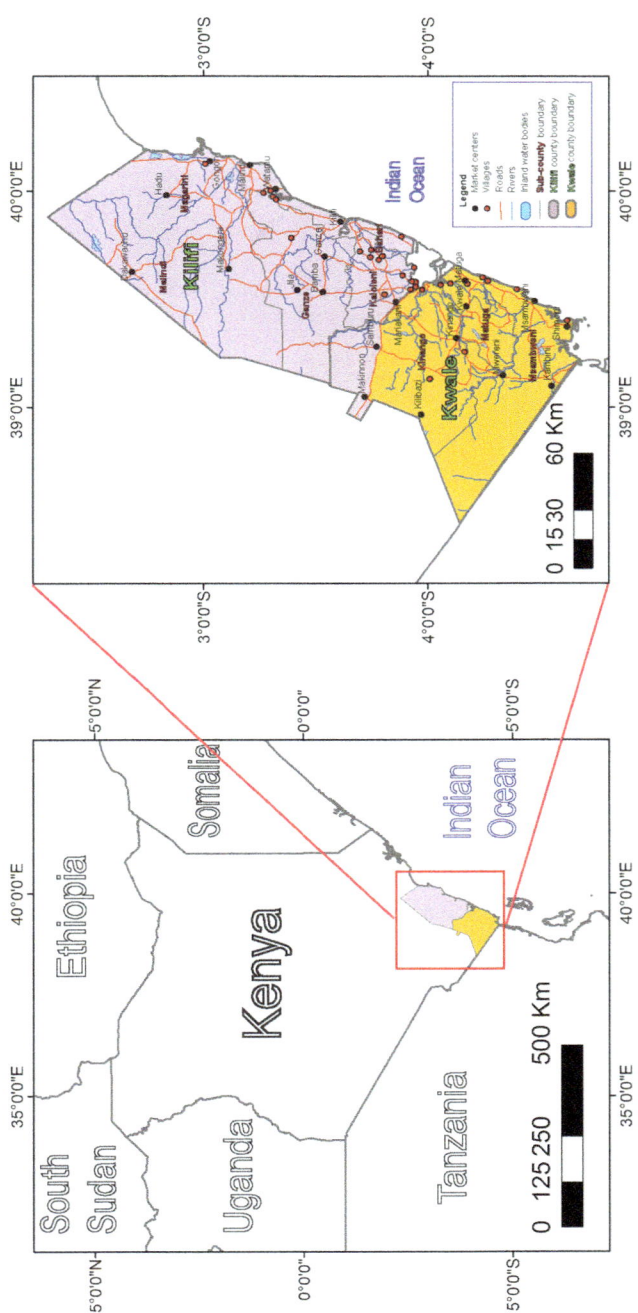

Fig. 3.1 Location of study sites

Table 3.1 Characteristics of study communities

Community	Characteristics
Giriama	Located in Kilifi County
	Located in the semi-arid and arid zones of the coastal lowlands
	Mixed farming is the main livelihood activity in the semi-arid areas, while livestock rearing is the main activity in the arid areas
	The road network is fairly good, but other infrastructure development is still low (e.g. low access to electricity)
Chonyi	Located in Kilifi County
	Located on a wet fertile ridge
	Crop farming (both annual and perennial crops) is the main livelihood activity, followed by small-scale businesses and tourism
Rabai	Located in Kilifi County
	Located in a hilly area with average rainfall
	Mixed farming and livestock rearing are the main livelihood activities
	Infrastructure development is relatively good with most households having access to electricity, tarmac roads and piped water
	Good access to regional and local markets
Digo	Located in Kwale County
	Landscape is characterized by a monsoon climate and coastal plains, flood plateaus, coastal uplands and Nyika plateaus with rivers and streams
	Livestock keeping and agricultural production are the main livelihood activities
	The rainforests and national parks offer great potential for tourism
Duruma	Located in Kwale County
	Located in the semi-arid Nyika plateau, which receives much lower rainfall compared to the rest of Kwale County
	High poverty levels due to high rates of unemployment, school dropout and early marriages
	Livestock rearing is the main livelihood activity
	Crop farming is mainly undertaken for subsistence

Table 3.2 Data collection methods and sample sizes

Site	Number of villages	Number of respondents		
		Household questionnaires	Focus group discussions (FGD)	Key informant Interviews
Giriama	7	35	1	10
Chonyi	7	35	1	10
Rabai	7	35	1	10
Digo	6	30	1	10
Duruma	4	20	1	10
Total	**31**	**155**	**5**	**50**

The household survey covered broad themes such as (a) past and current sources of income, (b) climatic patterns and their effect on crop production, (c) farming practices, (d) agrobiodiversity conservation, (e) social networks, and (f) biocultural innovations. FGDs were used both to validate the information collected through household surveys and gather new information especially on biocultural practices and innovations. A checklist was used to guide the discussions on the following broad thematic areas: community livelihoods, agrobiodiversity, social networks and biocultural practices and innovations.

The criteria for selecting FGDs participants were (a) gender composition (at least a third of the participants were women); (b) people with special knowledge related to agrobiodiversity such as herbalists; (c) traditional and spiritual leaders; (d) rainmakers and local farmers growing traditional crop varieties; (e) community members with special societal functions such as members of *Kaya council of elders*; and (f) representatives of various community groups such as farmers' groups and village banking groups. We purposively selected key informants from community members with exceptional knowledge of traditional practices and systems such as *Kaya* elders, herbalists, farmers and community leaders.

3.3 Results

3.3.1 Livelihood Activities and Food Security

The household survey revealed that crop production was the most important livelihood activity across the five communities, with 51% of households identifying it as their main livelihood activity. Duruma community was the exception as livestock rearing was the main livelihood activity (the other communities also undertake livestock rearing but on a smaller scale). Other livelihood activities included small business ownership (31%) and labour in urban areas (18%).

The contribution of crop production to household food security was highest in Giriama (33%) and lowest in Duruma community (7%). Key informant interviews attributed these patterns in Giriama to the availability of large tracks of arable land and several ILK-based practices and innovations that reduce the vulnerability of crop production in the face of climatic change. On the contrary, the Duruma area is semi-arid and hence has low agricultural productivity, leading to community dependence on livestock.

Most of the surveyed households (96%) in the five communities reported a general decline in crop yield (by 31.2%) and reduced resistance to pests and diseases (by 16.1%) of the staple food crops such as maize, cassava, cowpeas and green grams between 2003 and 2013. This was possibly due to declining and erratic rainfall patterns and the high incidence of pests and diseases, which have significantly affected food crop production, and hence food security. Livestock production declined by 21% since 2003, with the number of livestock reducing mainly due to the high incidence of pests and diseases and the frequent and prolonged droughts that have hampered the availability of pasture.

3.3.2   Adaptation Strategies for Food Security

The different communities have developed various adaptation strategies that are largely shaped by the specific agro-ecological, geographical and climatic characteristics of the respective areas (Table 3.3). For example, households in the Duruma community (which occupies a semi-arid area with perennial water problems (Sect. 3.2.1) have developed an innovative method of excavating water pans within homesteads to provide drinking water for livestock and reduce the spread of livestock diseases from communal watering points. In addition, many households in the Duruma community undertake early planting before the onset of rainfall in order to efficiently utilize the little rainfall that is available for crop production in this semi-arid environment (Table 3.3).

Many households in the Rabai community rely heavily on spiritual prayers and sacrifices in order to avert climate-related extreme events. They also preserve seeds of local cultivars in home-based seed banks for use during the subsequent planting seasons (Table 3.3). Key informants attributed these practices to the strong cultural values and traditional resource governance system (*Kaya council of elders*) implemented in the community.

As Chonyi is located in a semi-arid area, all members of the community cultivate extensively drought-tolerant crops such as cassava (Table 3.3). Furthermore, 92.6% of the surveyed households plant early-maturing crop varieties in order to overcome possible challenges associated with reduced and unreliable rainfall. Similar to other communities, some households also engage in seed preservation and early planting (Table 3.3).

In the Digo community, which is located in a hilly moist forest ecosystem, conservation tillage and afforestation are the main adaptation strategies, which are undertaken by 66.7% and 21.1%, respectively, of the surveyed households (Table 3.3). Conservation tillage is mainly performed during drought periods, as a strategy to utilize the available water sustainably. Afforestation entails the

Table 3.3 Main adaptation strategies for food security in the study communities

| | Response (%) | | | | |
Adaptation strategy	Rabai	Digo	Chonyi	Giriama	Duruma
Use traditional bio-pesticides	23.7	8.8	4.6	33.5	29.4
Plant crops early	4.9	0.0	19.5	0.0	75.6
Preserve seeds in home-based banks	37.0	5.2	14.3	23.4	20.1
Plant early-maturing and drought-tolerant crops	0.0	0.0	92.6	0.0	7.4
Cultivate mainly drought-tolerant crops	0.0	0.0	100	0.0	0.0
Undertake irrigation farming	33.3	66.7	0.0	0.0	0.0
Offer prayers and sacrifices	88.2	2.9	0.0	8.8	0.0
Plant high-quality seeds	0.0	0.0	0.0	100	0.0
Diversify income through small-scale businesses	55.0	5.0	15.0	15.0	10.0
Afforest degraded ecosystems	0.7	21.1	2.0	35.4	40.8
Excavate water pans	0.0	0.0	0.0	0.0	100.0

integration of nitrogen-fixing agroforestry and fodder tree species and is widely applied for increased crop and animal yields respectively.

In the Giriama community located in semi-arid and arid areas, the local community mainly relies on planting high-quality local cultivar seeds (usually preserved by the community) to ensure high crop productivity (Table 3.3). The Giriama community is considered to be very innovative in this regard and has developed bio-pesticides comprising of traditional herbal plants such as *Encephalartos hildebrandtii*, *Dialium orientale*, *Landolphia kirkii*, *Brachystegia spiciformis*, *Dalbergia melanoxylon*, *Afzelia quanzensis*, *Brachylaena huillensis*, *Vepris glomerata* and *Manilkara sansibarensis*. These bio-pesticides are used to control pests and diseases and ensure high crop productivity.

3.3.3 Strategies for Agrobiodiversity Conservation

The findings suggest that the five communities grow more local cultivars than hybrids for the two staple crops: maize and cassava. However the cultivation of local cultivars has been decreasing over time as the cultivation of hybrid varieties becomes more prevalent (Fig. 3.2). On average, the number of households growing local cultivars for staple food crops has decreased significantly, from 100% in 1982 to about 61.7% in 2012.

There was very little loss of crop varieties between 1982 and 1992 due to the strong cultural attachment to traditional varieties and traditional governance structures in the past. Furthermore, the number of introduced and hybrid crop varieties at that time was also very low. However, over the next 15 years, there was a massive decline in the variety of local cultivars, which peaked in 2004. Key informants attributed this to the erosion of local culture, as literacy rates increased, traditional

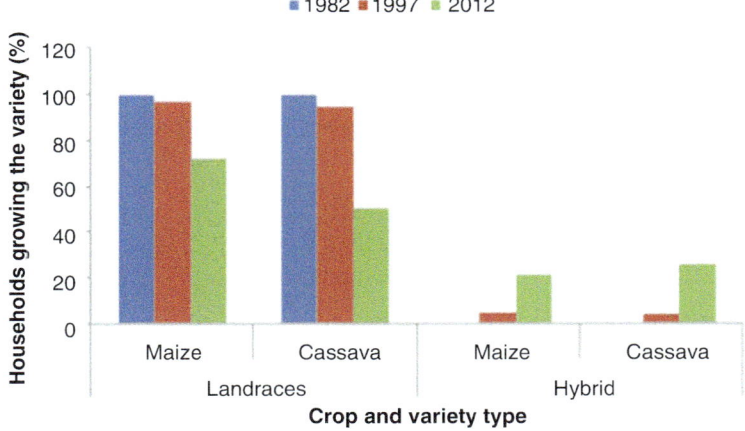

Fig. 3.2 Cultivation patterns of local cultivars and hybrids

governance systems weakened, and the adoption of improved and hybrid crop varieties increased due to intensive extension efforts that promoted their adoption at the time and free issuance of hybrid seeds by government.

Respondents from the Digo community reported the highest number of introduced new crop varieties in the last 30 years (18 varieties). To a large extent, this introduction was driven by (a) increased tourism in the area, which offers a ready market for the crops and cash to buy hybrid seeds, (b) migrant farmers from other regions and (c) favourable weather conditions in the area, which influenced the introduction of fast-growing and high-yielding crop varieties. Other communities also reported the introduction of new crop varieties but albeit at a lower extent: Chonyi (14 varieties), Giriama (11 varieties), Rabai (9 varieties) and Duruma (2 varieties).

FGD attributed the introduction patterns in Chonyi, Giriama and Rabai to their proximity to major markets in Kilifi (Chonyi and Giriama) and Mombasa (Rabai), where there is a high demand for food crops since they are major trading towns in the Coast region. Households in these communities tended to adopt crop varieties that grow and mature faster, in order to meet the ever-increasing market demand from the urban areas. The Duruma community recorded the lowest number of introduced crop varieties, as households only adopted crop varieties that could tolerate the prolonged dry spells in the area, as it is relatively dryer compared to the other sites.

On the contrary, the period between 2008 and 2012 was characterized by increases in the number of local crop cultivars grown (Fig. 3.2). This was due to increasing incidences of pests and diseases, frequent droughts and the high agricultural input requirements for the adopted hybrids and introduced varieties (e.g. fertilizer, pesticide). Farmers resumed growing more traditional varieties due to their resistance to pests and diseases and drought tolerance. These trends indicate that despite the previous decline in the production of local cultivars, the local farmers have re-introduced them due to their drought tolerance. The re-introduction of local cultivar crop varieties has also resulted to the wider re-adoption of traditional farming practices aiming to enhance farm productivity and conserve agrobiodiversity. Some of the traditional farming practices include the diversification of planted crop varieties and the self-saving of local cultivar seeds in home-based seed banks for use in the next planting season.

The analysis of the different seed-saving strategies indicates that the saving of reliable and high-quality seeds can contribute to the conservation of agrobiodiversity through the continued availability of quality germplasm for the propagation of plant genetic resources. Seed-saving is undertaken for many different food crops and often plays an important role in enhancing local food security. In particular, the household surveys established that 71% of the surveyed households across the five communities select and save high-quality planting material from the previous harvests for planting during the subsequent seasons. Seed varieties of local cultivars are mainly sourced from self-saved seed reserves, with Chonyi and Rabai having the highest number of farmers saving seeds for the next planting season at 89.9% and 78.6%, respectively. However, there was little exchange of seeds, both between farmers within the same community and between farmers from different communities (0.1%

Table 3.4 Gender roles in seed selection

Site	Households (%)	
	Men	Women
Rabai	19.5	80.5
Digo	41.8	58.2
Chonyi	55.8	44.2
Giriama	4.6	95.4
Duruma	25.2	74.8
Mean	**29.4**	**70.6**

Table 3.5 Desirable characteristics of staple food crop varieties

Desirable characteristics	Variety type (%)				
	Local cultivar	Self-improved	Community improved	Introduced improved	Hybrid
Resistance to drought and pests	91.7	4.9	0.10	1.4	2.1
Early maturation	80.0	1.3	1.3	7.5	10.0
High yield	80.6	0.9	0.0	12.1	6.5
Low rainfall requirement	100.0	0.0	0.0	0.0	0.0
Seed size	0.0	0.0	0.0	20.0	80.0
Desirable taste	82.6	8.7	0.0	0.0	8.7
Easy storage and availability	100.0	0.0	0.0	0.0	0.0
Easy to pound	100.0	0.0	0.0	0.0	0.0

and 1.0%, respectively). The high level of seed self-saving in Chonyi and Rabai points to the fact that many farmers are growing local cultivars, as the hybrid and introduced varieties cannot be self-saved. FGDs attributed the low level of seed exchange between farmers and communities to the eroding social cohesion due to the declining interaction within and between communities, which hinders their ability to freely exchange ideas and vital commodities such as seeds and foodstuff.

We analysed gender roles in the selection of seeds and found that women undertake most of the seed selection for local cultivars across the five communities (70.6%), with men playing a minor role (29.4%) (Table 3.4). However, gender roles in seed selection vary among the five communities, with men playing a major role in some communities such as Digo and Chonyi (Table 3.4).

Most of the desirable characteristics of saved seeds of staple food crops (i.e. maize, cassava, pulses) in the study communities are found in local cultivars (Table 3.5). Seed size is the only major desirable characteristic associated with introduced and hybrid crop varieties (Table 3.5). When selecting seeds of local cultivars, women tend to consider superior characteristics such as their ability to tolerate drought, pests and diseases, as well as taste and ease of seed storage (Table 3.5). According to key informants, and particularly the *Kaya* elders, this is

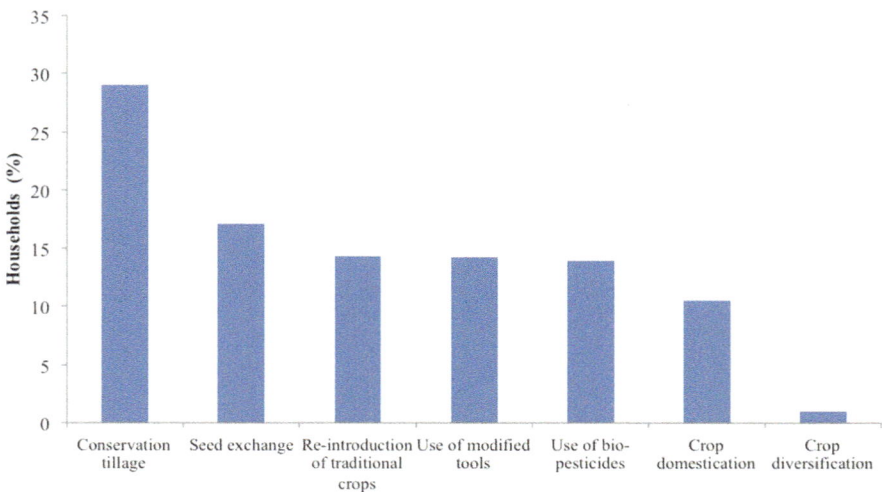

Fig. 3.3 Adoption of ILK-based practices and innovations

an indication that women are the custodians of traditional farming practices in the study communities.

3.3.4 ILK-Based Practices and Innovations

Conservation tillage[2] is the most widely adopted ILK-based practice among households in all study communities (29.0% of households) (Fig. 3.3). This was followed by seed exchange among farmers (17.1%), re-introduction of traditional crops (14.3%), use of modified tools (14.2%), use of bio-pesticide (13.9%), crop domestication (10.5%) and crop diversification (1.0%) (Fig. 3.3). There are numerous locally driven technological, social and institutional practices and innovations for food security in the study communities (Table 3.6). The diversification of crop varieties to reduce the risk of crop failure (43%) and the domestication of wild plants for income, medicine and food security (35%) are the most widely adopted practice in the five communities. The main social practices and innovations include the formation of communal farming and marketing groups including barter trade groups (30%) and the revival of customary laws and practices to preserve traditional values and crop diversity (25%) (Table 3.6). The latter includes local rules that regulate the utilization of wild food and medicinal plants and rules that encourage the

---

[2]Conservation tillage is a more sustainable cultivation system that combines minimal soil disturbance, mulching and crop rotation, in order to improve soil properties and other biotic factors.

Table 3.6 ILK-based practices and innovations in the study communities

Type	Innovation/practice	Community
Technological	Diversified varieties of the same crop in the same plot in a single season	Giriama
	Conservation tillage	Giriama
	Plant large areas with drought-tolerant crops	Giriama, Chonyi
	Treat livestock wounds and diseases using assorted herbal plants	Giriama, Duruma
	Domesticate wild medicinal and food plants	All
	Dig livestock water pans in homesteads	Duruma
Market/institutional/ social	Develop cultural villages to showcase culture and promote ecotourism	Rabai, Digo
Market/social	Form communal farming and marketing groups focusing on barter trade	All
Institutional/social	Revive and preserve customary laws and practices	Rabai, Giriama

cultivation of local cultivars for key crops such as maize, cowpeas, millet and sorghum for use in traditional ceremonies.

Poverty coupled with the increasing cost of living is the main economic factor driving the increasing adoption of ILK-based practices and innovations. Village banking is the most widely embraced livelihood-related innovation for diversifying household income sources and acting as a livelihood buffer during emergencies (61% of the surveyed households). Farmers from different communities also jointly form marketing groups (10%) in order to exchange commodities and ideas, access regional markets and negotiate competition for sale of their agricultural products. In addition, 5% of the surveyed households undertake value addition activities for traditional food crops and products (e.g. cassava, sweet potatoes, fruits, handcrafts). The domestication of wild medicinal and food plants is also being practiced by 14% of households, mainly for the conservation of important plant species (e.g. *Tamarindus indica*, *Adansonia digitata*, *Ziziphus mauritiana*, *Acacia mellifera*, *Salvadora persica*), for food security and for generating additional income from the sale of food and herbal products.

3.4 Discussion

3.4.1 Livelihoods and Food Security

Food crop production is the most important livelihood activity in most study communities, followed by small businesses and labour in urban areas (Sect. 3.3.1). The main exception is the Duruma community, where pastoralism is the dominant livelihood activity mainly due to the semi-arid characteristics of the Kinango

Sub-county where the community is located (Sect. 3.2.1). These findings reflect other similar studies, which have established that in semi-arid areas in Kenya (e.g. Coastal and Eastern lowlands), food crops generally contribute less to total household income and food security (10–43%), when compared to areas of higher agricultural potential (Kibaara et al. 2009). However, such livelihood trends might compromise local food production (and possibly food security) in the long run, as more young people that would have otherwise provided farm labour are increasingly migrating to urban centres.

Between 2003 and 2012, there was also a reported decline in staple food crop yields and resistance to pests/diseases, impacting negatively on food security (Sect. 3.3.1). The increased prevalence of pests and diseases, coupled with prolonged droughts, also reportedly has had a negative effect on livestock production, further compromising food security (Sect. 3.3.1). This reflects other studies in rural Kenya, which have reported the declining production of livestock and food crops in semi-arid areas over the last decade due to climatic factors (KARI 2007; Awuor 2009; Onyutha 2018). This situation has possibly forced local communities to develop alternative livelihood support mechanisms. However, in semi-arid areas, rural communities tend to have a lower preparedness to climate change and an overdependence on natural resources (and particularly agriculture) (Waha et al. 2013; Van Ittersum et al. 2016), which constitutes the focal areas for improving food security and rural incomes (Kibaara et al. 2009; Ongugo et al. 2014; Rippke et al. 2016) (Sect. 3.4.2). This is in line with the growing literature, which indicates that climate change will significantly affect agricultural production and (consequently) food security for millions of people in SSA and other parts of the world (Brown and Funk 2008; Lobell et al. 2008; Challinor et al. 2014; Rippke et al. 2016) (see also Chap. 1 Vol. 1; Chap. 2 Vol. 2).

3.4.2 Adaptation Strategies for Food Security

The study communities have developed several local adaptation strategies to respond to the impacts of climate change on declining crop productivity and food security (Sect. 3.3.2). These adaptation strategies vary across communities, even if communities are located in areas with similar agro-ecological conditions. Some of the factors that seem to dictate the type of the adopted adaptation strategies include the socio-economic characteristics of the community and the strength of traditional governance institutions and cultural practices (Sect. 3.3.2). For instance, the Duruma community digs livestock watering pans to overcome perennial water shortage in their semi-arid environment, while the Rabai community relies heavily on prayers and sacrifices to avert natural disasters, reflecting their strong cultural belief system (Sect. 3.3.2). It has been argued that the strong cultural values and traditional resource governance system of the Rabai community connect its members to their spiritual world often using spiritual prayers and sacrifices to avert climate related disasters (Ongugo et al. 2014).

Overall, the most widely implemented adaptation strategies to enhance food security access in the five communities included the (a) utilization of traditional bio-pesticides to control crop and animal pests and diseases; (b) preservation of seeds in home-based seed banks; (c) cultivation of crops that mature early and tolerate drought; and (d) cultivation of large amounts of drought-tolerant crops (Sect. 3.3.2). Such adaptation approaches are quite common in many different agrarian contexts of SSA (Dinar et al. 2008; Vincent et al. 2013; Spindell-Berck et al. 2019) (see also Chap. 10 Vol. 1).

The cultivation of local crop varieties that are drought, pest and disease-resistant is one of the most widespread and effective adaptation strategies (Sect. 3.3.3), which has also been observed in other studies in coastal Kenya (Onyango 2016; Makoti and Waswa 2015). For example, 90% of the households in the coast region of Kenya tend to grow drought-tolerant crops (Makoti and Waswa 2015), with these indigenous crop varieties faring relatively better in low fertility soils with low rainfall and offering a diversified farming production model (Onyango 2016).

However, despite their local importance, the cultivation of local crop varieties has been decreasing over time, concomitant with the increase in the cultivation of hybrid varieties (Sect. 3.3.3). This was attributed to the aggressive promotion of hybrid crop varieties by agricultural extension agents and is likely to contribute to the long-term loss of genetic diversity as farmers increasingly adopt just a few fast-maturing and high-yielding hybrid varieties (see also Chap. 10 Vol. 1). However, due to their unique traits, local cultivars can act as safety nets during periods of food insufficiency, which makes their preservation particularly important (Sect. 3.3.3). The preservation of local cultivars in communal seed banks and the domestication of wild plants have immensely contributed to agrobiodiversity conservation. Women tend to play a more important role than men in the selection of the seeds of the local cultivars and can thus be considered as custodians of traditional farming practices (Sect. 3.3.3). This reflects well other studies, which have established how women contribute substantially to the conservation of agricultural biodiversity in Kenya (Morandi 2015).

3.4.3 Capitalizing and Maintaining ILK Practices and Innovations

The integration of ILK practices into climate change mitigation and adaptation strategies has been emphasized in the Kenyan National Climate Change Response Strategy, as a central element of the current efforts to enhance local adaptation, crop productivity and food security (Republic of Kenya, 2013c). More broadly, many large-scale scientific assessments have re-affirmed the importance of ILK practices for ensuring climate change adaptation and food security in agrarian contexts of SSA (IPBES 2018) (Chap. 1 Vol. 1; Chap. 9 Vol. 2).

Our study identified a wide range of traditional farming practices that are currently utilized in the study communities and rooted on ILK that has withstood the test of time (Sect. 3.3.4). Similar environmentally sound and low-cost ILK-based practices and innovations to those used by the Mijikenda community in coastal Kenya have been identified in many other studies (Ongugo et al. 2014; Wekesa et al. 2017). Indeed ILK practices such as integrated farming methods (e.g. mulching, mixed cropping, crop rotation), use of bio-cosmological indicators to identify planting seasons, selection of high-yielding and locally adapted crop varieties, soil fertility management, integrated pest management and crop preservation and storage are common in many rural areas in Kenya and other parts of SSA (Altieri and Nicholls 2013; Ponge 2013) (see Chap. 10 Vol. 1).

Several factors support ILK-based practices and innovations in the study communities including individual innovators, traditional institutions, social organizations and networks, as well as economic and cultural factors (Sect. 3.3.4). Many studies in Kenya have revealed that indeed community interactions through social networks, traditional ceremonies and traditional governance systems are important for preserving and promoting ILK-based practices and innovations (Ongugo et al. 2014; Wekesa et al. 2017).

For example, individual farmers often develop local innovations anchored on ILK and freely share them with other farmers through community-based groups and organizations such as women and farmer groups. Such groups contribute manifold to social cohesion and the exchange of information and planting materials, which further support ILK innovations.

At the same time, traditional institutions such as the *Kaya council of elders* play a key role in supporting and preserving ILK-based practices and innovations, in tandem with other relevant cultural practices. For example, the restricted access to *Kaya* forests has contributed to biodiversity conservation, as some biodiversity-rich areas are sacred and thus not exploited for timber and fuelwood (Chap. 7 Vol. 1). Furthermore, local cultivars are used in traditional rituals, while coping strategies, agricultural practices, seeds and weather predictions are openly shared among community members during traditional ceremonies. In particular, most of the traditional ceremonies related to prayers, sacrifice offerings, healing and initiation are strongly associated with natural resources. For example, traditional prayers and sacrifices aimed at appeasing the spiritual world require the use of grains of local cultivar varieties (e.g. mustard, millet, sorghum, maize) and indigenous animal breeds (e.g. cattle, sheep and chicken) and are conducted in the *Kaya* forests. Similarly, most traditional healing ceremonies entail the use of various plant parts, which are effectively conserved due to their strong cultural significance.

Similarly, the Mijikenda communities have established cultural villages as opportunities for showcasing their cultural ceremonies, rituals and practices related to agrobiodiversity conservation. The cultural villages essentially bring community members together to showcase their cultural heritage, both as a means of preserving it and generating alternative income through cultural tourism. For the former, the different traditional ceremonies and festivals provide a platform to share information among community members and are therefore important in supporting and maintaining ILK-based innovations and practices. For the latter, traditional artefacts (e.g. baskets, cooking ware) and domesticated wild plants can be sold to tourists directly (e.g. as fruits) or after value addition (e.g. as herbal medicine) to diversify income.

3.4.4 Policy Implications and Recommendations

The study highlighted the important role of ILK-based practices and innovations for enabling adaptation to climate change and ensuring food security. In this respect a major recommendation of this chapter is to develop and implement appropriate policies and strategies to safeguard ILK and associated local practices and innovations at the local and national levels.

At the community level, there is a need to strengthen traditional institutions, collective landscape management and related governance systems among the Mijikenda to halt both ILK erosion and biodiversity loss. This can be achieved by strengthening the conservation of the sacred *Kaya* forests as biocultural heritage territories through integrated landscape management approach. This can have important co-benefits between different Sustainable Development Goals (SDGs) such as SDG2 (Zero Hunger), SDG13 (Climate Action) and SDG15 (Life on Land).

At the national level, the Science, Technology and Innovation Act 2013 provides for protection of all innovations. However, as the process of patenting traditional innovations is too complex, innovations anchored on modern science are still given preference. Thus the process of patenting should be made less complex to incorporate ILK-based innovations, as a possible means of unlocking a closer collaboration between ILK holders and research institutions, which is still lacking as in many SSA contexts (IPBES 2018) (see also Chap. 8 Vol. 1). Among multiple other SDGs, this can have ripple effects for SDG8 (Descent Work and Economic Growth) and SDG9 (Industry, Innovation and Infrastructure).

Finally, the conservation of local cultivars is still a major challenge due to capacity and resource constraints. Although farmers recognize the suitability and the benefits of local cultivars, the extension service providers are oriented towards conventional/modern hybrid crop varieties that are promoted at the expense of local cultivars (see also Chap. 10 Vol. 1). Thus, there is need to sensitize extension service providers to recognize the benefits of ILK practices with regard to the conservation and planting of local cultivars. Building appropriate portfolios of local and external cultivars could better ensure local food security in the face of climatic variability (Chaps. 6, 10 Vol. 1).

3.5 Conclusions

This chapter explored livelihood and food security patterns in five local communities in the coastal region of Kenya. Crop production was the main source of livelihood and income for most communities, but it declined between 2003 and 2012. Climate change most likely caused this low crop productivity through the combined effects of unpredictable rainfall, prolonged dry periods and increased incidences of weeds, pests and diseases. As a result, food security was compromised in many of the study communities during the past years.

To cope with the decreasing crop and livestock productivity, the five Mijikenda communities have developed a number of practices and innovations anchored on ILK, which vary between communities depending on their cultural values and socio-economic and ecological conditions. Most of the communities widely use these ILK-based practices and innovations to improve crop productivity and ensure food security in the face of climate change, including (a) crop diversification, (b) early planting and adoption of drought-tolerant and fast-growing local cultivars, (c) crop rotation, (d) conservation tillage, (e) domestication of food and medicinal plants from *Kaya* forests and (f) use of bio-pesticides.

The local communities use their cultural values, social networks, customary resource management practices and traditional governance systems to ensure the preservation of these ILK practices. Considering that the ILK-based innovations and practices discussed in this chapter can be an effective tool to achieve community resilience to climate change, urgent action is needed to integrate them meaningfully into relevant policies and climate change adaptation strategies at the local, national and international levels.

Acknowledgement The authors acknowledge the financial support from the European Union that funded Smallholder Innovations for Resilience (SIFOR) Project and the technical and logistical support from Kenya Forestry Research Institute towards undertaking the study.

References

Altieri AM, Nicholls CI (2013) The adaptation and mitigation potential of traditional agriculture in a changing climate. Clim Change:1–13. https://doi.org/10.1007/s10584-013-0909-y

Awuor C (2009) Increasing drought in arid and semi-arid Kenya. In: Ensor J, Berger R (eds) Understanding climate change adaptation: lessons from community-based approaches. Practical Action Publishing, Rugby, pp 101–114

Bremner J (2012) Population and food security: Africa's challenge. Policy Brief, Population Reference Bureau

Brown ME, Funk CC (2008) Food security under climate change. Science 319:580–581

Challinor AJ, Watson J, Lobell DB, Howden S, Smith D, Chhetri N (2014) A meta-analysis of crop yield under climate change and adaptation. Nat Clim Chang 4:287–291. https://doi.org/10.1038/nclimate2153

Collier P, Conway G, Venables T (2008) Climate change in Africa. Oxf Rev Econ Policy 24 (2):337–353

Dinar A, Hassam R, Mendelssohn R, Benhin J (2008) Climate change and agriculture in africa, Impact assessment and adaptation strategies. Earthscan, London

Enete AA (2009) Middlemen and smallholder farmers in cassava marketing in Africa. Tropicultura 27:40–44

FAO (2009) How to feed the World in 2050. Food and Agriculture Organization of the United Nations (FAO), Rome

FAO (2011) The state of the World's land and water resources for food and agriculture: managing systems at risk. Earthscan, New York

FAO (2015) The state of food insecurity in the world. In: Meeting the 2015 international hunger targets: taking stock of uneven progress. Rome, Italy

Hall C, Dawson TP, Macdiarmid JI, Matthews RB, Smith P (2017) The impact of population growth and climate change on food security in Africa: looking ahead to 2050. Int J Agric Sustain 15:124–135

Intergovernmental Panel on Climate Change (IPCC) (2014a) Climate change 2014: impacts, adaptation, and vulnerability. Part A: global and sectoral aspects. Contribution of Working Group II (WG2) to the Fifth Assessment Report (AR5) of the Intergovernmental Panel on Climate Change (IPCC), Cambridge University Press

Intergovernmental Panel on Climate Change (2014b) Climate change 2014: synthesis report. Contribution of Working Groups 1, 11, and 111 to the Fifth Assessment Report of the IPCC. IPCC, Geneva, 155pp

IPBES (2018) The IPBES regional assessment report on biodiversity and ecosystem services for Africa. Secretariat of the intergovernmental Science-Policy Platform on Biodiversity and Ecosystem Services (IPBES), Bonn

Kenya Agricultural Research Institute (KARI) (2007) Land use practices in Mbeere District: biophysical and socio economic challenges, Copping strategies and opportunities: a baseline survey report. KARI, Nairobi

Kenya Forestry Research Institute (KEFRI) (2016) Capability mapping for growing high value tree species in the coast region of Kenya; a guide for farmers, Forest managers, extension agents and investors. Kenya Forestry Research Institute, Nairobi. 74pp

Kibaara B, Ariga J, Olwande J, Jayne TS (2009) Trends in Kenyan agricultural productivity: 1997–2007. Tegemeo Institute of Agricultural Policy and Development, Egerton University

Lobell DB, Burke MB, Tebaldi C, Mastrandrea MD, Falcon WP, Naylor RL (2008) Prioritizing climate change adaptation needs for food security in 2030. Science 319:607–610

Makoti A, Waswa F (2015) Rural communities coping strategies with drought- driven food insecurity in Kwale County, Kenya. J Food Secur 3(3):87–93

Morandi F (2015) Women are custodians of biodiversity, vital to the world. Bioversity International, Nairobi

Mutegi J, Ameru J, Harawa R, Kiwia A, Njue A (2018) Soil health and climate change: implications for food security in sub-Saharan Africa. International Society for Development and Sustainability

Ongugo P, Wekesa C, Ongugo R, Abdallah A, Akinyi L, Pakia M (2014) Smallholder innovation for resilience: qualitative baseline study, Mijikenda Community, Kenyan Coast. International Institute for Environment and Development (IIED), London

Onyango AO (2016) Finger millet: food security crop in the arid and semi-arid lands of Kenya. World Environ 6(2):62–70

Onyutha C (2018) African food insecurity in a changing climate: the roles of science and policy. Food Energy Secur e00160. https://doi.org/10.1002/fes3.160

Ponge A (2013) Integrating indigenous knowledge for food security. Perspectives from the Millenium Village Project at Bar-Sauri in Nyanza Province in Kenya. Institute of Policy Analysis and Research- Kenya and Institute of Education, University of London

Poya A, Yogesh V, Knill P, Foy T, Harrold M, Steele P, Tanner T, Hirsch D, Oosterman M, Rooimans J, Debois M, Sperling F (2002) Poverty and climate change reducing the vulnerability of the poor through adaptation. In: Draft presented at 8th conference of parties to UNFCC, New Delhi

Prasad YG, Maheswari M, Dixit S, Srinivasarao C, Sikka AK, Venkateswarlu B, Sudhakar N, Prabhu Kumar S, Singh AK, Gogoi AK, Singh AK, Singh YV, Mishra A (2014) Smart practices and technologies for climate resilient agriculture. Central Research Institute for Dryland Agriculture (ICAR), Hyderabad, p 76

Republic of Kenya (2009) National population and housing census. Kenya National Bureau of Statistics, Nairobi

Republic of Kenya (2013a) Kilifi County first county integrated development plan 2013–2017. Government of Kenya, Nairobi

Republic of Kenya (2013b) Kwale County first county integrated development plan 2013–2017. Government of Kenya, Nairobi

Republic of Kenya (2013c) National climate change response strategy. Government of Kenya, Nairobi

Rippke U, Ramirez-Villegas J, Jarvis A, Vermeulen SJ, Parker L, Mer F, Howden M (2016) Timescales of transformational climate change adaptation in sub-Saharan African agriculture. Nat Clim Chang 6:605–609. https://doi.org/10.1038/nclimate2947

Rosenzweig C, Hillel D (2008) Climate variability and the global harvest: impacts of El Niño and other oscillations on agro-ecosystems. Oxford University Press, 280 pp

Roué M, Césard N, Adou Y, Yao C, Oteng-Yeboah A (eds) (2016) Indigenous and local knowledge of biodiversity and ecosystem services in Africa, Knowledges of nature 8. UNESCO, Paris

Spindell-Berck C, Berck P, Di Falco S (eds) (2019) Agricultural adaptation to climate change in Africa: food security in a changing environment. Routledge, London

Van Ittersum MK, van Bussel LGJ, Wolf J, Grassini P, van Wart J et al (2016) Can sub-Saharan Africa feed itself? PNAS 113(52):14964–14969

Vincent K, Cull T, Chanika D, Hamazakaza P, Joubert A, Macome E, Mutonhodza-Davies C (2013) Farmers' responses to climate variability and change in southern Africa–is it coping or adaptation? Clim Dev 5(3):194–205

Waha K, Müller C, Bondeau A, Dietrich JP, Kurukulasuriya P, Heinke J, Lotze-Campena H (2013) Adaptation to climate change through the choice of cropping system and sowing date in sub-Saharan Africa. Glob Environ Chang 23:130–143. https://doi.org/10.1016/j.gloenvcha.2012.11.001

Wekesa C, Ndalilo L (2018) Sustainable use of biodiversity in socio-ecological production landscapes and seascapes (SEPLS) and its contribution to effective area-based conservation: the case of Kaya forests on the Kenyan Coast. United Nations University Institute for the Advanced Study of Sustainability, Tokyo

Wekesa C, Ndalilo L, Ongugo P, Leley N, Swiderska K (2015) Traditional knowledge based innovations for adaptation and resilience to climate change: the case of coastal Kenya. In: XIV World Forestry Congress, Durban, South Africa, 7–11 September 2015

Wekesa C, Ndalilo L, Swiderska K (2016) Kaya forests: role in climate change adaptation among the Mijikenda community. In: Socio-ecological production landscapes and seascapes (SEPLS) in Africa, UNU-IAS & IR3S/UTIAS, United Nations University Institute for the Advanced Study of Sustainability Tokyo

Wekesa C, Ongugo P, Ndalilo L, Amur A, Mwalewa S, Swiderska K (2017) Smallholder farming systems in coastal Kenya: key trends and innovations for resilience. International Institute for Environment and Development (IIED), London

Chapter 4
Reframing the Challenges and Opportunities for Improved Sanitation Services in Eastern Africa Through Sustainability Science

Sara Gabrielsson, Angela Huston, and Susan Gaskin

4.1 Introduction

The need for access to sufficient, clean and reliable drinking water has long been recognized as a major development goal (WHO/UNICEF 2017). However the long-term public health benefits of clean water provision will only be sustained if hygienic sanitation conditions are present (Bartram and Cairncross 2010). The availability of water, sanitation and hygiene (WASH) services is essential for a healthy and dignified life, but these services are astonishingly still unavailable to a third of the global population (WHO/UNICEF 2017).

The main purpose of WASH programmes is to separate humans from contact with faeces (and associated pathogens) as a means of preventing disease transmission through faecal-oral pathways.[1] However, recent sanitation statistics indicate that out of the 962 million people living in Sub-Saharan Africa (SSA), as many as 220 million (23%) still practice open defecation, 300 million (31%) rely on

---

[1]WASH-related diarrhoea is the most widespread and dangerous of these diseases, estimated to have caused 842,000 deaths in 2012 alone (Pruss-Ustun et al. 2014).

S. Gabrielsson (✉)
Lund University Centre for Sustainability Studies, Lund, Sweden
e-mail: sara.gabrielsson@lucsus.lu.se

A. Huston
McGill University, Montreal, QC, Canada

IRC International Water and Sanitation Centre, The Hague, The Netherlands
e-mail: huston@ircwash.org

S. Gaskin
McGill University, Montreal, QC, Canada
e-mail: susan.gaskin@mcgill.ca

© Springer Nature Singapore Pte Ltd. 2020
A. Gasparatos et al. (eds.), *Sustainability Challenges in Sub-Saharan Africa II*,
Science for Sustainable Societies, https://doi.org/10.1007/978-981-15-5358-5_4

unimproved sanitation facilities, 172 million have limited sanitation services (18%) and only 270 million (28%) have access to basic sanitation (e.g. an improved pit latrine not shared with other households) (WHO/UNICEF 2017). In addition to being a major health risk, inadequate sanitation has a high negative impact on earnings, with the national economic benefits of proper sanitation investments being well-documented (WSP 2012; UN-Water 2008). The World Bank Water and Sanitation Program has estimated that poor sanitation has major economic costs to 18 SSA countries (USD 5.5 billion per year), the greatest proportion of which is associated with premature death due to diarrheal diseases (WSP 2012).

While there are abundant approaches and frameworks for implementing sanitation services, there is still scarce evidence of the long-term success of sanitation interventions in SSA (Davis 2016). The international development community and national governments have not been able to meet the sanitation demands of the rapidly growing population across SSA, despite billions of dollars invested over the past decade (Davis 2016; WHO/UNICEF 2015, 2017) (Chaps. 1, 5 Vol. 1). Although the Gross Domestic Product (GDP) is increasing rapidly in many SSA countries, the continent is simultaneously coping with the negative impacts of climate change, high rates of population growth, massive migration into urban areas and the expansion of informal settlements (see Chap. 1, Vol. 1). This has resulted in increasing economic inequality that has left millions of people without the basic human right to sanitation (Cross and Coombes 2014; Oates et al. 2014).

Eastern Africa is a region of SSA that exemplifies the challenges and problems associated with the lack of successful implementation of WASH programmes. Despite large increases in the national GDP of Kenya, Tanzania and Uganda (eightfold, tenfold and six-fold, respectively, since 1990), the sanitation sector has not kept pace in terms of service delivery improvements and remains chronically underfunded (see Chaps. 1, 5, Vol. 1). In fact the fraction of the population with access to sanitation has only slightly increased from 25% to 30% in Kenya, from 7% to 16% in Tanzania and from 6% to 13% in Uganda between 2000 and 2015 (World Bank 2017). UN Water found that 80% of SSA countries report insufficient financing for the sanitation sector, which perpetuates the tendency to seek external *solutions* to the sanitation challenge, instead of developing new and robust local financing schemes and owner-operator-regulator relationships (UN Water 2014 cited by Davis 2016) (Chaps. 1, 5 Vol. 1). Although national-level policies widely recognize sanitation as the responsibility of the government (UN Water 2014), the international development and non-profit communities play a large role in both financing and implementing sanitation solutions in eastern Africa (WHO 2017). The extensive, but often inconsistent, investment from external support agencies has usurped the responsibility of the governments and has allowed national and local sanitation systems to remain weak (Ekane et al. 2014). Sanitation is often bundled with (but usually as a second priority to) drinking water and broader WASH sector activities, while donor investment remains fragmented, inconsistent and unsupported by national policy (Galli et al. 2014; Ekane et al. 2016) (Chap. 5 Vol. 1).

The above suggest that the proper implementation and wide-scale adoption of WASH activities are a major sustainability challenge in eastern Africa. Indeed, between 2000 and 2015, the Millennium Development Goals (MDGs) were driving the international development agenda, with MDG 7c aiming to halve the number of people globally without access to an improved water source and to sanitation facilities (Hickling 2014). In 2015, it was evident that most SSA countries, including Kenya, Tanzania and Uganda, had failed to meet their WASH targets, with the largest gap being for basic sanitation (WHO/UNICEF 2017). While this failure is due in part to the chronic underfunding of the sanitation sector, it is also due to the lack of coordinated investment efforts, from both SSA governments and donors, and the lack of national policy frameworks (Galli et al. 2014; Ekane et al. 2016) (Chap. 5, Vol. 1). Even though, the commitment to improve sanitation coverage was an important step of MDG targets, the actual headline indicators focused on infrastructure construction without promoting sufficient investment in supporting systems promoting sanitation use and providing system maintenance (Davis 2015). The MDGs also limited their scope to the toilet itself and did not consider the entire *sanitation chain*,[2] which must also be addressed to achieve the desired long-term health benefits (Galli et al. 2014; Mulumba et al. 2014). Furthermore, the MDGs did not target household and community hygiene practices that must be understood and addressed at the same time as technological sanitation solutions are provided, in order to effectively break the faecal-oral disease pathway (Tilley et al. 2014).

The newly adopted Sustainable Development Goals (SDGs), through SDG 6, aim to overcome the shortcomings of MDG 7c by including targets for universal coverage, faecal sludge management and wastewater treatment (UNICEF 2017). These are ambitious targets that will motivate significant investment. However, even with the currently high levels of investments and good understanding of the challenges (Chap. 5, Vol. 1), these targets are unlikely to be met in most SSA countries (WHO/UNICEF 2017) without a radical paradigm shift in how we *view* and *do* sanitation.

Sustainability Science may help to reframe the sanitation challenge in SSA by offering insights about the requirements necessary to implement and maintain sustainable sanitation services. Sustainability Science is an emerging discipline with a vibrant research community that brings together scholarship and practice from different perspectives (e.g. global and local, north and south) and disciplines from the natural sciences, social sciences, engineering and medicine (Clark and Dickson 2003; Kates 2011; Ziegler and Ott 2011). Sustainability Science aims at finding solutions to complex problems, such as those characterized by numerous feedback loops and interactions with other sectors, which require behaviour change and have high damage potential, urgency and no obvious optimal solution (Rittel and Webber 1973; Wiek et al. 2011) (Chap. 1, Vol. 1). Solutions are sought by attempting to generate, integrate and link use-inspired knowledge and channel it into transformative action through participatory, deliberative and adaptive techniques

---

[2]The *sanitation chain* refers to the series of processes necessary in order to safely manage human waste. The steps are capture, containment, transport, treatment and disposal/reuse (Galli et al. 2014).

(Kates et al. 2001; Bäckstrand 2003; Clark and Dickson 2003; Komiyama and Takeuchi 2006; van Kerkhoff and Lebel 2006; Sarewitz et al. 2010; Jerneck et al. 2011; Wiek et al. 2011; Talwar et al. 2011) (Chap. 8 Vol. 2). The strength of this approach lies in redefining the functions, mandating and scoping of scientific inquiry and understanding human-environment systems as integrated (i.e. coupled) rather than separable or even separate (Clark 2007; Kates et al. 2001).

Sustainability Science adopts a systems thinking perspective. It is *problem-driven* and *solution-oriented*, making it conducive to understanding and responding to the complexity posed by poor sanitation in SSA (Huston and Moriarty 2018; Neely 2019; Andersson et al. 2016). Using this approach for the sanitation challenge first requires the identification and description of the entire sanitation chain in a particular geographic setting, i.e. how does the sanitation system function and how do related interventions perform, while considering the different value-laden goals and objectives (Schultz et al. 2008). For example, it could be possible to investigate the balancing and reinforcing feedback between the generation, removal, disposal and potential reuse of faeces and urine and the motivation for (and impacts of) different sanitation interventions on humans and ecosystems (Andersson et al. 2016). Such an inquiry would require extensive place-based knowledge and the use of theories, methods and tools from an array of disciplines (Jerneck et al. 2011). This would need to be produced in a *transdisciplinary* manner with the involvement of wider societal actors such as NGOs, the private sector, national/local government and local communities (Galli et al. 2014) (Chap. 8 Vol. 2).

Once the sanitation system is understood, practical solutions to the underlying sustainability challenges must be sought. Solution-oriented principles first require questioning the sustainability of existing solutions and then exploring alternative pathways through strategic and operational questions to identify which transition pathways are viable (Loorbach 2010; Jerneck et al. 2011). In the context of sanitation implementation in SSA, this step could include a critical analysis of the applicability and sustainability of water-based sanitation systems (also called wet sanitation),[3] considering their reliance on reliable piped water supply and sewer networks. Wet sanitation may not be appropriate considering the expected water scarcity due to climate change in much of SSA (see Chap. 1, Vol. 1). In addition, only 4.6% of the SSA population has access to sewers (Oates et al. 2014). This begs at least an investigation into alternative sanitation options as a means of meeting SDG6 (Clean Water and Sanitation), as well as other goals related to sanitation such SDG3 (Good Health and Wellbeing). Furthermore, Sustainability Science approaches could be used to explore the *synergistic effects* of turning human waste into valuable products (e.g. biogas, fertilizers, animal feeds) and how this could

---

[3]Wet sanitation refers to a system of capture and transport of excreta that uses water as a carrying medium. This is the Victorian era model used in western style flush toilets and sewer networks. Dry sanitation does not use water and hence does not require sewers but needs an alternative transport and treatment method for the more solid medium, composed primarily of faeces.

contribute to transformational change through the environmental and income generating benefits it could offer (Diener et al. 2014).

The aim of this chapter is twofold and is anchored in Sustainability Science thinking (Sect. 4.2). The first is to provide a heuristic analysis of how the sanitation problem (and its solutions) is commonly conceived and how this perpetuates cyclical failure in the SSA context (Sect. 4.3.1). The second is to offer empirical examples from eastern Africa of some practical solutions that are breaking out of this failure cycle in eastern Africa by adopting new and innovative approaches to improve and sustain sanitation service operation and maintenance (Sect. 4.3.2). We then discuss the governance implications of reframing the sanitation challenge in the region and make policy recommendations for the future (Sect. 4.4).

4.2 Methodology

The lack of priority for (and the competitive nature of) funding for sanitation in SSA reduces the opportunity to talk openly about failure and explains why so few failed projects have been reported and properly documented (Davis 2016). Not only does this diminish the potential learning from failure, but it also encourages stakeholders to understate and oversimplify its causes. Our methodological approach is anchored in Sustainability Science, and in particular the principle of systems thinking and is twofold: critical and exploratory. It is critical in its attempt to understand the drivers of systemic failure from a sector perspective (Sect. 4.3.1) and exploratory as it outlines how specific characteristics can allow sanitation systems to break out of this failure cycle (Sect. 4.3.2).

Our initial analysis of the drivers of systemic failure in the sanitation sector (Sect. 4.3.1) incorporates information from academic and grey literature, as well as the authors' own experience from practical sanitation work and research in the region during the last 5 years (2013–2018). Rather than doing detailed analyses of 'failed' sanitation projects and programmes across eastern Africa, our focus is to identify and discuss crosscutting aspects of the cyclic failure in the sector that persist across rural and urban contexts across the region.

An overview of the sanitation context in each of the three study countries (i.e. Kenya, Tanzania, Uganda) was obtained through seven transect walks combined with informal interviews with different actors (Table 4.1). These walks took place in one rural village (Lumuli, Tanzania), two peri-urban settings (Naivasha, Kenya and Chuka, Kenya), and four informal urban settlements (Keko in Dar es Salaam, Tanzania; Kibera in Nairobi, Kenya; Kiwanja Ndege in Naivasha, Kenya and Kabalagala in Kampala, Uganda) (Fig. 4.1).

We then analyze data from six implemented WASH schemes in the three countries (Box 4.1), to identify the characteristics conducive to breaking out of the failure cycle (Sect. 4.3.2). A snowball sampling method was used to identify the study schemes that are a representative sample of a specific type of projects (see below) rather than an exhaustive list of sanitation schemes in the region. In

Table 4.1 Characteristics of key informant interviews and focus group discussions

Respondent	Organization	Location	Data collection method
Research officer	Umande Trust	Kibera, Nairobi, Kenya	Key informant interview
Project manager	Umande Trust	Kibera, Nairobi, Kenya	Key informant interview
Deputy director	Umande Trust	Kibera, Nairobi, Kenya	Key informant interview
Soweto high-rise savings group	Umande Trust	Kibera, Nairobi, Kenya	Focus group discussion
Teacher, St. Christina School	Umande Trust	Kibera, Nairobi, Kenya	Informal interview
MUVI self-help group members	Umande Trust	Kibera, Nairobi, Kenya	Focus group discussion
Government relations specialist, Management team member	Sanergy	Makuru, Nairobi, Kenya	Key informant interview
Engineer	Sanergy	Makuru, Nairobi, Kenya	Key informant interview
Director	Centre for Community Initiatives	Keko, Dar es Salaam, TZ	Key informant interview
Sanitation engineer	Centre for Community Initiatives	Keko, Dar es Salaam, TZ	Key informant interview
Tumaini Letu group	Centre for Community Initiatives	Keko, Dar es Salaam, TZ	Focus group discussion
Co-founder	Sanivation	Naivasha, Kenya	Key informant interview
Energy production, team member	Sanivation	Naivasha, Kenya	Key informant interview
Programme assistant, decent living project	Environment Alert	Kabalagala, Uganda	Key informant interview

particular, the sanitation schemes were selected from communities that are currently underserved by sanitation services, which had adopted an innovative on-site technology in either a rural or an urban setting. They are self-financed (i.e. not through charity) and are either based on profit-making business models or are community-owned and managed schemes. We identify (a) the characteristics that make them different to the current business-as-usual sanitation approaches, (b) how their adaptive systems incorporate 'soft' elements and (c) how broader systems strengthening can build and support an environment conducive to sustainability.

Fig. 4.1 Study sites in Kenya, Tanzania and Uganda

Box 4.1 Study Sanitation Schemes

Umande Trust is a rights-based agency that works in informal settlements ('slum' communities) in and around Nairobi, Kenya. Umande uses a multi-level approach that focuses on delivering a 'product'(e.g. access to urban water, bio-sanitation, solid waste management services) by creating a raft of community-led processes to support it. Such processes are, for example,

(continued)

Box 4.1 (continued)

partnerships for change, integrated urban environmental planning, sanitation governance, human rights and urban services financing. The Umande Trust team is comprised of community organizers, academics, geospatial analysts, urban planners, human rights advocates, civil engineers, social scientists, environmental scientists as well as gender, youth and enterprise development resource persons. For more information refer to www.umande.org.

Centre for Community Initiatives (CCI) is a national support NGO formed in Tanzania. Its aim is to provide technical and financial assistance to local communities in informal settlements. CCI focuses on building resilient communities and supporting them to meet their needs. Their work could entail the provision of direct help to local communities by installing sanitation infrastructure or providing complementary support from the community such as establishing savings schemes, community resource mobilization and organization, enumeration and mapping support, exchange visits, partnership support, technical assistance, capacity building, leadership and management support, outreach, advocacy, action-oriented research and documentation. For more information refer to www.ccitanzania.org.

Sanergy is a social enterprise that provides low-cost hygienic sanitation facilities that are rapidly installed and designed to function in dense informal settlements in Nairobi, Kenya. Sanergy employs a franchise business model that provides business training and microcredit loan opportunities to small-scale sanitation engineers, who then maintain, operate and expand access to toilets in high-demand areas. Sanergy toilets are urine-diverting dry toilets, capturing the waste for reuse in agriculture and energy production. For more information refer to www.sanergy.com.

Sanivation is a social enterprise that instals container-based toilets in homes in local communities near Naivasha, Kenya. The toilets are installed for free and Sanivation charges a small monthly fee to empty them, transforming the waste into a clean burning alternative to charcoal. The enterprise focuses strongly on providing a *service* rather than simply a toilet or charcoal alternative. Thus they focus on the wants and needs of local communities, while at the same time addressing the entire sanitation chain in an effort to reduce faecal contamination hazards in urbanizing communities. For more information refer to www.sanivation.com.

The Decent Living Project in Kampala, Uganda, aims to improve the lives of residents in informal settlements through the development of WASH services for local needs. The project takes a three-pronged approach combining (a) advocacy for WASH needs and services, (b) construction of facilities with local artisans, and (c) the development of business enterprise to support a range of sanitation and water-related business models. Environment Alert, the implementing NGO, works with local entrepreneurs and youth groups to

(continued)

Box 4.1 (continued)
identify and understand local needs. It then supports the development of business models to fill these gaps in an equitable manner (e.g. brickmaking for improving facilities, waste treatment for use in urban agriculture). For more information refer to http://envalert.org/phase-2%E2%80%B2-decent-living-dl/.

These selected sanitation projects share three similarities: (a) they are deeply embedded in the local context; (b) they have a balanced approach to meeting human needs and assuring environmental sustainability, thereby taking a *service delivery approach*; and (c) they recognize the need to move away from aid-based approaches and towards financially viable sanitation solutions. By using these three aspects, we demonstrate how these organizations start from a complex and dynamic understanding of the current situation, which allows them to consider alternative future scenarios, rather than being limited to the business-as-usual approach.

The sanitation schemes were analyzed using qualitative data collected from in-person key informant interviews with staff, project site visits and focus group discussions with community members participating in the sanitation schemes (Table 4.1). The empirical data was collected during fieldwork in Tanzania (January to March 2015 and July of 2016), Kenya (June of 2015 and 2016) and Uganda (June of 2016). Additionally we conducted a desk review of the available documentation about each organization's approach and on published peer-reviewed articles about their progress (e.g. O'Keefe et al. 2015; Otsuki 2016).

The empirical data was analyzed and synthesized using a Sustainability Science approach, which is founded in an iterative learning process, to identify systemic properties that improve or inhibit sustainability. We should note that this analysis focuses on the qualitative aspects of sanitation provision in eastern Africa, rather than on providing a comprehensive technical and financial analysis of the different approaches. Through this rapid assessment approach, we identify key aspects that affect the technical and financial viability of the study sanitation schemes. In a sense, we seek to identify how novel ideas are implemented in innovative ways, exhibit institutional learning and adaptive capacity and have the potential for long-term sustainability. Rather than using a formal definition of *success*, we describe *promising* approaches based on their transformational change potential, in-built flexibility and suitability for scaling up.

Lastly, it must be recognized that the authors were born, raised and educated in the lobal North. Therefore, despite our profound recognition of the western bias as a contributing factor for systemic failure (Sect. 4.3.1.2), it is not removed from our research and perspectives, despite having spent a significant amount of time living and working in SSA (periodic stays of several months since 2006). We seek to mitigate the influence of this bias by undertaking a context-laden and systemic approach to this analysis.

4.3   Results and Discussion

4.3.1   Factors of Cyclic Failure

Through a critical analysis of the business-as-usual approach to sanitation using a sustainability thinking lens, we identify persistent shortcomings and trends associated with inadequate solutions (Sects. 4.3.1.1, 4.3.1.2, 4.3.1.3 and 4.3.1.4). Figure 4.2 outlines the relationship between the main factors and challenges described in Sect. 4.3.1 and demonstrates how they create a reinforcing cycle of failure for the business-as-usual sanitation sector in eastern Africa.

4.3.1.1   Lack of Systems-Based Thinking

When linear or single-issue thinking is applied, only one possible outcome is considered for a given intervention. A sanitation chain that recognizes the interlinkages between different processes from the capture to the safe disposal remains inadequate if the people, hygiene behaviours and capacities at each step of the chain are not considered. The construction, maintenance and sustained use of latrines are separate but related issues that must all be addressed in order to achieve the positive health outcomes associated with improving WASH access. Often the focus is on either technology or behaviour change, without adequate consideration of

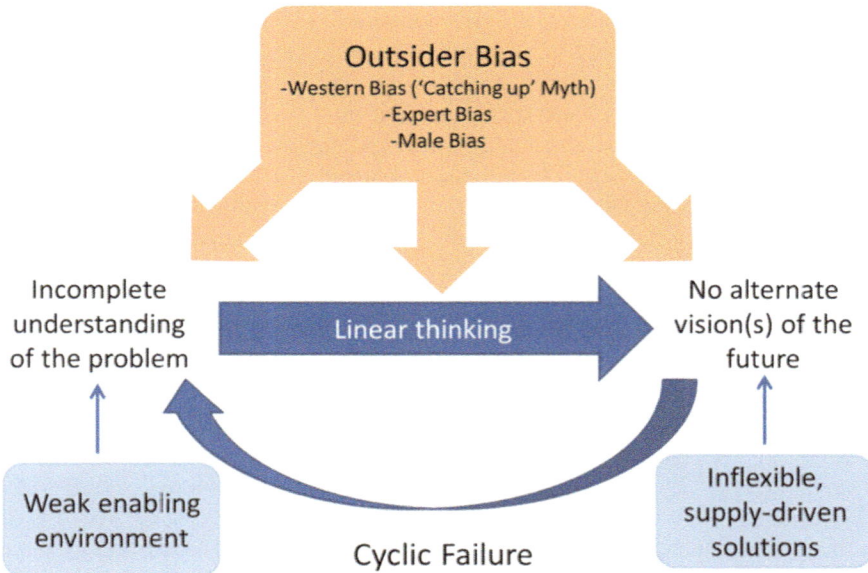

Fig. 4.2 Cyclic failure of the business-as-usual model for sanitation solutions in eastern Africa

what types of technology and what social conditions can jointly promote the sustained use of sanitation facilities. In other cases, facilities are developed without an adequate consideration of the land tenure, environmental conditions or legal and regulatory constraints that will affect the success of the intervention (Sepalla 2002; Huston and Moriarty 2018).

A *systems approach* acknowledges that the overall functioning of systems is more than just the sum of their parts (Stroh 2015). Such an approach demands a good understanding of the complex interactions within a system prior to developing interventions that aim to shift its dynamics (Stroh 2015). Systems-based solutions are holistic rather than symptom-based, they are achieved more slowly and their endpoint is more variable than conventional (single-issue) solutions (Galli et al. 2014). However, they are also more flexible because the dynamic and unpredictable nature of the system is recognized. In the context of sanitation, the sustainability of sanitation service provision depends on the interrelated factors of robustness of the economic conditions, effective governance, supportive social systems and sufficient natural resources (Galli et al. 2014).

4.3.1.2   Outsider Biases

The problems caused by short-term thinking (Sect. 4.3.1.1) are exacerbated when solutions are designed by outsiders having little or no long-term experience in the implementing context and who are unaware of (and unable to predict) the possible outcomes of a given intervention.

An example from the sanitation sector is community-led total sanitation (CLTS), which seeks to change sanitation habits by triggering shame in communities that practice open defecation and thus catalyze the construction and sustained use of latrines (Chambers and Myers 2016). This method was developed in Asia and has been applied broadly across much of SSA with little consideration of the different but related needs (Davis 2016). While CLTS will produce a peak in toilet construction and use, inadequate consideration of the specific needs for expertise, spare parts, environmentally appropriate technologies and context-specific public health behaviour training will limit its sustainability over the longer term and hinder the achievement of the desired health impacts over time (Davis 2016).

Sustainability Science calls for an increased understanding of the perspectives of the many stakeholders and their capacity to fulfil their roles and responsibilities (Jerneck et al. 2011). In western societies, where sanitation systems are well established, sanitation is usually the responsibility of the state (municipality) with its highly trained functionaries. Its operation and maintenance is financed through taxes or user fees, while capital projects are financed through loans and partnerships with upper levels of government. The in-house user interface with the sanitation system, the toilet, is bought by homeowners but mandated and regulated by public building codes and health standards. These multiple stakeholders (and their interactive roles for developing and maintaining a sanitation system) suggest that deep knowledge of the context, social behaviours and political economy is necessary in order to understand and remove the multiple barriers to sanitation success. In this sense, path dependence refers to the constraint that the identified set of options/solutions for any given sustainability challenge is limited by past decisions and that visions of success can emerge from other contexts and under different conditions (Rip and Kemp 1998).

Below we discuss three outsider biases, which can be a key cause of the cyclic failure of sanitation systems, contributing to the lack of sustainable sanitation services in SSA: (a) western bias, (b) expert bias and (c) male bias.

Western bias is the 'myth of catching up development' (Mies 1998), which assumes that the development of all countries will evolve along the same path to reach the same endpoint. For sanitation, the progression is to advance up the 'sanitation ladder' from open defecation, via pit latrines and pour flush toilets connected to septic tanks to toilets connected to a sewerage system (WHO/UNICEF 2015). Many large-scale projects are locked into the bias of water-based sewerage (Tilley 2008) and a financial and technological dependency that may not be appropriate or relevant in the eastern African context. Currently only 20% of the population in eastern Africa use sewers (WHO/UNICEF 2017),[4] and current strategic plans and legislative frameworks do not account for the remaining 80% who depend on non-sewered sanitation infrastructure (Ekane et al. 2016). The bias towards the western standard of a water-based sanitation system fails to recognize the unsustainability of this technology for water-stressed regions, such as eastern Africa (Penner 2010). The pursuit of this ideal hampers the organic development of locally appropriate sanitation systems, and instils a sense of inferiority or 'backwardness' when a country fails to advance up the linear sanitation ladder (Penner 2010). Questioning the environmental sustainability and financial viability of the top rungs of the ladder, some scholars are now calling for a revision of the concept of the sanitation ladder to add alternative benchmarks for improved WASH services, which recognize other more sustainable and/or appropriate sanitation technologies for countries of the Global South (Kvarnström et al. 2011).

Expert bias is due to outside 'experts' from the Global North or South (often sanitation engineers), who advise local communities on the construction of a predetermined sanitation technology. Challenges may arise when the outside experts fail to understand the local cultural context and behaviour or practices of the non-'experts', particularly for local decision-making processes during the planning and execution of activities. In one example from Iringa, Tanzania, unnecessary conflicts arose when the sanitation engineer from India declared to the village leadership (exerting his power as an expert) that he knew the local needs better than the community who was to use the WASH system. This eroded the established trust between the outsiders and the community, further complicating the implementation of the system, which ultimately led to the abandonment of the project (Project Manager, personal communication, July 4, 2014).

---

[4]Data from WHO/UNICEF 2017 baseline. Mean estimate for Sub-Saharan Africa as a whole

Male bias reflects the domination of men in the sanitation sector across all levels of implementation (Seager 2010). This poses a major challenge for sanitation sustainability. While women and girls are disproportionately affected by inadequate WASH services due to their biological needs, roles as caretakers of domestic tasks and established societal taboos; they have the least power to change this situation (Taylor 2009; Gabrielsson and Ramasar 2013). Males hold the decision-making power in eastern Africa at all levels, ranging from the individual household to the sub-village, village, district, region and government levels (Gabrielsson and Ramasar 2013). Unless the men in power fully understand and prioritize the importance of sanitation at all these levels, the sustainable implementation of sanitation systems will be difficult to achieve (Seager 2010). A typical example of this lack of priority is the fact that the majority (63%) of school toilets in Tanzania lack facilities to dispose of menstrual hygiene care products, forcing schoolgirls to dump them inside the latrines (causing clogs and overfilling) or bring the soiled and smelly pads back home (NIMR 2016).

There are many specific consequences of these outsider biases, but the overall effect is the development of poorly designed, disjointed and misaligned sanitation strategies based on short-term goals (Davis 2016). In addition, many sanitation strategies disregard or neglect the needs, priorities, voices and participation of the most vulnerable community members in sanitation projects and related national policies (Tsinda et al. 2013). The pathways leading from sanitation-related problems to the identified solutions are either inadequate and/or inflexible (linear) and therefore cannot access alternate visions of the future (Galli et al. 2014).

Such inflexible approaches and outsider biases contribute to dire sustainability challenges for sanitation efforts in eastern Africa and may partly explain why current solutions to 'fix' the problem(s) are not working. Sections 4.3.1.3 and 4.3.1.4 explore using a sustainability science lens to more broadly assess the enabling environment (e.g. the political economy and governance framework), in which sanitation interventions take place.

4.3.1.3 Weak Governance and Inadequate Long-Term Financing

Sanitation interventions and services require the involvement of individuals, households, local communities (and/or schools), operators and multiple levels of government (Galli et al. 2014). Many actors must be engaged to coordinate between sectors, yet sanitation often remains a low priority even for actors legally responsible for it (Ekane et al. 2016). For example, in Tanzania, the responsibility for sanitation is divided between the Ministry of Health, Community Development, Gender, Elderly and Children, the Ministry of Water and Innovation and the Ministry of Education, Science and Technology. This contributes to fractured governance and financing, as sanitation implementation problems are often dismissed as 'somebody else's problem' (Kimwaga et al. 2013; Ekane et al. 2014).

As already discussed in Sect. 4.1, due to the fragmented financing landscape, many eastern African governments look externally for possible solutions and support (Mulumba et al. 2014). However often such interventions are plagued with issues that reduce their long-term viability and sustainability. For example, the typical 3–5-year project funding cycles of NGOs, outside initiatives and even government programmes limits long-term planning for sustainability. In a survey of 48 US-funded WASH NGOs, 89% and 96% reported limited timeframes and lack of funding for long-term monitoring, respectively, as key hindrances to their ability to contribute to sustainable water and sanitation services (Davis 2015). Too often, we observe 'better-than-nothing' solutions that improve the situation for a short period and then fail, which are repeated and subsequently marked as successes (Jenkins and Sugden 2006). For example, continuing with the CLTS example (Sect. 4.3.1.1), communities were successfully triggered to build their own latrines; however, many were poorly constructed and could not withstand seasonal floods (Davis 2016). After as little as one season, many of the new latrines became open holes filled with human waste, presenting safety hazards to the community. Furthermore, without resources to rebuild (although some communities may be triggered to build better sanitation options) and the means to act, they develop a feeling of lesser dignity (Sanitation Engineer, CCI, personal communication, March 30, 2016). Inability to recognize the systemic inadequacy of linear and piecemeal solutions can lead to cyclic failure and prevent the development of alternative and more sustainable sanitation solutions (Waterkeyn and Waterkeyn 2013; Strande and Brdjanovic 2014).

Measuring performance accurately and beyond the initial project period is critical for a data-driven learning cycle to increase sustainability (Sparkman 2012). All-or-nothing indicators that report only on the presence of infrastructure and the progression of service provision (linearly) up the sanitation ladder neglect the complexities of sanitation provision. Building on lessons learned from CLTS, we observe that merely counting the number of activities or events triggered is insufficient to track and understand actual improvement (Sparkman 2012). For sustained success, it is important to monitor if (and how) actual sanitation practices are implemented over time. For example, many sanitation schemes that have used CLTS to trigger change in sanitation behaviour rarely provide dedicated funding and time for monitoring and reporting (Davis 2016). This makes it difficult to assess the overall impacts and performance of such schemes.

Monitoring can also lead to perverse incentives. For example, the strong focus towards meeting the MDG targets drove sector initiatives aimed at improving their national statistics in the fastest way possible, instead of working holistically to reach the most vulnerable and achieving sustained progress (Fukuda-Parr et al. 2014). For example, in peri-urban areas of Nairobi, Kenya, we witnessed how subsidies were used to replace existing latrines with 'improved latrines' in order to improve MDG statistics. However, a local staff member of an involved NGO suggested that these funds may have been better used to address the currently inadequate waste transportation and treatment options.

4.3.1.4   Focus on Supply-Driven Solutions

Sustainability science and systems thinking provide a lens to investigate not only the challenges of the broader enabling environment but also those of the sanitation interventions themselves. Strategies that aim to *solve* sanitation sustainability challenges in Kenya, Tanzania and Uganda continue to be dominated by supply-driven solutions (Nyonyintono and Musembi 2011). This is reinforced by the outsider biases (Sect. 4.3.1.2), where funding, expertise and technology from donor countries might drive incoming *solutions*, rather than by the local needs. Such *solutions*, developed with an incomplete understanding of the sanitation system and local context, share certain characteristics that may help explain the limited success of sanitation interventions in the past.

Supply-driven sanitation solutions are typically biased towards the use of hardware. They tend to focus more on the design and construction of sanitation technologies (e.g. toilets), rather than their adoption, sustained use and contribution towards change in hygiene behaviour (Strande and Brdjanovic 2014; Andersson et al. 2016). Generally, a bias towards hardware solutions also implies that the proposed sanitation technology is imposed by the implementing organization, rather than selected at the local level. For example, a study in Rwanda found, when revisiting households who had received advanced sanitation technologies in their homes (urine diversion dry toilets), that many were not in use or were used improperly, thus negating any potential benefits (Ekane et al. 2016). Large investments in technology can also lock the users into a specific technology pathway, limiting their avenues for adopting alternative and new sanitation technologies and behaviours. It may also reduce options to use locally available and more affordable construction materials (Rip and Kemp 1998; Kvarnström et al. 2011).

This bias was present in the implementation of an integrated water and sanitation scheme by an Indian-based NGO in rural Iringa, Tanzania. The organization insisted on using porcelain-made squatting slabs for their pour flush toilets, to replicate the system they implement in India. However, as Tanzania lacks a porcelain factory, porcelain sanitation ware had to be imported and transported by trucks to the interior of the country. Furthermore, good quality PVC pipes to distribute water or sewage were also imported. As a result, the costs for this seemingly low-cost sanitation scheme were higher in eastern Africa than in South Asia, where such construction materials are locally produced. This extra cost in Tanzania had to be borne by either the organization or the users. If there is no financial mechanism and strategy to enable users to save for this investment or pay for it incrementally over time, it becomes impossible for users to pay. In the Iringa case, the Indian organization did not account for these high material costs at the start of the project, nor did it make any attempts to enable villagers to pay for the porcelain ware. Costs therefore outgrew the project budget, and funds had to be diverted from other planned activities, such as masonry assistance, which had to be paid for by the villagers. Ultimately, this was one of the main reasons why many members of the local community opted not to participate in the scheme.

In addition to potentially limiting the availability of (and accessibility to) afford-able construction materials, a hardware bias also runs the risk of being culturally inappropriate. A typical example would be to insist on building dry toilets in Muslim communities where anal cleansing using water is the norm (Nawab et al. 2006). Another example would be to build only communal toilets in areas where female mobility is constrained and their safety may be at risk, thus limiting their access to (and use of) WASH facilities (Nallari 2015).

Many supply-driven solutions are also market-based and hence managed by private sector stakeholders. Private sector investments offer some promising oppor-tunities, but a disadvantage of such solutions is that both the responsibility for their management and their costs are borne by individuals, rather than the broader community (Ekane et al. 2014). For example, during the implementation of Eco-San (ecological sanitation) toilets in Uganda, households were asked to pur-chase on-site treatment technologies. While this is promising for waste containment, marketing to households shifts the responsibility to the individual and allows the state to neglect its role in developing services for its citizens (Huston et al. 2019). Market-based sanitation initiatives, therefore, allow the government to neglect its responsibility in the sanitation service chain. In Tanzania, as in much of eastern Africa, the regulatory environment is under-resourced and thus market-driven solu-tions run the risk of "enabling" the private sector to exploit the citizens and neglect those most marginalized (Ekane et al. 2014). Omitting the most vulnerable segments of society may not only limit the potential reach of the sanitation services to the unserved but also fail to reach the adopted SDG 6 target of ensuring sanitation for all by 2030.

4.3.2 Breaking the Cycle of Sanitation Failure

Sections 4.3.2.1, 4.3.2.2, 4.3.2.3 and 4.3.2.4 outline the characteristics of a small selection of examples of *promising* approaches, which are not yet defined as sanitation *successes*. These interventions holistically address several different aspects that cause cyclic failure within the sector; nothing however is fail-proof. Their iterative learning-based approach makes them robust and resilient. Particularly successful aspects of their model demonstrate the use of critical Sustainability Science thinking to overcome the challenges described in the previous section.

4.3.2.1 Promote Place-Based Solutions

A key aspect of breaking the cycle of failure is to develop sanitation solutions that are appropriate for the physical, socio-economic and cultural context within which they will be deployed (Tilley et al. 2014). While this is common sense, experience on the ground suggests that sanitation solutions rarely fully fit the characteristics of the area, unless local communities and experts are closely involved in the implementation process (Mbaria 2014).

CCI in Tanzania now involves local communities and experts in identifying and developing interventions after initial failures to scale up the use of specific and favoured technologies. They acknowledge that when entering a new context, a problem that may initially look similar to a previous one, is in fact likely to be unique and may require a different approach. There is no 'one-size-fits-all' solution. This strategy seeks to avoid falling into the trap of promoting sanitation solutions that are eventually abandoned or fall into disrepair shortly after deployment (Davis 2015). At the start of each intervention, CCI spends a considerable amount of time to understand the community in which they are planning to work and then engages directly with individuals and groups within the community to develop the most appropriate solutions. For example, in Keko in Dar es Salaam, the groundwater table is very high, so a water-based system or even an improved pit latrine is not appropriate. With the local community participating in close consultation, CCI designed and constructed a urine diversion toilet with three holes that accommodates both the physical constraints of the area (i.e. the high water table), as well as the cultural issues (e.g. provision for anal cleansing). Another local community wanted to explore ways to reduce the need for costly pit emptying for household latrines. Families unable to afford emptying services often experience seasonal overflow that causes environmental pollution and possibly contaminates surface/groundwater sources (Strande and Brdjanovic 2014). Working with CCI, they adapted a version of a tiger toilet, an on-site system using worms to process faeces (Furlong 2016), as a means of reducing the volume of waste, with the added benefit of reducing smell and being able to reuse the vermicompost for local horticultural production. The combination of demand for the service and the involvement of the community in adapting the technology led to its appreciation, sustained adoption and use.

Sanivation and Sanergy also have built-in flexibility in their place-based sanitation solutions in Kenya. In their case, this flexibility lies in the use of local materials for the construction and manufacturing of their services and products. Sanivation relies on simple and locally available machinery to manufacture bio-charcoal derived from human waste, making it possible to hire local operators without the need for intensive training. This makes the technology scalable and reduces the overall production expenditures, keeping the price of the bio-charcoal lower than its wood-based alternative. This provides consumers with a significant incentive to switch their domestic fuel use (usually fuelwood or charcoal) to bio-charcoal with the added benefit of reducing deforestation and reducing related greenhouse gas emissions (Felix 2015) (see also Chap. 5, Vol. 2).

Sanergy also uses local materials to construct their Fresh Life Toilets. These are prefabricated at the Sanergy headquarters in Nairobi and are then assembled on-site in 2 days. The urine diversion dry toilets can also be disassembled into their cement block components. As a result there is flexibility in determining deployment location, as they are easy to transport to otherwise difficult to reach areas, characterized by high population density, erratic house planning and lack of access for cars and trucks. The quick assembly time also reduces costs and the risk of theft of the materials during the construction process. Moreover, the design of the collection buckets (for both urine and faeces) inside the toilets and the meticulously planned daily collection of the waste keeps the toilets from overflowing. This improves cleanliness and facilitates maintenance for the franchise operators. A few specific components must still be imported; however, Sanergy is working to achieve sufficient scale so that it is feasible to establish manufacturing of all parts locally in the Nairobi area (Engineer, Sanergy, personal communication, May 10, 2016).

4.3.2.2   Situate Sanitation Within Broader Governance Systems

Sanitation intervention implementers must not only focus on the infrastructure but also on the social and cultural context of the targeted communities. This means that, in addition to completing the targeted intervention, support is provided to local systems for monitoring, regulation and maintenance, as well as for developing service demand (Moriarty et al. 2013). This dual focus can better situate sanitation interventions within the larger system, within which the targeted local communities work, live and thrive. Sanitation interventions can be perceived as entry point activities (or stepping-stones), to achieve broader sustainable development goals. Several of the organizations whose approach towards sanitation delivery is more successful are also involved in issues beyond sanitation. They usually link sanitation delivery to other important sustainability issues related to agriculture, energy, gender empowerment and livelihood/income diversification (CCI, year; Umande Trust, year; Floret 2017).

For example, the bio-centres of Umande Trust provide both public access to pour flush toilets and facilities for hand-washing and showering, as well as spaces for cooking, banking, community meetings, housing and the development of local business enterprises. The bio-centres, therefore, offer local communities a place to access affordable sanitation, hygiene services and cooking facilities fueled from the biogas generated from the human waste. The local community is responsible for managing and maintaining the bio-centre and in the process has an opportunity to develop financial literacy, engage in leadership training and have alternative income sources. These co-benefits enhance the feeling of ownership, increase capacity building and, in particular, build trust among diverse stakeholders. The integration of these communal activities helps overcome community conflicts and is a powerful tool for gender equality, as it can enhance the voices and decision-making power of women (Floret 2017).

Similarly, the Community WASH Centres in Kampala initiated by Environment Alert (in partnership with WaterAid) are closely managed by a caretaker from within the local community. As a result, users gain access to not only a toilet but also a clean shower and a reliable service for refilling drinking water containers.

4.3.2.3 Foster Multi-Stakeholder Collaboration and Coordination

As discussed in Sect. 4.3.1, in order to enhance the effectiveness of sanitation interventions, it is important to engage with many different stakeholders on different levels. Sanitation provision in urban slums, in particular, exemplifies the need for creative collaboration between different actors, who interact in the geographically and financially constrained environment of informal settlements (Galli et al. 2014).

The Umande Trust developed its first community water and sanitation biogas centre in Kibera slum (Nairobi) in 2004, during a period when the Kenyan government did not recognize such community facilities as safe or viable sanitation options. Umande realized that the sanitation technology options outlined in policies at that time (e.g. household latrines and septic tanks) were not feasible options for residents of Kibera. It immediately began advocating for both improved sanitation from a rights-based approach, as well as to gain legitimacy for their technological solution as a sanitation option that could meet the needs of the most vulnerable residents of the slum.

Umande simultaneously built up the business skills of the groups operating the toilets and advocated to be recognized as a formal stakeholder for urban sanitation in Nairobi in order to increase their voice and influence within the sector. This was essential to ensure recognition of their community sanitation centres as safe and viable options both to prevent any future conflict with the government and to prepare the ground for possible future collaborations with the government.

Sanergy followed an entirely different approach to sanitation in informal settlements, prioritizing change at the policy level. They employed more than six full-time staff members to build a relationship with different levels of government through involvement in ministry working groups and municipal planning teams. These staff members are advocates for policy change, who promote the harmonization of Sanergy's activities and targets with those of the government. They also work to support the development of capacity in the government for the regulation of their services, as a step towards a more sustainable and scalable model of service provision (Government Relations Specialist, Sanergy, personal communication, May 10, 2016).

CCI, in addition to working closely with community members and local organizations, works in partnership with Dar es Salaam's public water and sewerage utility company DAWASA on several projects. CCI is also a member and avid participant in activities and conferences organized by Slum Dwellers International.[5]

Ultimately, to effectively develop and sustain sanitation services, there is a need to engage individuals from different backgrounds to use systems thinking for long-term planning, identify the core issues and implement transformative change (Sect. 4.3.1). In this sense, while institutional development is important, human capacity to

---

[5] Slum Dwellers International is a network of community-based organizations that advocates for the human right to land and to basic services in informal settlements by sharing lessons from other organizations working in similar contexts.

fill and use even the most effective institutional, governance and financial structures is also crucial. When capacity and resources are continuously supplied from outside the local/national context, then outsider bias persists (Sect. 4.3.1.2), and there is insufficient investment in building local human capacity, where it is ultimately needed to sustain positive change. For example, a study on the human resource needs to meet Tanzanian water and sanitation MDG targets found a shortage of 4501 water supply and sanitation engineers, 447 social development professionals and 7589 operations and management professionals (Kimwaga et al. 2013). The development of the first local PhD programme in sanitation (in 2016) is a promising start but demonstrates the lack of priority of the government for capacity building in the sector. It also highlights the monumental challenges ahead to meet future sanitation demands in the country.

4.3.2.4 Identify and Leverage Alternative Funding Mechanisms

The public financing gap for sanitation in eastern Africa, combined with the unpredictability of donation-based finance models, suggests the need for innovative funding mechanisms to increase the financial viability of sanitation service delivery models (Abeysuriya et al. 2015). The western model for sanitation is defined by government laws and regulations and is financed by a robust taxation and public financing system to operate and maintain the infrastructure (and its management) for the transport and treatment of waste, while the user invests directly in toilets in the home (Sect. 4.3.1.2). In the long term, it is the responsibility of the government to provide basic water and sanitation services. However, the severe limitation of government budgets and capacity in eastern Africa (due in part to the high rate of population growth and urbanization) means that, in the interim, demand for the service is met by a market of sanitation service providers, who are independent organizations, private businesses and social enterprises (McFarlane et al. 2014).

Sanergy's approach is robust, in the sense that its innovative resource recovery technology establishes a value chain that integrates the demand for toilets, need for employment, development of business opportunities, production of organic fertilizers and provision of a source of low-cost energy. By using a franchise model of individually owned public pay-per-use toilets, Sanergy remains scalable and adaptable to the diverse and changing needs in densely populated urban settlements. Microcredit loans are available for new franchise owners, who are supported with training in business management and accounting. Toilet owners pay a monthly fee to Sanergy, who in return hires individuals to clean and empty the urine diversion dry toilets on a daily basis and makes a profit by converting the 'waste' into fertilizer and bioenergy. Thus, both the business model and the technology are suitable for dense informal settlements and are flexible and adaptable to the changing urban landscape. Sanergy has received extensive financial support from donors, and this initial investment was important for overcoming the hurdles of developing an innovative start-up business. Its social enterprise model aims to achieve financial independence for both the company and the franchise owners. Rather than only providing the sanitation infrastructure, the *service delivery* model considers the complex system of the operating environment and can become an established private service provider for unplanned settlements that cannot be served by municipal governments.

Both CCI and Umande Trust, much like Sanergy, have developed pay-per-use systems for shared toilets that generate revenues to meet operational and maintenance costs. The concept of paying to use a toilet is not new, but there are many aspects to be considered to increase the sustainability and financial viability of this model (Arimah 1996). Umande Trust also recognizes the dangers of walking outside at night carrying cash to use a toilet. To reduce this disincentive to use the toilet, payment can be made with a personalized no-cash punch card that reduces the risk of robbery. In addition, they choose only to work with pre-existing community groups as managers for new facilities to reduce the potential conflicts between group members jointly managing the community WASH business.

These three organizations have adopted an innovative financial model to support sanitation service delivery. In addition, they invest in the people needed to operate them. Rather than (or parallel to) direct financial support, they facilitate skills training, entrepreneurial coaching, leadership development and business management. As the deputy director of Umande stated 'We don't build toilets we build communities' (Deputy Director, Umande, Personal communication, July 10, 2015).

4.3.2.5 Enhance Value-Addition and Co-benefits

One common thread in the holistic approaches to sanitation discussed above is the recognition that human faeces can potentially be a valuable resource rather than merely a waste flow whose environmental impacts have to be mitigated. Human waste can be transformed into organic fertilizer, animal feed or an energy source (Drechsel et al. 2011).

For example, CCI and Sanergy convert human urine and faecal waste into fertilizers to be used for agricultural purposes. Such fertilizers are in high demand in eastern Africa, because soil fertility is low and chemical fertilizers are expensive (Diener et al. 2014; Andersson 2015). Sanivation develops bio-charcoal and Umande biogas derived from human waste, both of which can be used as a domestic fuel for cooking. As co-products of the sanitation service, such fuels can provide an added income stream and also reduce the demand for conventional cooking fuels such as charcoal and fuelwood which are linked to ecosystem degradation and reduction of time availability for women and girls (Drechsel et al. 2011, Diener et al. 2014; Semiyaga et al. 2015) (see also Chaps. 2, 7 Vol. 1; Chap. 5 Vol. 2).

The above are good examples of Sustainability Science thinking, where the adoption of a systems thinking, problem-focused and solution-oriented mindset can provide solutions to persistent social and environmental sustainability challenges. These *solutions* are deeply embedded in the needs of the local communities and recognize that the sanitation service provision system is comprised of many different actors and users. Flexibility and adaptability are key elements of these operational models as they invest both in people and in infrastructure. Such resource recovery systems can contribute to the vision of a future where human waste becomes part of a larger resource recovery value chain and coordinated action between multiple actors can drive behaviour change (Tilley et al. 2014).

4.4   Policy and Practice Implications and Recommendations

This chapter is situated at the interface of multiple SDDs such SDG6 (Clean Water and Sanitation), SDG3 (Good Health and Wellbeing), SDG8 (Decent Work and Economic Growth) and SDG9 (Industry, Innovation, and Infrastructure), among others. Given the complexity and heterogeneous nature of sanitation challenges in eastern Africa, and its growing population, urbanization and environmental concerns, we suggest several recommendations including to:

- Embrace a diverse set of sanitation approaches to fit the broad spectrum of contexts
- Engage and include stakeholders from the outset of sanitation interventions
- Develop and invest in sector coordination and learning platforms
- Experiment with and promote successful sanitation options
- Build capacity among sanitation users and decision-makers
- Develop innovative funding mechanisms
- Improve and invest in proper monitoring and evaluation processes

A diverse set of approaches is required to fit the broad spectrum of sanitation contexts in SSA, which would need to be supported by a comprehensive policy framework. The examples discussed in Sect. 4.3.2 show how the awareness and use of place-based knowledge, flexible financial mechanisms, systems thinking and value addition in the sanitation chain are all important factors that can enhance the sustained adoption and overall sustainability of relevant interventions. There is no one-size-fits-all approach. New technology policies should encourage customized designs that are appropriate to the diverse urban and rural landscapes, rather than favour one sanitation technology over another. Appropriate policies can guide private sector investment, and effective regulatory mechanisms should be in place for these actors at both the national and local level. The policy framework for sanitation interventions needs to be integrated properly into the broader set of other relevant policies and reflect the diverse needs and contexts of individual countries and/or cities.

Stakeholder engagement and inclusivity in the development and implementation of sanitation interventions will support the development of interventions, which fits the context and will be adopted by the community. Stakeholder inclusion should not only be a consultative process but can take the form of equal partnerships and/or involvement in leadership positions. By including diverse voices that counterbalance the three types of outside bias (i.e. expert, male, western) (Sect. 4.3.1.2), it could become possible to better understand how infrastructure, management and awareness-raising approaches can be integrated to fit local needs. Civil society organizations (CSOs) can document the realities and current inadequacies in the sanitation sector, as well as comment on (and advocate for) sanitation options that are more suited to local needs. As many CSOs face severe resource limitations, partnerships with research institutions, private enterprises and other parties can support such grassroots advocacy.

Development and investment in sector coordination and learning platforms can address the fragmented nature of sanitation-related governance at the national level. This can be achieved by strengthening cooperation and coordination between national agencies and ministries responsible for (or with synergies to) sanitation and by developing sector learning platforms and reviewing existing mechanisms to facilitate joint discussions and planning of critical sanitation issues. These same mechanisms can offer opportunities not only to identify common challenges and inadequacies in current sanitation practices but also to highlight successes and share learning from promising examples, such as the case studies outlined in Sect. 4.3.2. This would require the documentation of successes, failures and lessons learnt from current initiatives. Certainly, this can only happen when multiple actors can recognize the limitations of the existing sanitation service delivery models, particularly in informal settlements in growing urban areas. Such findings can then be used to inform and update national policies. In addition, sector learning platforms such as sanitation working groups and learning teams can encourage collaborative partnerships between research institutions, NGOs, government and private companies to explore and finance new innovative sanitation pathways. Financial support for convening learning platforms can be built in or annexed to donor-financed projects.

Experimentation and promotion of successful sanitation options can enable the development of various sanitation interventions that are or can be adapted to the broad spectrum of sanitation contexts in SSA. A critical policy recommendation in this regard is the implementation of small-scale trials for a range of different sanitation interventions, followed by a comprehensive analysis of the context in which they are implemented, their overall performance and the factors contributing to success/failure. The examples in this chapter show promise for sustainability and for scaling up. However, this will require an enabling environment to achieve scaling up and long-term sustainability in the form of government recognition/oversight and creative mechanisms for long-term financing.

Building local capacity is particularly important in the context of sanitation in SSA. A key recommendation for donors and international actors would be to couple all of their sanitation projects and programmes with genuine efforts to integrate them within (and support) local systems and government initiatives. Local expertise within Kenya, Tanzania and Uganda need to be better recognized and expanded in order to channel resources more effectively and support the scaling up of locally appropriate solutions. The current resource gap should be targeted not only with solution-oriented projects that treat the 'symptoms' of the inadequate sanitation system but also with long-term investments to foster local knowledge and build capacity within the sector. These areas include, but are not limited to integrated sanitation management, environmental engineering, hygiene education, menstrual hygiene management, community leadership and facilitation, faecal sludge management, resource recovery technology development, sanitation financing, sanitation marketing and sanitation business development.

Innovative funding mechanisms would be required to bridge the funding gap. Despite the continued and significant support of external agencies for sanitation in eastern Africa (Sect. 4.1), the constraints associated with short-term funding cycles (e.g. the need to show immediate and measurable results) will most certainly persist even for the most well-intentioned donors. However, by adopting a creative, flexible and adaptive approach can help pair shorter-term interventions with sustainable local systems building through the combination of different types of investments. The promising sanitation examples presented in Sect. 4.3.2 have all identified ways to overcome serious economic constraints by establishing sustainable business models. However, all of the studied projects still required (and received) seed funding or even long-term donor support during their development phase. In a sense, it was once the economic constraint was removed that innovation became possible. Thus, donors and investors should consider offering financing in the form of start-up grants to promising projects and loans to more diverse implementers. In order to promote innovation in the sector, this early funding can be supplemented with financial training for small- and medium-sized private companies interested in starting a sanitation enterprise. To ensure their financial viability, sanitation service schemes must consider different mechanisms for revenue generation, including user contributions or service payments for the building and maintenance of the actual sanitation services.

Improvements and investments in proper monitoring and evaluation processes will, in the longer term, contribute to meeting the SDG headline indicators. Holistic monitoring approaches are required that reflect the messy nature of progress and implementation in complex environments and the multiple factors needed to catalyze positive change and increase the likelihood of sustainability. Various 'sustainability indicators' are currently available for evaluating the likelihood of whether an intervention or a single part of broader infrastructure will last over time (Schweitzer et al. 2014). For example, some donors, such as the Netherlands Ministry of Foreign Affairs (DGIS), have been enforcing a clause that requires all of their funding contract implementers to perform a sustainability assessment using appropriate indicators to guarantee that funded projects will remain functional for a minimum of 10 years after initial completion (Ward 2017).

4.5 Conclusions

By using Sustainability Science as a guiding lens, this chapter attempted to reframe the sanitation challenge in eastern Africa. The focus was not only on the factors that perpetuate sanitation failure but also on the characteristics and competencies conducive to the development of sustainable sanitation systems in the region. In a nutshell, sanitation *solutions* that start from an incomplete understanding of the problem and the broader context tend to rely on path-dependant and often supply-driven strategies. Such solutions are unable to break out of a cycle of failure to succeed in the complex environment within which they are implemented. The development of more holistic and sustainable approaches is often curtailed due to the lack of learning from failure and the inadequate space for iteration and adaptation based on inclusive perspectives, which perpetuates the outsider bias. This, in turn, curtails the development of an enabling environment and makes it difficult to seek support and operate within existing governance structures.

However, through the analysis of six promising sanitation projects, we have shown how various actors in the WASH sector in eastern Africa have managed to break this cycle of failure and develop alternate sanitation pathways that fit the geographical, cultural and financial realities of each project context. These sanitation approaches, although different, all demonstrate context adaptability and compatibility, mechanisms that ensure financial viability, technologies that are culturally appropriate and an emphasis on environmental sustainability through resource recovery and closed-loop thinking.

These examples illustrate that breaking the cycle of failure is possible if there is a proper enabling environment. This enabling environment can foster the key competencies needed to respond to complex sustainability challenges and must be (a) *descriptive* in how specific needs and linkages to other systems are identified; (b) *critical* of universally accepted sanitation solutions; (c) *cooperative* in the design, implementation, management and monitoring of activities, and (d) *visionary* through the inclusion of new ways of handling waste and turning it into value for the benefit of people as well as the environment.

While the six study projects show promise for scalability, they are still small-scale relative to the scope of the sanitation challenge in the region. However, they show that alternatives to the business-as-usual approaches to sanitation service delivery are both feasible and desirable. We can learn from both their strengths and their limitations when investing in new ideas and alternative sanitation futures.

That said, the scale of the sanitation challenge is enormous in SSA, and it must be tackled by (1) embracing a diverse set of approaches to fit the broad spectrum of contexts, (2) engaging and including stakeholders from the outset, (3) developing and investing in sector coordination and learning platforms, (4) experimenting with and promoting successful options, (5) building capacity, (6) developing innovative funding mechanisms and (7) improving and investing in proper monitoring/evaluation processes. This recognizes that challenges are transdisciplinary and multiscale, affected by governance, finance and sector coordination.

As a final note, by demonstrating the multiple benefits of improved sanitation on health, dignity, livelihood/income diversification and gender empowerment, we have shown that sanitation interventions can contribute to numerous SDGs. If the systemic linkages and mutual benefits discussed throughout this chapter are recognized beyond the community level to become anchored in government policies and funding priorities, then they could create the right enabling environment at the regional, national and subnational level. However, in order to achieve universal sanitation coverage, a radical paradigm shift anchored on Sustainability Science principles would be required in how we *think about* and *do* sanitation. Only then will we be able to learn from past failures and build local capacity to enable investments in futures that we have not yet imagined.

Acknowledgements The authors would like to extend their deepest gratitude to the community members and sanitation organizations who voluntarily contributed precious time to share valuable insights about their lives and sanitation activities in their respective communities. Thanks also goes to the Swedish research council for sustainable development, *FORMAS*, who enabled this research through Dr. Gabrielsson's PostDoc project, 'Water for life and dignity: A study on the sustainability and scalability of community-based water supply and sanitation systems in Tanzania'.

References

Abeysuriya K, Kome A, Willetts J (2015) Enabling investment in urban sanitation services through the sustainable full cost recovery principle. In: 38th WEDC international conference. WEDC

Andersson E (2015) Turning waste into value: using human urine to enrich soils for sustainable food production in Uganda. J Clean Prod 96:290–298

Andersson K, Rosemarin A, Lamizana B, Kvarnström E, McConville J, Seidu R, Dickin S, Trimmer C (2016) Sanitation, wastewater management and sustainability: from waste disposal to resource recovery. United Nations Environment Programme and Stockholm Environment Institute, Nairobi/Stockholm

Arimah BC (1996) Willingness to pay for improved environmental sanitation in a Nigerian City. J Environ Manag 48(2):127–138

Bäckstrand K (2003) Civic science for sustainability: reframing the role of experts, policy-makers and citizens in environmental governance. Glob Environ Polit 3(4):24–41

Bartram J, Cairncross S (2010) Hygiene, sanitation, and water: forgotten foundations of health. PLoS Med 7(11):e1000367

Chambers R, Myers J (2016) Norms, knowledge and usage. Frontiers of CLTS: innovations and insights issue 7. IDS, Brighton

Clark WC (2007) Sustainability science: a room of its own. PNAS 104(6):1737

Clark WC, Dickson NM (2003) Sustainability science: the emerging research program. Proc Natl Acad Sci 100(14):8059–8061

Cross P, Coombes Y (eds) (2014) Sanitation and hygiene in Africa where do we stand? Analysis from the AfricaSan conference, Kigali, Rwanda. IWA Publishing, London

Davis S (2015) Do donor restrictions affect sustainability of water and sanitation interventions? Results from a Pilot Survey. Improve International

Davis S (2016) Seeking sanitation success: phase I. report. Published by catholic relief services

Diener S, Semiyaga S, Niwagaba CB, Muspratt AM, Gning JB, Mbéguéré M et al (2014) A value proposition: resource recovery from faecal sludge – can it be the driver for improved sanitation? Resour Conserv Recycl 88:32–38

Drechsel P et al (2011) Recovery and reuse of resources: enhancing urban resilience in low-income countries. Urban Agriculture Magazine no. 25. RUAF 10 years

Ekane N, Nykvist B, Kjellén M, Noel S, Weitz N (2014) Multi-level sanitation governance: understanding and overcoming challenges in the sanitation sector in sub-Saharan Africa. Waterlines 33(3):242–256

Ekane N, Weitz N, Nykvist B, Nordqvist P, Noel S (2016) Comparative assessment of sanitation and hygiene policies and institutional frameworks in Rwanda, Uganda and Tanzania. Stockholm Environment Institute (SEI) Working Paper, 5

Felix M (2015) Future prospect and sustainability of wood fuel resources in Tanzania. Renew Sust Energ Rev 51:856–862

Floret M (2017) Building capacity of urban poor CBOs in Dar es Salaam to set up ecological sanitation business, final evaluation report, Dar es Salaam, Tanzania

Fukuda-Parr S, Yamin AE, Greenstein J (2014) The power of numbers: a critical review of millennium development goal targets for human development and human rights. J Human Dev Capab 15:2–3, 105–117. https://doi.org/10.1080/19452829.2013.864622

Furlong C (2016) Tiger worms: a win-win solution. In: Chambers R, Myers J (eds) Norms, knowledge and usage. Frontiers of CLTS: innovations and insights issue 7. IDS, Brighton

Gabrielsson S, Ramasar V (2013) Widows: agents of change in a climate of water uncertainty. J Clean Prod 60:34–42

Galli, G., Nothomb, C., and Baetings, E. (2014) Towards systemic change in urban sanitation. IRC, The Hague

Hickling A (2014) Status of sanitation and hygiene in Africa. In: Cross P, Coombes Y (eds) Sanitation and hygiene in Africa where do we stand? Analysis from the AfricaSan conference, Kigali, Rwanda. IWA Publishing, London

Huston A, Moriarty P (2018) Understanding the WASH system and its building blocks, Working paper series building strong WASH systems for the SDGs. IRC, The Hague

Huston A, Moriarty P, Lockwood H (2019) All systems go! Background note for the WASH systems symposium. IRC, The Hague

Jenkins MW, Sugden S (2006) Rethinking sanitation: lessons and innovation for sustainability and success in the new millennium. UNDP

Jerneck A, Olsson L, Ness B, Anderberg S, Baier M, Clark E et al (2011) Structuring sustainability science. Sustain Sci 6(1):69–82

Kates RW (2011) What kind of a science is sustainability science? PNAS 108:19449–19450

Kates R, Clark W, Corell R, Hall J, Jaeger C, Lowe I, McCarthy J, Schellnhuber H, Bolin B, Dickson N, Faucheux S, Gallopin G, Grubler A, Huntley B, Jager J, Jodha N, Kasperson R, Mabogunje A, Matson P, Mooney H, Moore B III, O'Riordan T, Svedlin U (2001) Sustainability science. Science 292(5517):641–642

Kimwaga R, Nobert J, Kongo V, Ngwisa M (2013) Meeting the water and sanitation MDGs: a study of human resource development requirements in Tanzania. Water Policy 15(S2):61–78

Komiyama H, Takeuchi K (2006) Sustainability science: building a new discipline. Sustain Sci 1:1–6

Kvarnström E, McConville J, Bracken P et al (2011) The sanitation ladder – a need for a revamp? J Water Sanit Hygiene Dev 1(1):3–12

Loorbach D (2010) Transition management for sustainable development: a prescriptive, complexity-based governance framework. Governance 23(1):161–183

Mbaria J (2014) Engendering positive change in slums – role of biocentres in the supply for affordable and sustainable water and sanitation services in informal settlements. Umande Trust, Nairobi

McFarlane C, Desai R, Graham S (2014) Informal urban sanitation: everyday life, poverty, and comparison. Ann Assoc Am Geogr 104(5):989–1011

Mies M (1998) Patriarchy and accumulation on a world scale: women in the international division of labour. Palgrave Macmillan, London

Moriarty P, Smits S, Butterworth J, Franceys R (2013) Trends in rural water supply: towards a service delivery approach. Water Altern 6(3):329

Mulumba J, Nothomb C, Potter A, Snel M (2014) Striking the balance: what is the role of the public sector in sanitation as a service and as a business? Waterlines 33(3):195–210

Nallari A (2015) "All we want are toilets inside our homes!" The critical role of sanitation in the lives of urban poor adolescent girls in Bengaluru, India. Environ Urban 27(1):73–88

Nawab B, Nyborg IL, Esser KB, Jenssen PD (2006) Cultural preferences in designing ecological sanitation systems in North West Frontier Province, Pakistan. J Environ Psychol 26(3):236–246

Neely K (2019) Systems thinking and transdisciplinarity in WASH. Systems thinking and WASH: tools and case studies for a sustainable water supply, 17

NIMR (2016) Baseline of school wash facilities in Tanzania. National Institute of Medical Research, Dar es Salaam

Nyonyintono Lubaale G, Musembi Musyoki S (2011) Pro-poor sanitation and hygiene in East Africa. Background paper presented at the East Africa practitioners workshop on pro poor urban sanitation and hygiene, LAICO Umubano Hotel, Kigali, Rwanda, March 2011

O'Keefe M, Lüthi C, Tumwebaze IK, Tobias R (2015) Opportunities and limits to market-driven sanitation services: evidence from urban informal settlements in East Africa. Environ Urban 27 (2):421–440

Oates N, Ross I, Calow R, Carter RT, Doczi J (2014) Adaptation to climate change in water, sanitation and hygiene – assessing risks and appraising options in Africa. Overseas Development Institute, London

Otsuki K (2016) Infrastructure in informal settlements: co-production of public services for inclusive governance. Local Environ 21(12):1557–1572

Penner B (2010) Flush with inequality: sanitation in South Africa. Places J. Available : https://placesjournal.org/article/flush-with-inequality-sanitation-in-south-africa/

Prüss-Ustün A, Bartram J, Clasen T, Colford JM, Cumming O, Curtis V et al (2014) Burden of disease from inadequate water, sanitation and hygiene in low-and middle-income settings: a retrospective analysis of data from 145 countries. Tropical Med Int Health 19(8):894–905

Rip A, Kemp R (1998) Technological change. Battelle Press, Columbus, pp 327–399

Rittel HWJ, Webber MM (1973) Dilemmas in a General theory of planning. Policy Sci 4:155–169. https://doi.org/10.1007/bf01405730

Sarewitz D, Kriebel D, Clapp R, Crumbley C, Hoppin P, Jacobs M (2010) The sustainable solutions agenda: the consortium for science, policy and outcomes. Lowell Center for Sustainable Production

Schultz J, Brand F, Kopfmuller J, Ott K (2008) Building a 'theory of sustainable development': two salient conceptions within the German discourse. Int J Environ Sustain Dev 7(4):465–482

Schweitzer R, Grayson C, Lockwood H (2014) Mapping of water, sanitation, and hygiene sustainability tools. Working paper 10, IRC/Aguaconsult/Triple-S

Seager J (2010) Gender and water: good rhetoric, but it doesn't "count". Geoforum 41(1):1–3

Semiyaga S et al (2015) Decentralised options for faecal sludge management in urban slum areas of sub-Saharan Africa: a review of technologies, practices, and end-uses. Resour Conserv Recycl 104:109–119

Seppälä OT (2002) Effective water and sanitation policy reform implementation: need for systemic approach and stakeholder participation. Water Policy 4(4):367–388

Sparkman D (2012) More than just counting toilets: the complexities of monitoring for sustainability in sanitation. Waterlines 31(4):260–271

Strande L, Brdjanovic D (eds) (2014) Faecal sludge management: systems approach for implementation and operation. IWA Publishing, London

Stroh D (2015) Systems thinking for social change: a practical guide to solving complex problems, avoiding unintended consequences, and achieving lasting results. Chelsea Green Publishing, Hartford

Talwar S, Wiek A, Robinson J (2011) User engagement in sustainability research. Sci Public Policy 38(5):379–390

Taylor B (2009) Addressing the sustainability crisis: lessons from research on managing rural water projects. WaterAid Tanzania [online] Available at: http://www.wateraid.org/~/media/Publica tions/sustainabilitycrisis-rural-water-management-tanzania.pdf

Tilley E (2008) Compendium of sanitation systems and technologies. Eawag

Tilley E, Strande L, Lüthi C, Mosler HJ, Udert KM, Gebauer H, Hering JG (2014) Looking beyond technology: an integrated approach to water, sanitation and hygiene in low income countries. Environ Sci Technol 48:9965–9970

Tsinda A, Abbott P, Pedley S, Charles K, Adogo J, Okurut K, Chenoweth J (2013) Challenges to achieving sustainable sanitation in informal settlements of Kigali, Rwanda. Int J Environ Res Public Health 10(12):6939–6954

UN Water (2008) Sanitation is an investment with high economic returns. Factsheet. http://www.gwopa.org/index.php/resource-library/3166-sanitation-is-an-investment-with-high-economic-returns

UN Water (2014) Investing in water and sanitation: increasing access, reducing inequalities. Special report for the SWA high-level meeting

UNICEF (2017) The 2030 agenda for sustainable development. www.unicef.org/agenda2030/

Van Kerkhoff L, Lebel L (2006) Linking knowledge and action for sustainable development. Ann Rev Environ Resour 31:445–477

Ward, R (2017) Sustainability clause, check and compact primer for programming and long term management. Prepared for WASH Alliance International. Available: https://wash-alliance.org/wp-content/uploads/sites/36/2017/03/201701_SCCC_Guide_V.9_FINAL.pdf

Waterkeyn JA, Waterkeyn AJ (2013) Creating a culture of health: hygiene behaviour change in community health clubs through knowledge and positive peer pressure. J Water Sanit Hygiene Dev 3(2):144–155

WHO (2017) UN-Water global analysis and assessment of sanitation and drinking-water (GLAAS) 2017 report: financing universal water, sanitation and hygiene under the sustainable development goals. World Health Organization, Geneva. Licence: CC BY-NC-SA 3.0 IGO

WHO/UNICEF (2015) Progress on sanitation and drinking water â 2015 update and MDG assessment

WHO/UNICEF (2017) Progress on drinking water, sanitation and hygiene: 2017 update and SDG baselines. World Health Organization and the United Nations Children's Fund, Geneva

Wiek A, Withycombe L, Redman CL (2011) Key competencies in sustainability: a reference framework for academic program development. Sustain Sci 6(2):203–218

World Bank (2017) GDP growth annual. http://data.worldbank.org/indicator

WSP (2012) Economic impacts of sanitation, world bank water and sanitation program. www.wsp.org/content/economic-impacts-sanitation

Ziegler R, Ott K (2011) The quality of sustainability science: a philosophical perspective. Sustain Sci Pract Policy 7(1):31–44

Part II
Southern Africa

Chapter 5
Ethanol as a Clean Cooking Alternative in Sub-Saharan Africa: Insights from Sugarcane Production and Ethanol Adoption Sites in Malawi and Mozambique

Anne Nyambane, Francis X. Johnson, Carla Romeu-Dalmau, Caroline Ochieng, Alexandros Gasparatos, Shakespear Mudombi, and Graham Paul von Maltitz

5.1 Introduction

Access to clean, reliable, affordable and sustainable energy is one of the greatest sustainability challenges currently facing sub-Saharan African (SSA) countries and has been encapsulated in Sustainable Development Goal 7 (SDG 7) (IEA 2014) (Chaps. 1, 2, 7 Vol. 1). It has been estimated that up to 730 million people in SSA (or 80% of the population) have no access to electricity or clean cooking fuels, relying instead on traditional solid biomass fuels such as charcoal and fuelwood for their domestic needs (IEA 2014; Zulu and Richardson 2013) (Chaps. 1–2 Vol. 1). In fact biomass fuels have traditionally dominated household energy needs, especially for cooking (Karanja and Gasparatos 2019; IEA 2014) (Chaps. 2, 7 Vol. 1). However, the sourcing and use of conventional biomass fuels such as fuelwood

A. Nyambane (✉)
Stockholm Environment Institute (SEI), Nairobi, Kenya

F. X. Johnson · C. Ochieng
Stockholm Environment Institute (SEI), Stockholm, Sweden
e-mail: francis.x.johnson@sei.org

C. Romeu-Dalmau
University of Oxford, Oxford, UK

A. Gasparatos
Institute for Future Initiatives (IFI), The University of Tokyo, Tokyo, Japan
e-mail: gasparatos@ifi.u-tokyo.ac.jp

S. Mudombi
Trade and Industrial Policy Strategies (TIPS), Pretoria, South Africa

G. P. von Maltitz
Centre for Science and Industrial Research (CSIR), Pretoria, South Africa

© Springer Nature Singapore Pte Ltd. 2020
A. Gasparatos et al. (eds.), *Sustainability Challenges in Sub-Saharan Africa II*,
Science for Sustainable Societies, https://doi.org/10.1007/978-981-15-5358-5_5

and charcoal for cooking has various sustainability impacts as discussed below (Chaps. 2, 7 Vol. 1).

On the one hand, the use of traditional biomass fuels is associated with substantial negative health outcomes due to indoor air pollution from smoke and other pollutants produced during fuel combustion (Fullerton et al. 2008; Amegah and Jaakkola 2016). Furthermore, fuelwood collection and cooking in inefficient stoves can increase the risk of injuries (Das et al. 2017) and time diversion from educational and income-generating activities (Edelstein et al. 2008; Fullerton et al. 2008; Langbein 2017). Often these impacts are gender-differentiated as women and girls spend disproportionately higher amount of time collecting fuelwood and cooking in poorly ventilated areas (Ezzati and Kammen 2002; Karanja and Gasparatos 2019). Furthermore, reliance on traditional biomass puts a major recurring burden on household budgets, especially in areas experiencing fuelwood scarcity (e.g. from overexploitation or climate change) or where fuelwood collection is not possible (e.g. in cities) (Openshaw 2010; Karanja and Gasparatos 2019).

Furthermore, some of the prevailing fuelwood harvesting practices have been linked to negative environmental outcomes related to deforestation, forest degradation, carbon stock loss and biodiversity loss (Chidumayo and Gumbo 2013; Hosonuma et al. 2012; IPBES 2018; Karanja and Gasparatos 2019) (Chaps. 1–2, 7, 9 Vol. 1). Broader landscape degradation from unsustainable fuelwood harvesting can increase the vulnerability of the rural poor through the loss and degradation of ecosystem services related to livelihoods and food security,[1] creating thus a vicious cycle of biomass dependence and poverty (Cerutti et al. 2015; Chidumayo and Gumbo 2013; Mugo and Ong 2006; IPBES 2018). At the same time, charcoal production (in inefficient kilns) and use (in inefficient stoves) has been associated with significant greenhouse gas (GHG) emissions (Chidumayo and Gumbo 2013; Okoko et al. 2017) (Chap. 2 Vol. 1). Dependence on charcoal and fuelwood has been estimated as accounting for 1.9–2.3% of global GHG emissions (Bailis et al. 2015), which is comparable to global emissions from the aviation sector.

Conversely the biomass fuel sector (and especially the charcoal sector) is a vital source of livelihoods especially for many of the rural poor that produce and sell charcoal either as a primary or a secondary income-generating activity (Jones et al. 2016; Smith et al. 2017; Woollen et al. 2016). In this sense, fuelwood (and the derived charcoal) are important provisioning ecosystem services, catering for multiple human needs in urban and rural contexts of the continent (Woollen et al. 2016; Zorrilla-Miras et al. 2018). Charcoal production sometimes is also a means of sustainably managing and/or eradicating invasive tree species such as *Prosopis juliflora*[2] (FAO 2018).

---

[1]Forest loss and degradation due to the unsustainable extraction of fuelwood has been associated with the loss of multiple provisioning, regulating, cultural and supporting ecosystem services in SSA (IPBES, 2018).

[2]*Prosopis* is an invasive species in Kenya, Somalia and Ethiopia that is known for degrading rangeland and thus affecting negatively the livelihoods of pastoralist communities (Zeray et al. 2017).

Eradicating the energy poverty associated with high reliance on traditional biomass fuels is a key for attaining sustainable development in SSA. For example, it has been suggested that the large-scale promotion and uptake of clean cooking options (and the simultaneous phasing out of traditional biomass fuels) can have multiple sustainability benefits (IEA et al. 2019; Maes and Verbist 2012)[3] (Chaps. 1–2 Vol. 1). There have been many pilot and large-scale programmes during the past decades in SSA, promoting improved biomass stoves and stoves using electricity, liquefied petroleum gas (LPG), biogas, ethanol and briquettes (World Bank 2017). At the same time, there have been similar attempts to promote more efficient charcoal production technologies (Adam 2009; Schure et al. 2019). If successful, such initiatives and efforts can contribute substantially to multiple SDGs such as SDG1 (No Poverty), SDG2 (Zero Hunger), SDG3 (Good Health and Wellbeing), SDG5 (Gender Equality), SDG7 (Affordable and Clean Energy), SDG12 (Responsible Consumption and Production), SDG13 (Climate Action) and SDG15 (Life on Land) among others (Chap. 2 Vol. 1).

However, reducing household dependence on traditional cooking fuels through the promotion and uptake of clean cooking options, despite its high positive sustainability outcomes, remains a major sustainability challenge in SSA (IEA 2014; World Bank 2014). Besides factors related to costs (e.g. high upfront stove cost, recurring fuel costs, repair/change costs) that often hinder the switch from traditional to modern fuels/stoves (World Bank 2014), there are several other funding constraints complicating the effective implementation of clean cooking programmes (Karanja et al. 2020). In many SSA contexts, there is a lack of readily available modern cooking options, motivation, and incentives for fuel switch, as well as financing and other support measures (Zhou et al. 2011; World Bank 2017). Furthermore, some stove designs fail to consider the needs and cultural preferences of stove users (e.g. space for multiple pots, ability to cook local dishes) (Karanja and Gasparatos 2019; Jürisoo et al. 2018). All these factors have contributed to the lack of large-scale adoption and sustained use of efficient biomass cookstoves in parts of the continent, even if their initial adoption was successful (EkouevI and Tuntivate 2012; Ruiz-Mercado et al. 2011; Karanja et al. 2020; Debbi et al. 2014).

Ethanol fuel and stoves is one of the clean cooking options promoted in some SSA countries such Ethiopia, Kenya and Mozambique (Chap. 2 Vol. 1). However most of these efforts have been rather small-scale and have not moved beyond the pilot stage (World Bank 2017; Rogers et al. 2013). On the other hand, there is a long tradition of sugarcane ethanol production for transport biofuels in countries such as Malawi, Zimbabwe and South Africa (Gasparatos et al. 2015). For example, Malawi has been blending ethanol in conventional gasoline since the 1980s (often at blends as high as 20%) and has been considered as one of the most successful countries in

---

[3]However, some negative trade-offs are also possible such as (a) the loss of rural income and employment opportunities in the charcoal sector and (b) low employment generation in the clean stove sector if the market is dominated by imported rather than domestically produced stoves (Karanja and Gasparatos 2019).

this regard globally (Johnson and Silveira 2014). However, despite its wide availability, bioethanol has not been adopted in the residential sector in Malawi except for some pilot projects.

A growing body of literature has been exploring both the factors affecting the adoption of ethanol stoves in SSA and the impacts associated with sugarcane ethanol production. For example, studies have used consumer choice techniques to uncover the key attributes inherent in the selection of ethanol stoves (Ozier et al. 2018; Takama et al. 2012). Some studies have also explored user perceptions of ethanol stoves compared to other readily available options (Benka-Coker et al. 2018; Mudombi et al. 2018a) or the actual indoor air pollution/emissions (Pope et al. 2017) and associated health effects (Alexander et al. 2018). Some recent studies have discussed the possible environmental effects of sugarcane production[4] on land use change (Beza and Assen 2017; Dlamini 2017; Romeu-Dalmau et al. 2018; Twongyirwe et al. 2018), biodiversity (Degefa and Saito 2017), water availability and quality (Hess et al. 2016; Ngcobo and Jewitt 2017; Kasambala Donga and Eklo 2018) and GHG emissions (Dunkelberg et al. 2014). Many studies have also explored impacts related to livelihoods/poverty (Manda et al. 2018; Mudombi et al. 2018b; Wendimu et al. 2016), food security (Dam Lam et al. 2017; Herrmann et al. 2018) and social conflicts (https://www.tandfonline.com/doi/full/10.1080/03057070.2016.1211401). As sugarcane can be viewed as a renewable resource in SSA, it could support energy access goals through the expansion of bioethanol use beyond the transport sector and into the residential sector (Johnson et al. 2017) (Chap. 2 Vol. 1).

The above suggests that when seeking to understand the potential and the sustainability of ethanol-based clean cooking options in SSA, it is important to understand broader factors affecting its production and adoption. This includes issues both related to the impacts of sugarcane production[4] and factors affecting the promotion, adoption and sustained use of clean cookstoves.

The aim of this chapter is to explore some key aspects of the production and adoption of ethanol as a clean cooking option in SSA. However, there is currently no country in SSA with a significant track record of both domestic ethanol production and large-scale adoption of ethanol stoves (World Bank 2017; Gasparatos et al. 2015). As this prohibits a comprehensive analysis across the entire value chain of ethanol for cooking, this chapter synthesizes insights from two different sites, one related to ethanol adoption (Maputo, Mozambique) and one related to sugarcane production (Dwangwa, Malawi). Collectively these two sites offer unique and distinctive experiences in ethanol production and use in the SSA context. In particular, we explore some of the factors that led to the rapid penetration and adoption of ethanol as a clean cooking fuel in Maputo, as well as some of the impacts of ethanol

---

[4]Sugarcane molasses is normally the most cost-effective feedstock for ethanol production in the SSA. Thus sugarcane production has particular relevance when considering the possible impacts of ethanol cooking options on the supply side (Johnson et al. 2017).

feedstock production in Dwangwa on selected ecosystem services. In this way, we capture supply and demand issues on a somewhat comprehensive manner.

Section 5.2 outlines the characteristics of the study sites and the methodology for assessing the impacts of sugarcane production (Dwangwa, Malawi) and user perceptions of ethanol stoves (Maputo, Mozambique). Section 5.3 contains the main results related to the adoption of ethanol stoves in Maputo (Sect. 5.3.1) and the ecosystem services impacts associated with land use change in Dwangwa (Sect. 5.3.2). Section 5.4 synthesizes this evidence and outlines policy implications and recommendations for enhancing the sustainability and viability of ethanol-based clean cooking interventions in SSA.

5.2 Methodology

5.2.1 Study Sites

5.2.1.1 Charcoal Sector in Mozambique and Malawi

The two study sites discussed in this chapter include a sugarcane production area (Dwangwa, Malawi, Sect. 5.2.1.3) and an ethanol stove adoption and use area (Maputo, Mozambique, Sect. 5.2.1.2) (shown in Fig. 5.1). Both Malawi and Mozambique are least developed countries characterized by low levels of development in terms of gross domestic product (GDP) and the Human Development Index (HDI), ranking among the lowest in the world.

In both Mozambique and Malawi, traditional biomass fuels dominate the household cooking sector, with only 4% and 3% of the population, respectively, having access to clean cooking in 2016 (World Bank 2020). In Mozambique, traditional biomass fuels such as fuelwood and charcoal dominate the cooking fuel market, accounting for as much as 59% and 23% of cooking fuel demand, respectively, nationally (but with large variations between rural and urban areas) (EUEI 2012). Urban households in Mozambique predominately use charcoal for cooking regardless of income level, which suggests the strong position of charcoal in the national energy system (Castán Broto et al. 2020; EUEI 2012). Similarly, the urban population of Malawi is highly reliant on charcoal and fuelwood for cooking (Government of Malawi 2009; Republic of Malawi 2012; Zulu 2010; Makonese et al. 2018). Studies have found that low-income urban households depend almost completely on charcoal for cooking (91.2%), with this dependence being lower for middle- (28.9%) and high-income households (10.2%) (Practical Action 2017).

Overall, the charcoal sector plays a prominent role in the national economies of both countries. In Mozambique, it is estimated that the charcoal sector employs between 136,000 and 214,000 people on a full-time basis (EUEI 2012). Charcoal production accounts for a large fraction of rural livelihoods, offering valuable income diversification especially considering that in most rural areas of the country formal employment and income opportunities are very scarce and infrequent

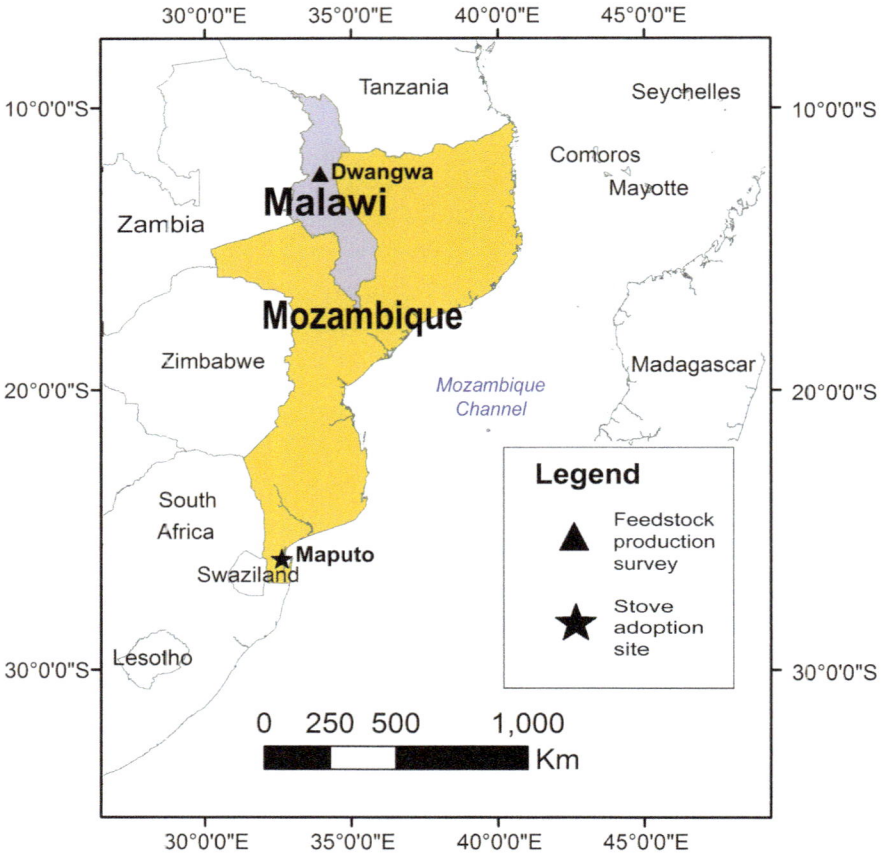

Fig. 5.1 Location of study sites

(Woollen et al. 2016; de Koning and Atanassov 2014; Jones et al. 2016). In Malawi, between 180,000 and 200,000 people are employed in the charcoal sector with an additional 928,000 people involved in the charcoal value chain (Kambewa et al. 2007). Similarly, the charcoal and fuelwood sectors are very important for rural livelihoods in Malawi, both in terms of absolute income and income diversification, considering the similar lack of formal income and employment options in most rural areas of the country (Smith et al. 2017; Kamanga et al. 2009).

At the same time, household expenditures on cooking fuel are quite high in both countries. In Mozambique, it is estimated that low-income households spend approximately 24% of their total income to purchase charcoal every month, while an average-income household spends about 15% of total income (Atanassov et al. 2012). Moreover, poor households essentially pay double the price for charcoal because they purchase it in small quantities, unlike high-income households that purchase in bulk and thus at lower prices (Atanassov et al. 2012). In Malawi, urban households also allocate a high proportion of their total income on energy, which ranges from 7.8% for high-income households to 13% for low-income households (Practical Action 2017).

Charcoal and fuelwood demand have been identified as a key driver of localized deforestation and ecosystem services degradation in both countries. Even though it is difficult to estimate exactly the actual deforestation rates in Mozambique (EUEI 2012), multiple studies have estimated the deforestation outcomes of charcoal production in many parts of Mozambique (Sedano et al. 2016, 2020; Silva et al. 2019). Similarly, charcoal and fuelwood production have been associated with significant deforestation in Malawi (Zulu 2010; Davies et al. 2010).

5.2.1.2   Maputo: Ethanol Stove Adoption Site

Maputo, the capital of Mozambique, has been the only major city in SSA to experience a large-scale commercialization of ethanol for cooking through a private sector initiative. In particular, CleanStar, a private company, promoted aggressively ethanol stoves and ethanol fuel as an alternative to charcoal, which dominates the cooking energy options in the city (Castán Broto et al. 2020; EUEI 2012) (Sect. 5.2.1.1). Even though the ethanol stove promotion built on a pre-existing initiative that commercialized ethanol for cooking in modest quantities, the large-scale pro-motion of ethanol effectively started in late 2012 (World Bank 2017). Ethanol stoves and fuel achieved rapid penetration shortly after the commencement of CleanStar activities through 160 ethanol distributors, reaching 34,000 consumers and a monthly consumption of 70,000–140,000 litres (Mudombi et al. 2018a). At the same time, CleanStar established an ethanol production facility in the city of Beira in central Mozambique using as feedstock cassava sourced from smallholder farmers from the north of the country (Bogdanski 2012; Costa 2019).

However, the ethanol production arm of CleanStar was discontinued in 2013 and its ethanol distribution component was transferred to a company called NDZiLO in June 2013 (Mudombi et al. 2018a). During CleanStar's restructuring, cassava ethanol production effectively ceased, thus cutting ethanol supply to stove users in Maputo (Costa 2019). To meet this supply gap, the company started importing ethanol from South Africa in the last months before its collapse. However, this imported fuel was cheaper but of lower quality, often causing the underperformance and malfunction of the canisters and eventually influencing several users to switch back to charcoal or LPG (Mudombi et al. 2018a). Following the eventual collapse of CleanStar, NDZiLO started importing high-quality ethanol fuel from South Africa in an effort to revive the ethanol stove sector. At the time of fieldwork in July 2015 (see Sect. 5.2.2.1), there was still a strong ethanol demand in Maputo, amounting to approximately 80,000 litres per month from 10,000 consumers.

Despite these difficulties sustaining CleanStar's activities, Maputo still arguably constitutes the largest consumer base for ethanol stoves in a major SSA city. In this respect Maputo offers a unique opportunity to explore the factors that constrained and facilitated the adoption of ethanol fuel and stoves, as an alternative to charcoal.

5.2.1.3 Dwangwa: Sugarcane Production Site

Malawi is the only SSA country to have fully integrated biofuels into its energy system (Sect. 5.1) (Chaps. 2–3 Vol. 1). Sugarcane ethanol has become a central element of the transport sector for more than 30 years (Johnson and Silveira 2014). In particular, sugarcane ethanol has been produced and blended with gasoline at proportions between 10 and 25% since the 1980s. This was essentially a response to the energy crises of the 1970s, which escalated the costs of importing refined oil products in this landlocked country. This has coincided with continuous efforts to integrate smallholders in the sugarcane value chain since the 1990s in order to enhance rural development (von Maltitz et al. 2019; Chinsinga 2017) (Chap. 3 Vol. 1). The sugarcane sector is also particularly important to the Malawian economy, accounting a substantial fraction of the national gross domestic product (GDP) (Chinangwa et al. 2017) (Chap. 3 Vol. 1).

Most of the sugarcane production is concentrated in the Dwangwa and Nchalo sugarcane belts (Chinangwa et al. 2017). The project site in Dwangwa contains a large-scale sugarcane plantation and a sugar mill operated by a multinational company (Illovo) since the late 1970s (Chinsinga 2017). Irrigated and rainfed sugarcane plots surround the core plantation and are managed by smallholders that sell sugarcane to the Illovo mill. The irrigated smallholders are part of the Dwangwa Cane Growers Limited (DCGL), while the rainfed smallholders are either independent or parts of smallholder associations (Chinsinga 2017; Gasparatos et al. 2018a). EthCo Malawi is a fully Malawi-owned company that operates an ethanol distillery, which uses the molasses by-product purchased from the Illovo mill. Due to the long history of sugarcane and ethanol production, Dwangwa offers a unique case study to explore some of the local impacts of bioethanol feedstock production in SSA.

5.2.2 Data Collection and Analysis

5.2.2.1 Maputo: Ethanol Stove Adoption Site

For the Maputo study, we use a qualitative research approach to elicit (a) user perceptions about ethanol stoves and fuel, compared to other options and (b) key lessons learned from the rapid and large-scale expansion of ethanol distribution, marketing and consumer use. For (a) we use a combination of a household survey with stove users and focus group discussions (FGDs) with ethanol users. For (b) we use in-depth interviews with an ethanol supplier (NDZiLO) and ethanol fuel and stove distributors.

For the household survey, we targeted 341 households that represent both users and non-users of ethanol stoves. The household survey was structured and mostly included closed-ended questions (with some open-ended questions). We selected respondents from neighbourhoods that had experienced a large-scale uptake of ethanol stoves and represented the predominant socio-economic background in Maputo (Mudombi et al. 2018a). We focused in the neighbourhoods of Benfica (N = 72), Chamanculo (N = 61), Hulene (N = 60), Mavalane (N = 69), Maxaquene (N = 65) and Urbanizacao (N = 14). For the household survey, we targeted the main female decision-maker within the household, which is usually the member most involved in food preparation and fuel/stove procurement (Gasparatos et al. 2018a). When the main female decision-maker was unavailable, we interviewed the spouse or another person that was involved in daily cooking activities (Gasparatos et al. 2018a). A full explanation of the survey protocol is included in Mudombi et al. (2018a).

Five FGDs were conducted with a total of 29 current and past ethanol users. The FGDs were mixed, with both men and women represented in each FGD. However, in contrast to the household survey, more men participated than women. This imbalance was due to the fact that the sampling list used to randomly recruit FGD participants was drawn from ethanol stove buyers from relevant shops. As men did the actual purchasing, it was challenging to invite the main female decision-maker from these randomly selected households. However, ground rules were laid out to ensure that all FGD participants contributed during the discussions.

Each FGD lasted between 60 and 90 min and contained open-ended questions about (a) participants' socio-economic characteristics; (b) reasons influencing the decision to purchase ethanol stove for cooking; (c) frequency of stove use; (d) type of food cooked using the ethanol stove; (e) convenience of purchasing ethanol fuel; (f) ethanol expenditures compared to the previous main cooking fuel; (g) advantages and disadvantages using ethanol for cooking compared to the previous main cooking fuel; (h) reasons for stopping ethanol use for cooking; (i) willingness to resume ethanol use for cooking if addressing the reason influencing the decision to stop ethanol use; and (j) any other relevant comments.

Finally, we conducted expert interviews with 25 ethanol fuel and ethanol stove distributors and one ethanol supplier (Zoe Enterprises). These expert interviews were conducted in person with each individual respondent and lasted between 30 and 60 min. They contained open-ended questions that aimed to elicit information about (a) ethanol fuel and ethanol stoves as a business venture compared to previous business ventures; (b) livelihood benefits compared to previous ventures; (c) challenges experienced during involvement in the ethanol sector and potential solutions; and (d) any other relevant comments.

Household survey data was analysed in Microsoft Excel and R version 3.2.2 (R Core Team 2015) using descriptive statistics (e.g. proportions, means). The FGDs and expert interviews were transcribed and analysed through qualitative content analysis techniques. The transcribed text was coded manually and classified into relevant categories with similar meanings and themes (Hsieh and Shannon 2005).

5.2.2.2 Dwangwa: Sugarcane Production Site

For the Dwangwa study, we aimed at identifying the possible impacts of land use change for sugarcane production on ecosystem services. We use a combination of analytical techniques including geospatial analysis, ecological surveys, soil analysis and household surveys to identify the effects of land use change due to sugarcane production on carbon sequestration (regulating service), woodland products (provisioning services) and recreation and religious values (cultural ecosystem services) (Sect. 5.3.2).

First, we track the land use change in the study site through geospatial analysis. We conduct this analysis for two periods, i.e. the years 1975 during the early stages of sugarcane production and 2015 when the household survey was undertaken (see below). We track land use change both within the area of sugarcane production and in the surrounding areas. Understanding the type of land that was converted for sugarcane production offers important insights about the type of affected ecosystem services. We use a selection of satellite images from the Landsat image archive and correct them following different processes. The final classification includes the following land cover types: (a) sugarcane, (b) high-density forest, (c) low-density forest, (d) bare land, (e) agriculture other than sugarcane, (f) grassland, (g) water and (h) cloud or shadow. We edit manually these different classes using our knowledge of the study site and test the inherent robustness of the classifications. A full explanation of the analytical process is included in Romeu-Dalmau et al. (2018).

Second, we estimate changes in carbon stocks due to the land use change associated with sugarcane production. Carbon storage is a major regulating ecosystem service associated with biofuel feedstock production (Gasparatos et al. 2018b). In particular, we estimate carbon stock change in the above-ground biomass, below-ground biomass and soil organic carbon (SOC), before and after land conversion for sugarcane production. We perform this analysis for each of the land uses identified above and estimate carbon stock change over a 20-year cycle.[5] Carbon stocks are estimated using primary and secondary data collected through:

- Standing biomass surveys in forest areas and soil sampling in forest, sugarcane and other agricultural areas (July 2015)
- Literature review about the standing biomass of sugarcane and other common agricultural crops in the study site (mainly maize)
- Literature review about allometric equations and other appropriate conversion factors

Subsequently, we use Monte Carlo simulation to assess the net changes in carbon stocks, as a means of incorporating the uncertainty associated with carbon stocks

---

[5]Sugarcane in Dwangwa is harvested annually during a 9-year cycle (on average). After the end of this cycle, the sugarcane plants are uprooted, and the land is left fallow for around 6 months. Subsequently, a new cycle starts with the planting of new sugarcane stems. As a result this 20-year cycle includes 19 years of sugarcane production (9 + 9 + 1 years) and 1 year where the land is fallow (6 + 6 months) (Romeu-Dalmau et al. 2018).

estimates. We follow the IPCC Guidelines for small data sets (IPCC 2006) and use R version 3.2.2 to perform the simulations (R Core Team 2015). A full explanation of the analytical process is included in Romeu-Dalmau et al. (2018).

Third, we use household surveys to identify which provisioning and cultural services local communities obtain from woodlands and their importance for the household. Following a literature review, we identify the main ecosystem services likely to be provided by the forest ecosystem in the study area (Gasparatos et al. 2018b). We then ask respondents two questions, with the first question eliciting whether respondents receive a given ecosystem service (Yes/No answer) and the second question eliciting the importance of this ecosystem service for the household. We survey households with differentiated levels of involvement in sugarcane production (i.e. plantation workers, sugarcane smallholders) and households not involved in sugarcane production (i.e. control groups). In particular, we sampled (a) formal plantation workers working for Illovo (N = 104); (b) irrigated sugarcane smallholders (N = 104); (c) rainfed sugarcane smallholders (N = 107); (d) a control group consisting of subsistence farmers living in the vicinity of the sugarcane-growing areas (N = 104); and (e) a control group consisting of subsistence farmers living approximately 50 km from the sugarcane belt (N = 99). Groups A–D live in the vicinity of the sugarcane plantation so it is safe to assume that they have been affected by the land use change, while Group E has not been affected by the land use change associated with sugarcane production. In order to ensure the effective randomization of respondents, we selected randomly the respondents of Groups A–C through lists obtained from Illovo, DCGL and the rainfed sugarcane grower associations (Sect. 5.2.1.3) and control groups through transect walks. Detailed information about the survey approach is included in Gasparatos et al. (2018a). The results are analysed through descriptive statistics (Sect. 5.3.2.2).

5.3 Results

5.3.1 Maputo: Ethanol Stove Adoption Site

5.3.1.1 Consumer Perceptions for Ethanol Fuel and Stoves

Out of the 341 surveyed households, 29% reported that they own (or previously owned) an ethanol stove. Of these, at the time of the survey, about 54% still used the ethanol stove regularly, 4% used it occasionally, and about 42% were no longer using it. Considering the entire household survey sample, the adoption profile is current users (17%), quitters (12%) and non-adopters (71%).

The different groups offered radically different reasons of why they adopted, not adopted and discontinued using the ethanol stoves (Table 5.1). The adopters usually cite the speed and convenience of the stove in terms of lighting and the lower smoke emissions. On the other hand, non-adopters and quitters mentioned the high costs

Table 5.1 Factors influencing the adoption and use of ethanol stoves and fuel according to the household survey

Category	Reason	Response (%)
Reasons for using	Faster to light and cook than charcoal stoves	68%
	Produce less smoke than charcoal stoves	21%
	More economical than charcoal stoves	7%
	Generally better than charcoal stoves	4%
Reasons for not using	Both the stove and the fuel are expensive	30.2%
	The stove is expensive	9.8%
	Bad reputation	8.5%
	Not appropriate to the needs of the household	8.5%
	The fuel is expensive	6.8%
	The stove is not trustworthy	6.8%
	Not interested in the stove	5.1%
	Own another stove	4.7%
	Lack of information and experience	3.8%
	The stove model/design is not good (e.g. the fuel evaporates)	3.8%
	Difficult to access the fuel	2.6%
	Safety considerations	2.1%
	The fuel smells and affects the taste of the food	1.7%
Reasons for discontinuing	Expensive	47.4%
	Stove broke down (already does not work)	12.3%
	The fuel is not accessible	12.3%
	Own another stove that works better	5.3%
	Disappointed with the product	5.3%
	Not suitable for the size of pots	3.5%
	Gave stove to another household	3.5%
	Slow for cooking	3.5%
	Does not have sufficient cooking heads	1.8%
	Stove was stolen	1.8%
	Safety considerations because of children in household	1.8%

Source: Mudombi et al. (2018a)

(both for stoves and fuel), the poor design that is prone to malfunction or unable to meet family needs and the lack of access to fuel.

FGD participants evoked similar reasons for adopting ethanol stoves and fuel and/or discontinuing use. In particular, former and current ethanol stove users pointed to the various factors influencing their decision to purchase ethanol stoves, with the most prominent being (a) the reasonable cost of the fuel and stove at the time of initial adoption, (b) convenience of use, (c) environmental friendliness, (d) lower smoke emissions, (e) higher safety, (f) cleanliness, (g) social exposure (i.e. influence from friends and neighbours) and (h) marketing strategies employed by the ethanol suppliers (Table 5.2). Most of the respondents indicated that the ethanol stove was easy to use, required little time to ignite and turn off and could be

Table 5.2 Factors influencing the adoption and use of ethanol stoves and fuel according to the focus group discussions

Category	Reason	Sample quotes
Reasons for adoption	Cost of stove and fuel	P3.12 "It is easy to use and cheaper than gas"
		P1.1 "Because I thought it was more economical than charcoal, but now it is no longer economical"
	Convenience of use	P5.29 "It is safe and fast and can be used anytime"
		P3.15 "The Ndzilo (bioethanol) stove can be used on rainy and windy days, any time and in every part of the house (living room, kitchen)"
	Environmental friendliness	P4.18 "I like the stove because it is portable and environmentally friendly."
		P4.21 "It does not pollute the environment"
	Safety and low smoke emissions	P 3.13 "When I saw the stove, I got interested because I heard that it was not dangerous for kids and it could be used indoors without smoke"
		P3.17 "It is not dangerous even kids can use it without any problem"
	Cleanliness	P3.10 "I like the way that the stove is used and it does not leave the pot dirty"
		P4.20 "It is easy to put off the fire and it does not mess the pots"
	Social exposure	P2.9 "I wanted to try because I had heard that it was good"
		P3.13 "When I saw the stove I got interested because I heard that it was not dangerous for kids and it could be used indoors without smoke"
	Effective marketing campaign	P1.5 "They made a lot of publicity on television and there were many stalls selling the product. When advertising is higher, there is a strong desire to buy and curiosity to try. Many people including myself got the Ndzilo at that time"
Reasons for discontinuation	Cost of fuel	P4.19 "Besides expanding stove selling, they must find a way to make the fuel economic by reducing the cost. The actual cost at the end of a month is high since the 5 litres last for only 3 days, so the cost gets to be higher than purchasing gas and charcoal"
		P1.1 "The cost of Ndzilo is high"
	Fuel availability	P2.6 "There was lack of Ndzilo fuel. I think it was in the past year [2014] and there was no explanation about the shortage of the product. I got worried because I had the stove at home and did not know what to do. When the product reappeared in the market the price had already changed; 5litres cost MZN 19 before, and after the shortage it costs MZN 220"
		P5.28 "There was a shortage of ethanol, I spent about 2 months without using Ndzilo"

(continued)

Table 5.2 (continued)

Category	Reason	Sample quotes
	Stove design	P4.21 "The flames from the stove are a bit far from the pot, so this wastes fuel"
		P5.26 "Ndzilo stove must be like paraffin stoves so that it could be easy to fill the tank"
	Quality of fuel	P5.29 "Ndzilo fuel had a bad smell"
		P4.21 "The quality of Ndzilo fuel was no longer the same"

Note: The respondent identifiers denote the number of the FGD (first number) and the specific respondent (second number)

used both inside and outside the house all year long. Social exposure was also listed as an important factor, as some respondents bought ethanol after seeing their neighbours using it.

However, despite incurring the initial stove purchase costs, some respondents eventually reduced the frequency of using the ethanol stove (or discontinued altogether its use). The most commonly mentioned reason was the high recurring fuel costs, which were in fact increasing over time (Table 5.2, Sect. 5.2.1.2). Similar to the survey, other reasons included the problematic stove design that was prone to malfunction (i.e. fuel tanks that rust easily) and the occasional lack of fuel availability (Table 5.2) (Sect. 5.2.1.2).

5.3.1.2   Factors Facilitating Rapid Ethanol Penetration

Expert interviews with personnel from Zoe Enterprises, a business venture in Maputo, provide further insights about the factors contributing to the initial rapid penetration of ethanol for household use in the city. These include the (a) enabling policy and institutional environment, (b) effective utilization of pre-existing market channels, (c) extensive awareness-raising campaigns and capacity-building efforts and (d) effective post-acquisition customer services and support.

Firstly, respondents suggested that the government of Mozambique had created a favourable environment for both local and international investors in the sector through the formulation of appropriate policies and regulatory framework. Furthermore, the government provided incentives and subsidies to bioethanol producers, which made ethanol production price-competitive for the household market. Central to all these was the approval of the National Policy and Strategy for Biofuels (NPSB), which provided clear guidelines to both the public and private sector to enhance participation in biofuel activities. The NPSB was aimed at reducing the dependence on imported fossil fuels and essentially created a local biofuel market, including for the household sector. Even though at the time of the survey, the cooking ethanol was imported from South Africa as CleanStar (and its smallholder-based cassava production and ethanol distillation) had ceased operations (Sect. 5.2.1.2), the government still subsidized the cost, hence reducing the overall cost assumed by customers. For example, according to the respondent from Zoe Enterprises:

> Initially, the government was not supportive as the only uses for ethanol they were familiar with, were as spirit in the health sector (which had no taxes) and as a beverage (which had 40% tax including import duty and VAT). However, after lobbying and explaining to the government that ethanol can be used for cooking in low-income households to substitute charcoal and firewood, we were able to convince them. Now we only pay 17% VAT, and no import duty. (Personal Communication, Manager, Zoe Enterprises, September 2015)

Secondly, it was also suggested that the effective capitalization of pre-existing market channels facilitated the rapid expansion of the ethanol sector in the city. Essentially, bioethanol was stored and distributed through a pre-existing network that commercialized imported ethanol gel fuel (Sect. 5.2.1.2), with the pre-existing consumers forming the initial consumer base of the ethanol. This allowed for the minimization of customer acquisition downtime. Furthermore, the ethanol fuel was retailed via existing outlets that were already known to the customer base, hence making easier the acquisition and improving the accessibility to the customers.[6] Moreover, selling both the stove and fuel in the same shop, and supply them through the same distributor reduced distribution costs and consumer effort in acquiring the stove and fuel. As was mentioned by an interviewee:

> Previously, Zoe Enterprises was selling gel fuel. We had our own business license, the premises, the staff, our gel fuel client base and our distribution networks. When we were approached by CleanStar Ventures who produced the ethanol, and which was going to be cheaper than the gel fuel we were distributing, we agreed [to work together]. They worked under our kind of umbrella. When they got their license, we started operating as CleanStar, with Zoe Enterprises becoming the department of sales and marketing. Our role was to produce the NDZilo fuel [a mixture of ethanol and other additives], and perform marketing activities, since we had links with the gel communities and distributing to our retailing networks. (Personal communication, Sales and Marketing Lead, Zoe Enterprises, September 2015)

Thirdly, a series of capacity-building efforts and awareness-raising campaigns through different channels further accelerated ethanol penetration in Maputo. Information sharing about the importance of the ethanol fuel and especially ethanol's economic, social and environmental benefits was a key element of these efforts. Relevant information about the ethanol stoves and fuel was disseminated through TV commercials, billboards and door-to-door visits by the sales team. These activities targeted a large segment of the local community and essentially familiarized many potential users with the new fuel. This ignited interest and catalysed the purchase of the stove and the fuel. As was mentioned by a respondent:

> We had a big team, so we coordinated, supervised, and worked six to seven days a week. We were in the community, we were in meetings. We started by entering one neighborhood, then we worked first with the chief who is the community leader who then convened meetings

---

[6]In some cases innovative partnerships further improved the visibility and accessibility of the ethanol fuel. For example, the ethanol distribution company partnered with a fish-retailing outlet, which sold bioethanol alongside its fish products.

with other smaller leaders. We had to do the demonstration, we had to explain to them and then we entered other neighborhoods and had our teams moving from place to place inside the community. We also did television advertisements and had billboards everywhere as we had more financial resources to support these activities. (Personal communication, Sales and Marketing Lead, Zoe Enterprises, September 2015)

Finally, the effective post-acquisition customer services and support was another major factor contributing to the quick uptake of ethanol fuel and stoves. By maintaining an updated record of their customers, ethanol stove retailers were able to follow up and receive user feedback. This included, for instance, reasons why ethanol purchase had declined over time. Oftentimes, as discussed above, this decline was influenced by poor stove design and fuel quality. This enabled the distributors to understand better the specific product-related problems and replace problematic stoves as needed. Such customer databases were important especially following collapse of the ethanol production (Sect. 5.2.1.2), as the distributors were able to contact directly ethanol stove users and explain the situation. Identifying and resolving consumer challenges early on helped in maintaining a demand for bioethanol despite the fuel supply challenges. This was described by an interviewee as follows:

We have a list of names and contact details of all the customers who have bought our ethanol stove. The information was mainly collected for the stove warranty and to get feedback from the customers. The information was used to contact the customers whose canister tanks had rusted for replacement and for informing them that NDZiLO was then available in the market. (Personal communication, Sales and Marketing Lead, Zoe Enterprises, September 2015)

5.3.2 Dwangwa: Sugarcane Production Site

5.3.2.1 Land Use Change

Irrigated sugarcane production in the Illovo plantation and the smallholders scheme (DCGL) have caused significant land use change (Fig. 5.2). In particular irrigated sugarcane cultivation led to the conversion of low-density forest, high-density forest and agricultural land dedicated to food crops and particularly maize, which is the staple crop in the area (Figs. 5.2 and 5.3a). Rainfed sugarcane production has also directly converted agricultural land, but it is difficult to quantify the actual magnitude of this land use change. The land use change observed in the surroundings area is minimal compared to the direct land use change occurring within the boundaries of irrigated sugarcane production (Fig. 5.3b). Although it is not possible to conclusively establish causality, the land use change observed in the surrounding area can be potentially linked to farm displacement elsewhere in the area and/or the attraction of population due to generation of direct and secondary employment opportunities.

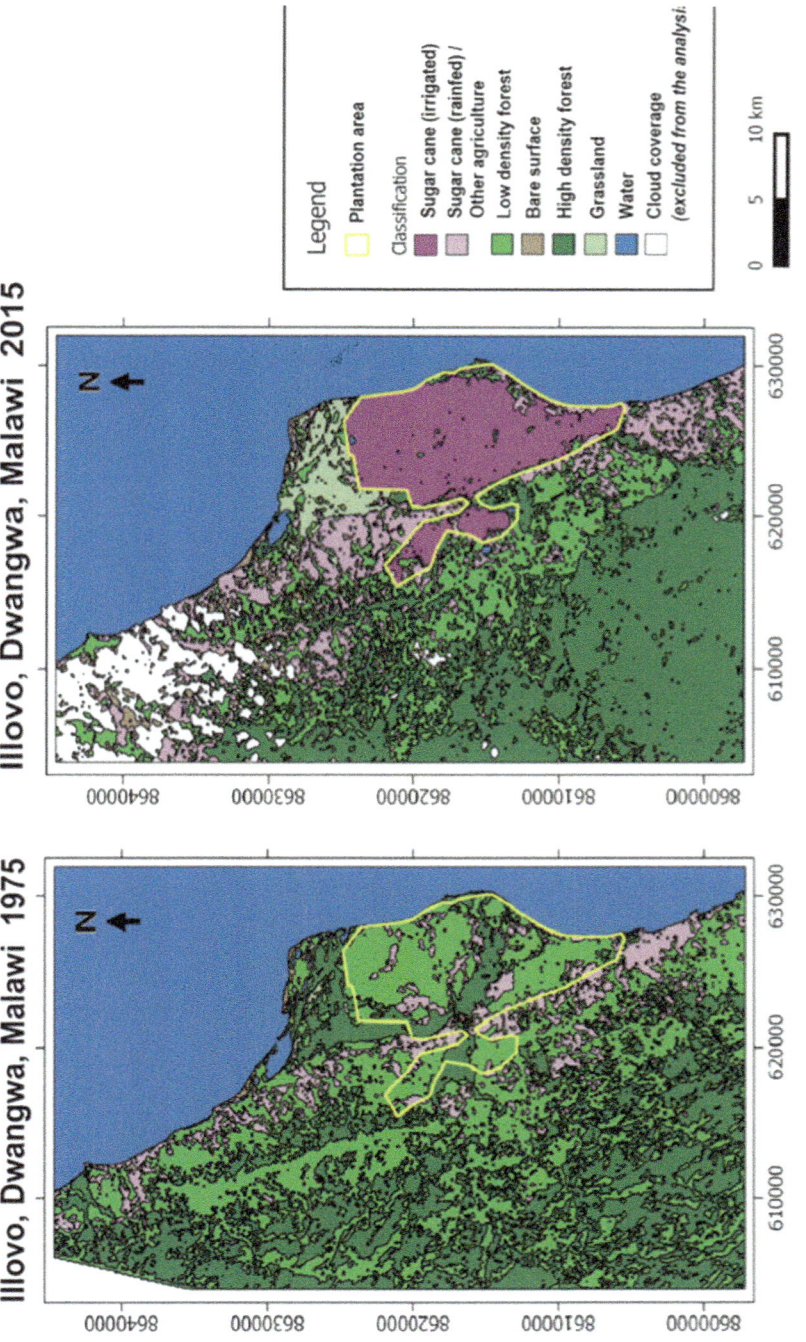

Fig. 5.2 Land use map of the Dwangwa area for 1975 and 2015. (Source: Romeu-Dalmau et al. 2018)

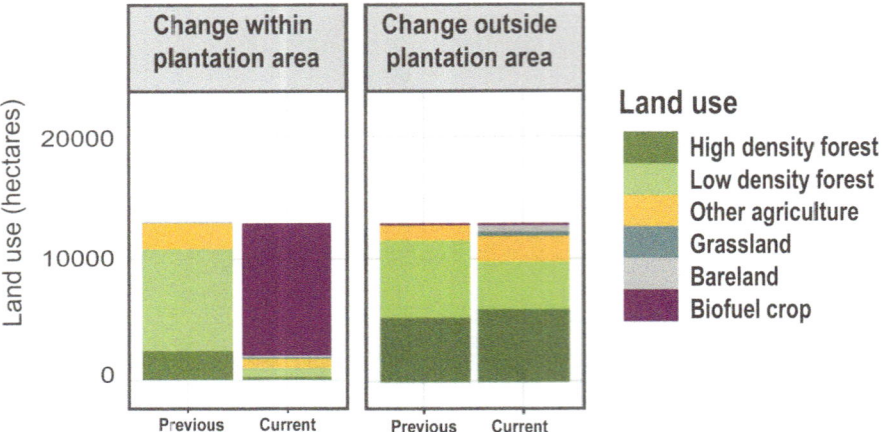

Fig. 5.3 Land use change within the irrigated sugarcane production area (**a**) and around it (**b**)
Note: Figure 5.3b illustrates the land use change that occurred in the surrounding areas of irrigated sugarcane production in a comparable area to Fig. 5.2b. (Source: Romeu-Dalmau et al. 2018)

Table 5.3 Carbon stocks in the different land uses (in tC/ha)

Land use	Carbon stocks (t/ha)			
	Above ground	Below ground	Soil organic carbon (SOC)	Total
High-density forest	42.9 ± 21.6	17.4 ± 7.1	54.1 ± 12.5	114 ± 25
Low-density forest	4.4 ± 3.8	2.6 ± 2.1	30.9 ± 6.6	38 ± 12
Sugarcane	20.9 ± 4.9	3.1 ± 0.7	40.5 ± 9.5	65 ± 10
Other agriculture	2.0 ± 1.2	0.4 ± 0.2	33.4 ± 12.0	36 ± 12
Grassland	1.4 ± 1.1	2.7 ± 2.1	39.8 ± 14.3	44 ± 15

5.3.2.2 Ecosystem Services

Sugarcane areas store on average 65 tC/ha, which is the second highest carbon storage amount among the different land uses (Table 5.3). Soil organic carbon (SOC) constitutes the largest carbon stock in each land use. Net carbon storage over a 20-year period is higher for sugarcane compared to surrounding land uses (Fig. 5.4). This suggests that sugarcane production can generate carbon stock gains in the study site, offering thus an important regulating service. This is possibly due to the fact that the densely planted sugarcane crops have higher standing biomass compared to the low standing biomass of surrounding agricultural and woodlands that are already partly degraded from fuelwood extraction (Sect. 5.2.1.3).

While these results show clear carbon sequestration benefits, the conversion of woodlands has also most likely led to the loss of forest-related provisioning ecosystem services such as fuelwood, medicinal plants, wild food and fodder for livestock. Although it is not possible to quantify the actual loss of these ecosystem services, the

Fig. 5.4 Net carbon stock change due to sugarcane conversion
Note: The box represents the interquartile range (IQR; difference between the 25th and 75th percentiles). The thick black line represents the median, and the top and bottom whiskers indicate the highest values within the upper range (75th percentile $+1.5 * IQR$) and the lowest values within the lower range (25th percentiles $-1.5 * IQR$). Source (Romeu-Dalmau et al. 2018)

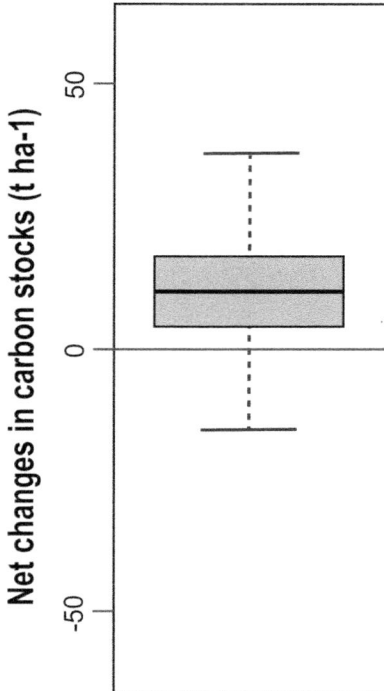

fact remains that these ecosystem services are important for the livelihoods of the local communities.

Based on the household survey, the two main provisioning services collected from forest were non-timber forest products (NTFPs) and fuelwood, with 62% and 48% of the responding households involved in their collection (Fig. 5.5). Fuelwood was collected throughout the year, while NTFPs and wild food were collected just a few times within the year. Other major provisioning services derived from woodlands include indigenous vegetables (21%) and wild fruits (20%). Fuelwood was reportedly considered to be of high importance to 94% of the surveyed households, with fodder and non-timber products also reportedly to be of high importance by 93% and 85% of the respondents, respectively (Table 5.4). A small proportion of respondents do not consider medicinal plants and honey to be of high importance for their households. On the other hand, woodlands do not seem to provide significant cultural ecosystem related to recreation and religious values. Respectively, 87% and 80% of the respondents mentioned that they do not derive such services from the landscape.

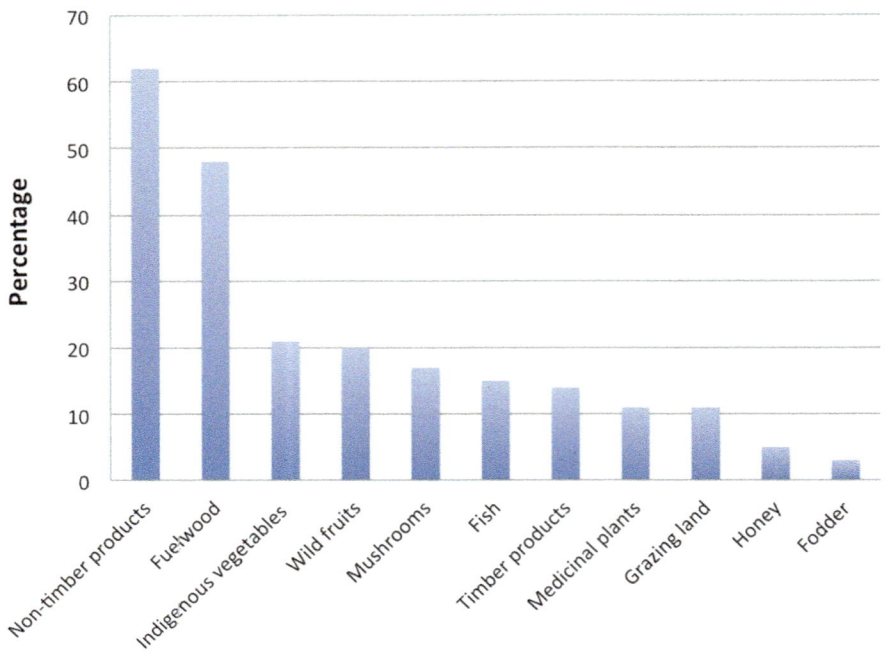

Fig. 5.5 Fraction of respondents obtaining provisioning ecosystem services from woodlands

Table 5.4 Self-reported importance of woodland ecosystem services for surveyed households

	No importance (%)	Low importance (%)	Medium importance (%)	High importance (%)	Number of respondents
Timber products	10	3	18	69	72
Non-timber products	1	2	12	85	323
Fuelwood	1	0	4	94	248
Medicinal plants	7	21	26	46	57
Honey	8	21	29	42	24
Wild fruits	5	13	29	54	104
Mushrooms	2	12	27	58	89
Fodder	19	13	19	50	16
Fish	4	5	37	53	75
Grazing land	5	2	0	93	57
Indigenous vegetables	4	7	27	63	107

5.4   Discussion

5.4.1   Synthesis of Findings

Section 5.3 outlined some of the critical issues associated with the demand and adoption of ethanol as a clean cooking alternative in Maputo (Mozambique) and the impacts of sugarcane production in Dwangwa (Malawi). In particular Sect. 5.3 explored issues of bioethanol marketing and use in Maputo, as well as the factors that influenced its successful and rapid penetration as a clean cooking fuel. In Malawi the analysis focused on the possible impacts of sugarcane production on specific ecosystem services.

When it comes to the feedstock production side, cassava-based ethanol production in Mozambique following the CleanStar model was rather complex, and due to multiple institutional, cost-related and logistical reasons, it eventually collapsed (Sect. 5.2.1.2) (Mudombi et al. 2018a; Costa 2019). On the contrary, sugarcane and ethanol production in Malawi are quite optimized, with a long history of achieving high levels of output at cost-competitive prices (IRENA 2016; Mitchell 2011). However, due to the large-scale production model adopted (Chap. 3 Vol. 1), sugarcane production in Dwangwa has caused extensive land use and land cover change, which has had variable effects on the provision of different ecosystem services and the wellbeing of local communities (Chinsinga 2017; Kiezebrink et al. 2015).

Land use change, and especially the loss of crop land, possibly had negative effects on provisioning services related to food crops and woodland products, which are important for the livelihoods of local communities (Sect. 5.3.2.2). However, other studies in the same area have suggested that the actual effects of sugarcane production on food crop production might have been less pronounced for small-holders, as improved access to fertilizers and other agricultural resources enabled higher yields, enabling thus, to some extent, the compensation of cropland loss (Herrmann et al. 2018) (see also Chap. 3 Vol. 1). For regulating services, land conversion for sugarcane production led to gains in carbon stocks (Fig. 5.4), which is rather unusual for biofuel projects that convert woodland (Achten and Verchot 2011). Most likely these carbon stock gains were due to the fact that the converted woodland areas were already rather degraded from extensive fuelwood collection (Romeu-Dalmau et al. 2018), which is prevalent throughout rural Malawi (Sect. 5.2.1.1). Landscape conversion did not seem to have had an appreciable effect on the provision of cultural ecosystem services, as few respondents indicated that they derive such services from woodland areas (Sect. 5.3.2.2). However, it should be noted that the pathways linking bioenergy production and cultural ecosystem services are rather complicated and indirect (Gasparatos et al. 2018b), usually depending on the actual context of bioenergy production and use (e.g. Ahmed et al. 2019; de Hoop 2018).

When it comes to the ethanol adoption side, our study finds that high costs have been the main reasons for not adopting or discontinuing ethanol use for cooking

(Tables 5.1 and 5.2). High costs have been identified as major constraints to stove adoption in many parts of SSA (Karanja et al. 2020; World Bank 2017; Rehfuess et al. 2014). Furthermore, users tend to prefer ethanol stoves for short cooking tasks such as boiling water for tea due to their convenience. Still, charcoal is preferred for longer and slower-cooking tasks, which further suggests that even for ethanol stove adopters, the comparatively higher operational cost factors limit its use to some degree (Mudombi et al. 2018a). Indeed many FGD participants suggested that the initial adoption of ethanol stoves was influenced by the reasonable initial costs, but this changed later on due to the collapse of domestic ethanol production, which in turn caused ethanol price hikes and decreases in quality (Table 5.2). Simple calculations based on the cost of energy supplied to the pot suggest that local ethanol prices need to drop to about 0.50 USD/L to make ethanol truly cost-competitive with charcoal, which is about half the current cost and rather close to production costs in major producing countries such as Brazil and the USA (Mudombi et al. 2018a).

As mentioned above the CleanStar ethanol production model was not optimized and solely geared towards the production of cheap ethanol, but attempted to incorporate many social aspects such as smallholder support during cassava farming (Costa 2019). Furthermore, it used a relative small-scale distillation facility that was located both far from the cassava production sites and the ethanol demand areas. The tendency to "weigh down" household fuel-switching efforts with multiple social benefits that may later complicate economic feasibility is among the reasons why some major cookstove programmes such as the Global Alliance for Clean Cookstoves tend to emphasize more commercial approaches involving building entrepreneurship and market demand (GACC 2014). However, such cost dynamics may change substantively if linked to large-scale production models such as the ones currently operational in Malawi (Puzzolo et al. 2019).

5.4.2 Policy Implications and Recommendations

The findings discussed in this chapter suggest major intersections between clean cooking and the SDGs. For example, the widespread adoption and sustained use of ethanol stoves and fuels essentially contributes directly to SDG7 on access to clean, affordable, reliable and sustainable energy (Sect. 5.1). However, many of the issues discussed in Sect. 5.3 associated with stove adoption and sugarcane production suggest strong intersections with multiple other SDGs such as SDG2 (Zero Hunger), SDG12 (Responsible Consumption and Production), SDG13 (Climate Action) and SDG15 (Life on Land), among others. This suggests that in order to maximize the positive impacts of ethanol stoves and fuel for contributing to progress to multiple SDGs, it would be important to both ensure the wide adoption and sustained use of ethanol stoves/fuel on the one hand, and the sustainable production of ethanol fuel on the other. Below we draw from the main results of this study some recommendations on how to enhance the potential of ethanol stoves and fuel to catalyse progress for meeting the SDGs.

First, to support the extensive development and widespread adoption of modern bioenergy options, SSA governments would need to design and implement comprehensive and cohesive policies that support various end uses, activities and actors along the bioethanol value chain (Chap. 2 Vol. 1). Comprehensive government support in terms of policies and institutional frameworks, subsidies and incentives are key for the establishment of biofuels investment and market (Jumbe et al. 2009; African Union 2013). This comprehensive support is clearly seen in Malawi through the long-term production of ethanol and its effective integration in the national energy system (Johnson and Silveira 2014). It was also partly seen in Mozambique in enabling both local and international actors to invest in biofuels sector (Schut et al. 2014). Such comprehensive support can in theory allow for cost reduction (see below) and the development of infrastructure for fuel/stove distribution, user education and awareness raising through diverse channels (Karanja et al. 2020). As discussed throughout this chapter, all these factors have played a major role in the long-term production of ethanol in Malawi and the rapid adoption of bioethanol in Maputo.

Second, it would be important to reduce the costs of ethanol production and provide incentives to consumers for stove uptake. It has been argued that in order to realize the wide commercialization and market penetration of clean fuels in SSA, it is crucial to understand the economics of biofuels industry (Amigun et al. 2008; Mitchell 2011). Regardless of whether ethanol is locally produced or imported, its costs as a household fuel must be competitive against other cooking fuels such as charcoal, LPG or electricity (World Bank 2017). However, the affordability of clean fuels such as bioethanol still remains a challenge, hindering their adoption and sustained use. This is further compounded by the fact that clean fuels such as ethanol are part of the formal sector and therefore regulated, compared to fuels such as charcoal and firewood, which are in the informal sector and hence unregulated (Smith et al. 2015; Ndegwa et al. 2016). A major issue directly related to the cost structure of ethanol is likely to be the taxes levied on ethanol. There are various convincing arguments that ethanol should be lightly taxed or not taxed at all when destined as a cooking fuel due to its health and environmental benefits compared to traditional cooking fuels (Sect. 5.1) (Chaps. 2, 7 Vol. 1) (Dalberg 2018; World Bank 2017). Governments can also support stove uptake by providing subsidies on stoves and fuels, therefore further enhancing its affordability to the consumers (Karanja et al. 2020).

Third, it would be important to develop and strengthen national and local bioethanol markets by putting in place institutional structures that provide a space for both the private and public actors to thrive along the bioethanol value chain (World Bank 2017; GACC 2014). The utilization and leveraging of existing market structures could be very important in fostering the scaling up of the adoption of a new product as the CleanStar experience has shown (Sect. 5.3.1). Governments can further support such efforts through research and development, building technical capacity in their respective ministries and raising awareness to potential users through targeted educational activities and campaigns (Karanja et al. 2020; Rehfuess et al. 2014). Further, and if appropriate, national government can facilitate the mandatory blending of ethanol into transportation fuels. This was done with success in Brazil and the USA and has not only created a viable biofuel market but has also driven down the cost of ethanol production (see comment above). This would simultaneously help reduce ethanol costs for cooking fuel and at the same time diversify the market options for producers, reducing thus their risks. However, embarking on a major transportation fuel ethanol programme would need sound research and justification. Furthermore, there would be a need for related guidelines and standards to ensure the effective consumer protection (Karanja et al. 2020).

Finally, depending on the location, scale of production and production practices, biofuel systems in SSA can have both positive and negative environmental, social and cultural impacts, including on ecosystem services (Gasparatos et al. 2015, 2018b). These impacts essentially influence the landscape sustainability and the wellbeing of local communities. It is thus important to adopt policies that acknowledge these aspects and possible trade-offs and provide guidelines on how to assess impacts and minimize negative trade-offs during the development of biofuel policies, programmes and projects. It should be noted that in SSA context, land use management strategies interact strongly with policies aimed at reducing dependence on traditional biomass and the significant GHG emissions associated with land use change (van de Ven et al. 2019). In this context, provisions that further support and strengthen the role of environmental impact assessments (EIAs) would be highly beneficial (OECD 2011), as EIAs are often the only avenue to hear the voices of local communities during the development of large agro-industrial projects in SSA (Ahmed et al. 2019).

5.5   Conclusions

Bioethanol, if priced correctly, has a great potential to substitute charcoal as a cooking fuel in SSA, therefore contributing directly to the achievement of SDG7. Its uptake can have multiple benefits related to health (SDG3), the stimulation of local industries and marketing chains (SDG9) and the reduction of GHG emissions (SDG 13) and deforestation and land degradation (SDG15). This is in addition to providing a clean fuel that appears to have a relatively high degree of user acceptance despite its comparatively higher cost. However, in order for ethanol cooking to evolve from a niche market within a fuel stacking usage context, there would be a need for substantive cost reductions to the point that it becomes comparable to charcoal. Alternatively (or better at the same time) the environmental and social costs of charcoal should be better reflected in local and national taxation regimes and energy and land use policies.

Considering the many possible environmental, social and economic benefits from moving away from charcoal to ethanol, there is a strong case to be made for national support for ethanol. Government support could be in the form of low taxation and the stimulus packages supporting low-cost ethanol production. The development of regional collaboration, trade and technology transfer could also support such markets, as exemplified by the fact that cost-effective supply was available in one country (Malawi), whereas robust demand existed in a neighbouring country (Mozambique).

Finally, there is real need to understand the pull factors that could accelerate the adoption and promote the sustained use of clean stoves and fuels in SSA. Addressing consumer affordability would be definitely critical towards this end. However, it would be equally important to understand some of the dynamics and trade-offs at both the fuel production and use levels and to identify ways to reduce the negative impacts and curb the barriers to adoption. Malawi and Mozambique are two countries that have complementary, albeit different, experiences in the production and adoption of ethanol fuel that can foster such learning for other SSA countries.

Acknowledgements The research was supported by the Ecosystem Services for Poverty Alleviation Programme (ESPA) (Grants: NE/L001373/1 and FELL-2014-107). The ESPA programme is funded by the UK Department for International Development (DFID), the Economic and Social Research Council (ESRC) and the Natural Environment Research Council (NERC). We acknowledge the support of Davies Luhanga, Manuel Chenene and Boris Attanasov for data collection in Malawi and Mozambique. Marcin Jarzebski and Jie Su helped with the editing of this chapter.

References

Achten WMJ, Verchot LV (2011) Implications of biodiesel-induced land-use changes for CO_2 emissions: case studies in tropical America, Africa, and Southeast Asia. Ecol Soc 16:14

Adam JC (2009) Improved and more environmentally friendly charcoal production system using a low-cost retort–kiln (Eco-charcoal). Renew Energy 34:1923–1925

African Union (2013) Africa bioenergy policy framework and guidelines: towards harmonizing sustainable bioenergy development in Africa. African Union, Addis Ababa

Ahmed A. Jarzebski MP, Gasparatos A (2019) Using the ecosystem service approach to determine whether jatropha projects were located in marginal lands in Ghana: Implications for site selection. Biomass and Bioenergy 114:112–124

Alexander DA, Northcross A, Karrison T, Morhasson-Bello O, Wilson N, Atalabi OM et al (2018) Pregnancy outcomes and ethanol cook stove intervention: a randomized-controlled trial in Ibadan, Nigeria. Environ Int 111:152–163

Amegah AK, Jaakkola JJK (2016) Household air pollution and the sustainable development goals. Bull World Health Organ 94:215–221

Amigun B, Sigamoney R, von Blottnitz H (2008) Commercialisation of biofuel industry in Africa: a review. Renew Sust Energ Rev 12(3):690–711

Atanassov B, Egas A, Falcão M, Fernandes A, Muhamane G (2012) Mozambique urban biomass energy analysis 2012. Mozambique Ministry of Energy, Maputo

Bailis R, Drigo R, Ghilardi A, Masera O (2015) The carbon footprint of traditional woodfuels. Nat Clim Chang 5:266–272

Benka-Coker ML, Tadele W, Milano A, Getaneh D, Stokes H (2018) A case study of the ethanol CleanCook stove intervention and potential scale-up in Ethiopia. Energy Sustain Dev 46:53–64

Beza SA, Assen MA (2017) Expansion of sugarcane monoculture: associated impacts and management measures in the semi-arid East African Rift Valley, Ethiopia. Environ Monit Assess 189(3):111

Bogdanski A (2012) Integrated food–energy systems for climate-smart agriculture. Agric Food Sec 1:9

Castán Broto V, Arthur MFSR, Guibrunet L (2020) Energy profiles among urban elite households in Mozambique: Explaining the persistence of charcoal in urban areas. Energy Research & Social Science 65:101478

Cerutti PO, Sola P, Chenevoy A, Iiyama M, Yila J, Zhou W et al (2015) The socioeconomic and environmental impacts of wood energy value chains in Sub-Saharan Africa: a systematic map protocol. Environ Evid 4:12

Chidumayo EN, Gumbo DJ (2013) The environmental impacts of charcoal production in tropical ecosystems of the world: a synthesis. Energy Sustain Dev 17(2):86–94

Chinangwa L, Gasparatos A, Saito O (2017) Forest conservation and the private sector in Malawi: the case of payment for ecosystem services schemes in the tobacco and sugarcane sectors. Sustain Sci 12:727–746

Chinsinga B (2017) The Green Belt initiative, politics and sugar production in Malawi. J South Afr Stud 43:501–515

Costa C (2019) The cassava value chain in Mozambique. World Bank, Washington DC

Dalberg (2018) Scaling up clean cooking in urban Kenya with LPG & bio-ethanol: a market and policy analysis. Dalberg, Geneva

Dam Lam R, Boafo YA, Degefa S, Gasparatos A, Saito O (2017) Assessing the food security outcomes of industrial crop expansion in smallholder settings: insights from cotton production in Northern Ghana and sugarcane production in Central Ethiopia. Sustain Sci 12(5):677–693

Das I, Jagger P, Yeatts K (2017) Biomass cooking fuels and health outcomes for women in Malawi. EcoHealth 14(1):7–19

Davies GM, Pollard L, Mwenda MD (2010) Perceptions of land-degradation, forest restoration and fire management: a case study from Malawi. Land Degrad Dev 21:546–556

de Hoop E (2018) Understanding marginal changes in ecosystem services from biodiesel feedstock production: a study of Hassan bio-fuel park, India. Biomass Bioenergy 114:55–62

de Koning P, Atanassov B (2014) Sustainable charcoal value chain Mozambique. Energy Engineers Solutions (EES), Maputo

Debbi S, Elisa P, Nigel B, Dan P, Eva R (2014) Factors influencing household uptake of improved solid fuel stoves in low- and middle-income countries: a qualitative systematic review. Int J Environ Res Pub Health 11:8228–8250

Degefa S, Saito O (2017) Assessing the impacts of large-scale agro-industrial sugarcane production on biodiversity: a case study of Wonji Shoa Sugar Estate, Ethiopia. Agriculture 7(12):99

Dlamini WM (2017) Mapping forest and woodland loss in Swaziland: 1990–2015. Remote Sens Appl Soc Environ 5:45–53

Dunkelberg E, Finkbeiner M, Hirschl B (2014) Sugarcane ethanol production in Malawi: measures to optimize the carbon footprint and to avoid indirect emissions. Biomass Bioenergy 71:37–45

Edelstein M, Pitchforth E, Asres G, Silverman M, Kulkarni N (2008) Awareness of health effects of cooking smoke among women in the Gondar region of Ethiopia: a pilot survey. BMC Int Health Hum Rights 8:10

Ekouevi K, Tuntivate V (2012) Household energy access for cooking and heating: lessons learned and the way forward. World Bank, Washington DC

EUEI (2012) Mozambique biomass energy strategy. European Union Energy Initiative Partnership Dialogue Facility (EUEI PDF), Eschborn

Ezzati M, Kammen DM (2002) The health impacts of exposure to indoor air pollution from solid fuels in developing countries: knowledge, gaps, and data needs. Environ Health Perspect 110:1057–1068

FAO (2018) Using Prosopis as an energy source for refugees and host communities in Djibouti, and controlling its rapid spread. Food and Agriculture Organization (FAO), Rome

Fullerton DG, Bruce N, Gordon SB (2008) Indoor air pollution from biomass fuel smoke is a major health concern in the developing world. Trans R Soc Trop Med Hyg 102(9):843–851

GACC (2014) Market Enabling Roadmap 2015-2017. Global Alliance for Clean Cookstoves (GACC), Washington DC

Gasparatos A, von Maltitz GP, Johnson FX, Lee L, Mathai M, Puppim de Oliveira JA, Willis KJ (2015) Biofuels in sub-Sahara Africa: drivers, impacts and priority policy areas. Renew Sust Energ Rev 45:879–901

Gasparatos A, von Maltitz G, Johnson FX, Romeu-Dalmau C, Jumbe C, Ochieng C, Mudombi S, Balde B, Luhanga D, Nyambane A, Lopes P, Jarzebski M, Willis KJ (2018a) Survey of local impacts of biofuel crop production and adoption of ethanol stoves in southern Africa. Nat Sci Data 5:180186

Gasparatos A, Romeu-Dalmau C, von Maltitz G, Johnson FX, Shackleton C, Jarzebski MP, Jumbe C, Ochieng C, Mudombi S, Nyambane A, Willis KJ (2018b) Mechanisms and indicators for assessing the impact of biofuel feedstock production on ecosystem services. Biomass Bioenergy 114:157–173

Government of Malawi (2009) National biomass energy strategy ministry of energy, Lilongwe

Herrmann R, Jumbe C, Bruentrup M, Osabuohien E (2018) Competition between biofuel feedstock and food production: empirical evidence from sugarcane outgrower settings in Malawi. Biomass Bioenergy 114:100–111

Hess TM, Sumberg J, Biggs T, Georgescu M, Haro-Monteagudo D, Jewitt G et al (2016) A sweet deal? Sugarcane, water and agricultural transformation in Sub-Saharan Africa. Glob Environ Chang 39:181–194

Hosonuma N, Herold M, De Sy V, De Fries RS, Brockhaus M, Verchot L et al (2012) An assessment of deforestation and forest degradation drivers in developing countries. Environ Res Lett 7(4):44009

Hsieh HF, Shannon SE (2005) Three approaches to qualitative content analysis. Qual Health Res 15 (9):1277–1288

IEA (2014) Africa energy outlook: a focus on energy prospects in Sub-Saharan Africa. International Energy Agency (IEA), Paris

IEA, IRENA, UNSD, WB, WHO (2019) Tracking SDG 7: the energy progress report 2019, Washington DC

IPCC (2006) IPCC guidelines for national greenhouse gas inventories. Intergovernmental Panel on Climate Change (IPCC), Bonn

IRENA (2016) Bioethanol in Africa: the case for technology transfer and south-south co-operation. International Renewable Energy Agency (IRENA), Abu Dhabi

Johnson FX, Silveira S (2014) Pioneer countries in the transition to alternative transport fuels: comparison of ethanol programmes and policies in Brazil, Malawi and Sweden. Environ Innov Soc Trans 11:1–24

Johnson FX, Leal MRLV, Nyambane A (2017) Chapter 15: Sugarcane as a renewable resource for sustainable futures. In: Rott P (ed) Achieving sustainable cultivation of sugarcane, vol 1. Burleigh-Dodds, Cambridge

Jones D, Ryan CM, Fisher J (2016) Charcoal as a diversification strategy: the flexible role of charcoal production in the livelihoods of smallholders in Central Mozambique. Energy Sustain Dev 32:14–21

Jumbe CBL, Msiska FBM, Madjera M (2009) Biofuels development in Sub-Saharan Africa: are the policies conducive? Energy Policy 37(11):4980–4986

Jürisoo M, Lambe F, Osborne M (2018) Beyond buying: the application of service design methodology to understand adoption of clean cookstoves in Kenya and Zambia. Energy Res Soc Sci 39:164–176

Kamanga P, Vedeld P, Sjaastad E (2009) Forest incomes and rural livelihoods in Chiradzulu district, Malawi. Ecol Econ 68:613–624

Kambewa P, Mataya S, Johnson (2007) Charcoal: the reality –a study of charcoal consumption, trade and production in Malawi. International Institute for Environment and Development (IIED), London

Karanja A, Gasparatos A (2019) Adoption and impacts of clean bioenergy stoves in Kenya. Renew Sust Energ Rev 102:285–306

Karanja A, Mburu F, Gasparatos A (2020) A multi-stakeholder perception analysis of the adoption, impacts and priority areas in the Kenyan clean cooking sector. Sustain Sci 15:333–351

Kasambala Donga T, Eklo OM (2018) Environmental load of pesticides used in conventional sugarcane production in Malawi. Crop Prot 108:71–77

Kiezebrink V, van der Wal S, Theuws M, Kachusa P (2015) Bittersweet: Sustainability issues in the sugar cane supply chain. SOMO, Amsterdam

Langbein J (2017) Firewood, smoke and respiratory diseases in developing countries—the neglected role of outdoor cooking. PLoS One 12(6)

Maes WH, Verbist B (2012) Increasing the sustainability of household cooking in developing countries: policy implications. Renew Sust Energ Rev 16:4204–4221

Makonese T, Ifegbesan AP, Rampedi IT (2018) Household cooking fuel use patterns and determinants across southern Africa: evidence from the demographic and health survey data. Energy Environ 29:29–48

Manda S, Tallontire A, Dougill AJ (2018) Outgrower schemes, livelihoods and response pathways on the Zambian "sugarbelt". Geoforum 97:119–130

Mitchell D (2011) Biofuels in Africa: opportunities, prospects, and challenges. World Bank, Washington DC

Mudombi S, Nyambane A, von Maltitz GP, Gasparatos A, Johnson FX, Chenene ML, Attanassov B (2018a) User perceptions about the adoption and use of ethanol fuel and cookstoves in Maputo, Mozambique. Energy Sustain Dev 44:97–108

Mudombi S, von Maltitz GP, Gasparatos A, Romeu-Dalmau C, Johnson FX, Jumbe C, Ochieng C, Luhanga D, Loes P, Balde BS, Willis KJ (2018b) Multi-dimensional poverty effects around operational biofuel projects in Malawi, Mozambique and Swaziland. Biomass Bioenergy 114:41–54

Mugo F, Ong C (2006) Lessons of eastern Africa's unsustainable charcoal business World Agroforestry Centre (ICRAF), Nairobi

Ndegwa G, Anhuf D, Nehren U, Ghilardi A, Iiyama M (2016) Charcoal contribution to wealth accumulation at different scales of production among the rural population of Mutomo District in Kenya. Energy Sustain Dev 33:167–175

Ngcobo S, Jewitt G (2017) Multiscale drivers of sugarcane expansion and impacts on water resources in Southern Africa. Environ Dev 24:63–76

OECD (2011) Strategic environmental assessment and biofuel development. Organization for Economic Cooperation and Development (OECD), Paris

Okoko A, Reinhard J, von Dach SW, Zah R, Kiteme B, Owuor S, Ehrensperger A (2017) The carbon footprints of alternative value chains for biomass energy for cooking in Kenya and Tanzania. Sustain Energy Technol Assess 22:124–133

Openshaw K (2010) Biomass energy: employment generation and its contribution to poverty alleviation. Biomass Bioenergy 34(3):365–378

Ozier A, Charron D, Chung S, Sarma V, Dutta A, Jagoe K et al (2018) Building a consumer market for ethanol-methanol cooking fuel in Lagos, Nigeria. Energy Sustain Dev 46:65–70

Pope D, Bruce N, Dherani M, Jagoe K, Rehfuess E (2017) Real-life effectiveness of "improved" stoves and clean fuels in reducing PM2.5 and CO: systematic review and meta-analysis. Environ Int 101:7–18

Practical Action (2017) Final quantitative report on the cost and efficiency of cooking fuels in Malawi. Practical Action, Lilongwe

Puzzolo E, Zerriffi H, Carter E, Clemens H, Stokes H, Jagger P, Rosenthal J, Petach H (2019) Supply considerations for scaling up clean cooking fuels for household energy in low- and middle-income countries. GeoHealth 3:370–390

R Core Team (2015) R: a language and environment for statistical computing#. R Foundation for Statistical Computing, Vienna. Retrieved from https://www.R-project.org/

Rehfuess EA, Puzzolo E, Stanistreet D, Pope D, Bruce NG (2014) Enablers and barriers to large-scale uptake of improved solid fuel stoves: a systematic review. Environ Health Perspect 122:120–130

Republic of Malawi (2012) Integrated household survey 2010–2011. National Statistical Office, Zomba

Rogers C, Sovacool BK, Clarke S (2013) Sweet nectar of the Gaia: lessons from Ethiopia's "Project Gaia". Energy Sustain Dev 17(3):245–251

Romeu-Dalmau C, Gasparatos A, von Maltitz G, Graham A, Almagro-Garcia J, Wilebore B, Willis KJ (2018) Impacts of land use change due to biofuel crops on climate regulation services: five case studies in Malawi, Mozambique and Swaziland. Biomass Bioenergy 114:30–40

Ruiz-Mercado I, Masera O, Zamora H, Smith KR (2011) Adoption and sustained use of improved cookstoves. Energy Policy 39(12):7557–7566

Schure J, Pinta F, Cerutti PO, Kasereka-Muvatsi L (2019) Efficiency of charcoal production in Sub-Saharan Africa: solutions beyond the kiln. Bois et Forêts des Tropiques 340:57–70

Schut M, Cunha Soares N, van de Ven G, Slingerland M (2014) Multi-actor governance of sustainable biofuels in developing countries: the case of Mozambique. Energy Policy 65:631–643

Sedano F, Silva JA, Machoco R, Meque CH, Sitoe A, Ribeiro N, Anderson K, Ombe ZA, Baule SH, Tucker CJ (2016) The impact of charcoal production on forest degradation: a case study in Tete, Mozambique. Environ Res Lett 11:094020

Sedano F, Lisboa SN, Duncanson L, Ribeiro N, Sitoe A, Sahajpal R, Hurtt G, Tucker CJ (2020) Monitoring forest degradation from charcoal production with historical Landsat imagery a case study in southern Mozambique. Environmental Research Letters 15:015001

Silva JA, Sedano F, Flanagan S, Ombe ZA, Machoco R, Meque CH, Sitoe A, Ribeiro N, Anderson K, Baule S, Hurtt G (2019) Charcoal-related forest degradation dynamics in dry African woodlands: evidence from Mozambique. Appl Geogr 107:72–81

Smith HE, Eigenbrod F, Kafumbata D, Hudson MD, Schreckenberg K (2015) Criminals by necessity: the risky life of charcoal transporters in Malawi. For Trees Livelihoods 24 (4):259–274

Smith HE, Hudson MD, Schreckenberg K (2017) Livelihood diversification: the role of charcoal production in southern Malawi. Energy Sustain Dev 36:22–36

Takama T, Tsephel S, Johnson FX (2012) Evaluating the relative strength of product-specific factors in fuel switching and stove choice decisions in Ethiopia. Discrete choice model of household preferences for clean cooking alternatives. Energy Econ 34:1763–1773

Twongyirwe R, Bithell M, Richards KS (2018) Revisiting the drivers of deforestation in the tropics: insights from local and key informant perceptions in western Uganda. J Rural Stud 63:105–119

van de Ven DJ, Sampedro J, Johnson FX, Bailis R, Forouli A, Nikas A et al (2019) Integrated policy assessment and optimisation over multiple sustainable development goals in Eastern Africa. Environ Res Lett 14:094001

von Maltitz GP, Henley G, Ogg M, Samboko PC, Gasparatos A, Ahmed A, Read M, Engelbrecht F (2019) Institutional arrangements of outgrower sugarcane production in southern Africa. Dev South Afr 36:175–197

Wendimu MA, Henningsen A, Gibbon P (2016) Sugarcane Outgrowers in Ethiopia: "forced" to remain poor? World Dev 83:84–97

Woollen E, Ryan CM, Baumert S, Vollmer F, Grundy I, Fisher J et al (2016) Charcoal production in the Mopane woodlands of Mozambique: what are the trade-offs with other ecosystem services? Philos Trans R Soc B Biol Sci 371(1703):20150315

World Bank (2014) Clean and improved cooking in Sub-Saharan Africa. World Bank, Washington DC

World Bank (2017) Scalable business models for alternative biomass cooking fuels and their potential in Sub-Saharan Africa. World Bank, Washington DC

World Bank (2020) Access to clean fuels and technologies for cooking (% of population). World Bank, Washington DC. Available at: https://data.worldbank.org/indicator/EG.CFT.ACCS.ZS?end=2016&start=2016&view=bar

Zeray N, Legesse B, Mohamed JH, Aredo MK (2017) Impacts of Prosopis juliflora invasion on livelihoods of pastoral and agro-pastoral households of Dire Dawa Administration, Ethiopia. Pastoralism 7:7

Zhou Z, Dionisio KL, Arku RE, Quaye A, Hughes AF, Vallarino J et al (2011) Household and community poverty, biomass use, and air pollution in Accra, Ghana. Proc Natl Acad Sci 108 (27):11028–11033

Zorrilla-Miras P, Mahamane M, Metzger MJ, Baumert S, Vollmer F, Luz AC et al (2018) Environmental conservation and social benefits of charcoal production in Mozambique. Ecol Econ 144:100–111

Zulu LC (2010) The forbidden fuel: charcoal, urban woodfuel demand and supply dynamics, community forest management and woodfuel policy in Malawi. Energy Policy 38:3717–3730

Zulu LC, Richardson RB (2013) Charcoal, livelihoods, and poverty reduction: evidence from sub-Saharan Africa. Energy Sustain Dev 17(2):127–137

Chapter 6
The Effect of Introduced *Opuntia* (Cactaceae) Species on Landscape Connectivity and Ecosystem Service Provision in Southern Madagascar

Rivolala Andriamparany, Jacob Lundberg, Markku Pyykönen, Sebastian Wurz, and Thomas Elmqvist

6.1 Introduction

The conversion of natural areas for other human uses such as agriculture and settlements is one of the primary drivers of global biodiversity loss (e.g. Schroter et al. 2005; Haddad et al. 2015; IPBES 2019). Biodiversity loss can have grave and complex consequences on ecosystem functions, the provision of ecosystem services, agricultural sustainability, and ultimately the livelihoods and wellbeing of local communities across the world (Kremen 2005; Kremen and Ostfeld 2005; Millennium Ecosystem Assessment 2005; Harrison et al. 2014; IPBES 2019) (see Chaps. 9–10 Vol. 1). Landscape fragmentation and habitat loss through land conversion can divide and separate landscape elements and the disrupt gene flow (Reh and Seitz 1990; Keller and Largiader 2003; Cushman 2006). The persistence of many species depends upon their dispersal ability in relation to the specific landscape configuration. Typically the separation of species populations and meta-populations occurs when the habitat of a species consists of patches that are much further apart than the distance over which these species are able to disperse (Gutiérrez and Harrison 1996).

Another major driver behind biodiversity loss is species introduction (Vitousek et al. 1997; Mooney and Hobbs 2000; Millennium Ecosystem Assessment 2005;

R. Andriamparany
Ambatovy Joint-Venture, Antananarivo, Madagascar

J. Lundberg · S. Wurz · T. Elmqvist (✉)
Stockholm Resilience Centre, Stockholm, Sweden
e-mail: jakob.lundberg@su.se; sebastian.wurtz@kunskapsskolan.se; Thomas.Elmqvist@su.se

M. Pyykönen
University of Gävle, Gävle, Sweden
e-mail: markku.pyykonen@hig.se

© Springer Nature Singapore Pte Ltd. 2020
A. Gasparatos et al. (eds.), *Sustainability Challenges in Sub-Saharan Africa II*,
Science for Sustainable Societies, https://doi.org/10.1007/978-981-15-5358-5_6

IPBES 2019). Some "alien" species can become invasive and outcompete or prey on more susceptible. native species (Mack et al. 2000; Tsutsui et al. 2000; Mooney and Cleland 2001; Sakai et al. 2001). There are numerous examples that illustrate the vulnerability of flora and fauna to alien invasive species, particularly on oceanic islands (e.g. Groombridge 1992; Russell et al. 2017).

Land use change and landscape fragmentation have become major drivers of ecosystem degradation and biodiversity loss in sub-Saharan Africa (SSA) (IPBES 2018; Laurance et al. 2017). Many of the major habitats in the region experience increasing pressure from human activity, especially due to agricultural expansion, overgrazing, and the overexploitation of natural products (Perrings and Halkos 2018; IPBES 2018) (see also Chaps. 7, 9–10 Vol. 1; Chap. 5 Vol. 2). Furthermore, invasive species increasingly have major ramifications for the loss of biodiversity and small-holder farmer livelihoods across the continent (Pratt et al. 2017; Boy and Witt 2013). Indeed, many highly biodiverse ecosystems across SSA face the combined effects of land use change, landscape fragmentation, and invasive alien species, leading to the loss of ecosystem services and the livelihoods of poor local communities that are reliant on these services (IPBES 2018).

Madagascar is such a case of a highly biodiverse area that faces multiple drivers of biodiversity loss. In particular, Madagascar is an oceanic island in SSA that is considered a global biodiversity hotspot due to the high prevalence of endemic species (Myers et al. 2000; Ganzhorn et al. 2001; MEEF Madagascar and UNEP 2014). However the unique biodiversity of Madagascar is under threat due to extensive land use change and the introduction of alien invasive species, among others (Kull et al. 2012; Marshall et al. 2018).

The human colonization of Madagascar is relatively recent, but human activity has had a substantial impact in the last 2000 years (Binggeli 2003). The introduction of agriculture and livestock has been the main cause of land conversion in Madagascar (Koechlin et al. 1974; Goodman and Patterson 1997). Numerous examples highlight the significant negative effects of introduced crop and livestock species on biodiversity and human livelihood in the island (Binggeli 2003; Kull et al. 2012). For example, the invasion of the parthenogenic marbled crayfish (*Procambarus virginalis*) in the central region of the island threatens freshwater biodiversity, rice cultivation, and fishing (Jones et al. 2008). In its centre-eastern humid forests, non-native invasive rats (*Rattus* spp.), with wide foraging niches, have a negative impact to the populations of native Malagasy rodents due to competition and disease transmission (Dammhahn et al. 2017).

However, some other introduced species have had a different effect on the native ecosystems and local communities. For example, the introduction of *Opuntia* cacti (Cactaceae) and predominantly *O. ficus-indica* (i.e. prickly pear or "Raketa" as locally known) to southern Madagascar has had a different trajectory. French settlers originally introduced the *Opuntia* in the eighteenth century as protection against raids from local people (Middleton 2002; Binggeli 2003). One important characteristic of the *Opuntia* cacti is that it grows extensively in degraded areas where people concentrate (Kaufmann 2004). Opuntia fruits are consumed both by local communities and animals (Leuteritz 2003), especially during the dry periods of the year or during droughts. *Opuntia* hedges may also increase landscape connectivity by providing corridor functions or even potentially serving as a habitat to species that are critical for seed dispersal and pollination (cf. Lundberg and Moberg 2003; Kruess and Tscharntke 1994). These multi-faceted contributions and significance of *Opuntia* for livelihoods in the southern Androy region is encapsulated in the proverb *Longo Tandroy sy Raiketa* – "The Tandroy and the prickly pear are relatives" (Decary 1930).

Considering the high population density and extensive land use change and landscape fragmentation in Madagascar, the conservation of its unique biodiversity is a mounting sustainability challenge for in the country. Conserving biodiversity outside of protected areas is a key sustainability priority but at the same time a big challenge considering the high prevalence of poverty in rural areas of Madagascar. Landscapes dominated by smallholder farms containing combinations of native and alien species can, in some cases, provide valuable habitat services to biodiversity (Kull et al. 2013) and should thus be considered as a possible option for biodiversity conservation outside of protected areas in the island. As outlined above, one alien species with possible biodiversity conservation benefits in Madagascar is the *Opuntia*.

The aim of this chapter is to provide a preliminary analysis of the potential of *Opuntia* hedges to connect forest patches and provide other ecosystem services in southern Madagascar. We use a combination of literature review, geospatial analysis, and ecological surveys to map the distribution of *Opuntia* hedges in a human-dominated landscape located in the southern part of the Androy region and identify bird and insect species that interact with these hedges.

Section 6.2 outlines the methodology used in this study. Section 6.3 highlights the main results of the geospatial analysis and the ecological surveys, identifying the extent to which *Opuntia* hedges can create corridors between forest fragments. Section 6.4 discusses some of the different benefits that *Opuntia* hedges can offer in relation to landscape connectivity and the provision of ecosystem services and makes the case that they have a large potential to enhance biodiversity conservation in human-dominated landscapes in the island.

6.2 Methodology

6.2.1 Study Site

The Androy region is situated in the southernmost part of Madagascar (see Fig. 6.1). The area is characterized by semi-arid climatic conditions with irregular rainfall averaging less than 800 mm per year. The annual rainfall declines from north to south and from northeast to southwest (Koechlin 1972). The dry season usually lasts 8 months (between March and October/November), but it can extend locally over several years. The wet season starts from December and lasts until March. Temperature ranges between 29 and 42 °C (EPP PADR 2006), with the mean temperature

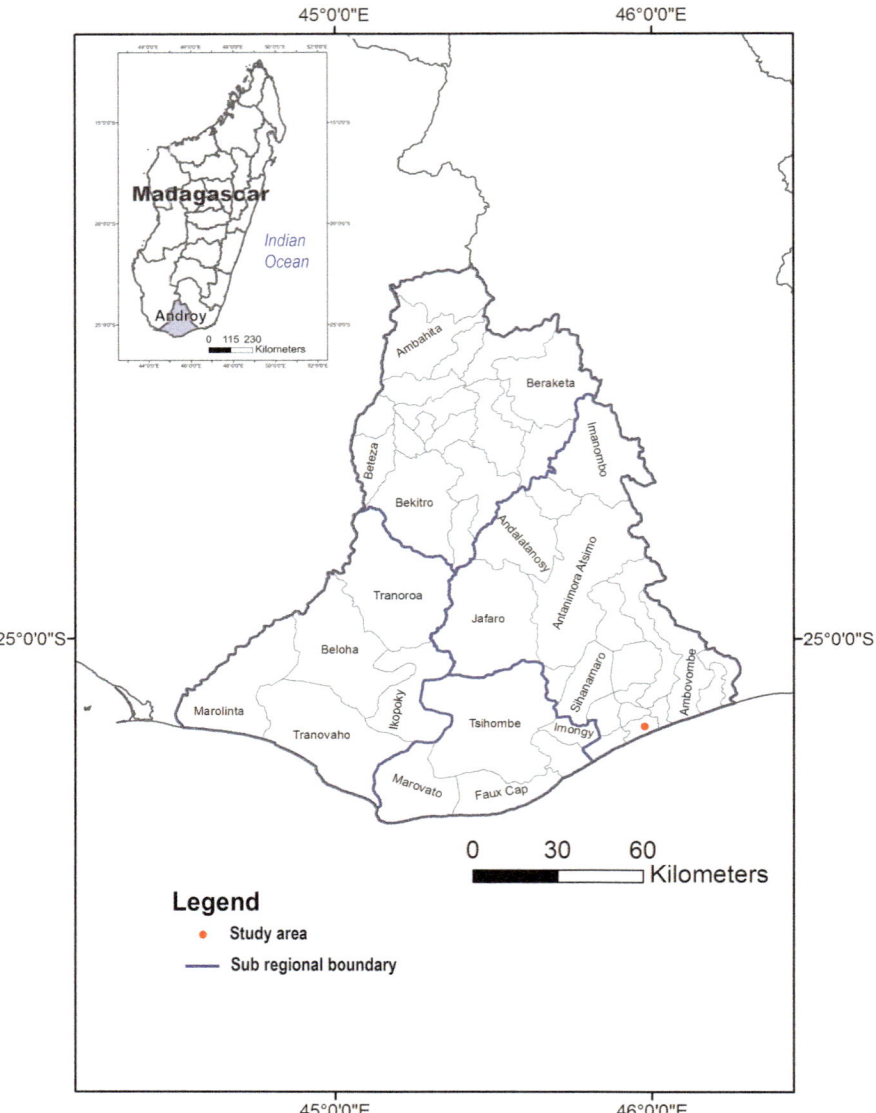

Fig. 6.1 Location of study site

generally being between 23 and 26 °C but the daily variability being as large as 22.5 °C during the cold season (in May–October).

The southernmost part of Androy region is a sandy area with paleodunes, which indicate a much more arid climate in the past, possibly during the last glacial maximum around 18.000 BP. The northern part is a hilly upland on Precambrian crystalline bedrock.

Settlements in the southern sandy areas have a long history and are dominated by the Tandroy people, which are primarily cattle herders. However, the cultivation of maize, beans, and sweet potatoes is essential for the local livelihoods, particularly in the south. The long dry periods and intense exposure to the predominant south-eastern wind often cause sand drifts, which cover the crops, and therefore often reduce crop yields. During these periods, the rural population often depends on the Raketa, which includes several varieties of *Opuntia* spp. (Cactaceae) and primarily *Opuntia ficus-indica*. Raketa fruits ripe in January and February, just before the first crop harvest. Thus the local communities use the fruits as an essential food supplement, when other food sources are absent or in short supply. The importance of these fruits for local food security increases during the dry years, when they become the main diet for humans (preferably the variety *raketasonjo*) and the cladodes of the cacti are used as livestock fodder.

For this study, we focused on an area of 9 km^2 (3x3 km, 900 ha) in the southern part of Madagascar. The study site is situated in the Androy region 17 km southwest of the district's main urban centre, Ambovombe (25°16'S, 45°59'E) (Fig. 6.2). The study site includes six heterogeneous forests that are larger than 1 hectare and are locally protected as sacred forests. The landscape is highly fragmented and human-dominated, with the forest patches surrounded by a matrix of small-scale agriculture and grazing land (Elmqvist et al. 2007, see Fig. 6.2). Most of the forest fragments are sacred forests protected through *fady*, a practice of taboo associated with local religious beliefs (Tengö et al. 2007). The area is located in an ecoregion that has among the highest levels of plant endemism in Madagascar, at both generic (48%) and species level (95%) (Koechlin 1972). Most of the endemic plants and animals are found primarily in these forest patches (Tengö et al. 2007).

6.2.2   Data Analysis

6.2.2.1   Remote Sensing Analysis

First we undertake a remote sensing analysis to identify the different forest patches and understand better landscape fragmentation. We use the IKONOS image 2003 November 05, which has a spatial resolution of 1x1m in the panchromatic band and 4x4m in the multispectral bands. Only the larger *Opuntia* hedges and stands are clearly visible in the multispectral images, whereas the smaller hedges are only visible in the panchromatic band. To detect vegetation from satellite imagery, we generate a standardized tNDVI index[1] in ERDAS Imagine (Leica Geosystems

---

[1]The index is calculated as tNDVI = Sqrt((Band four-Band three/Band four+Band three) + zero point five).

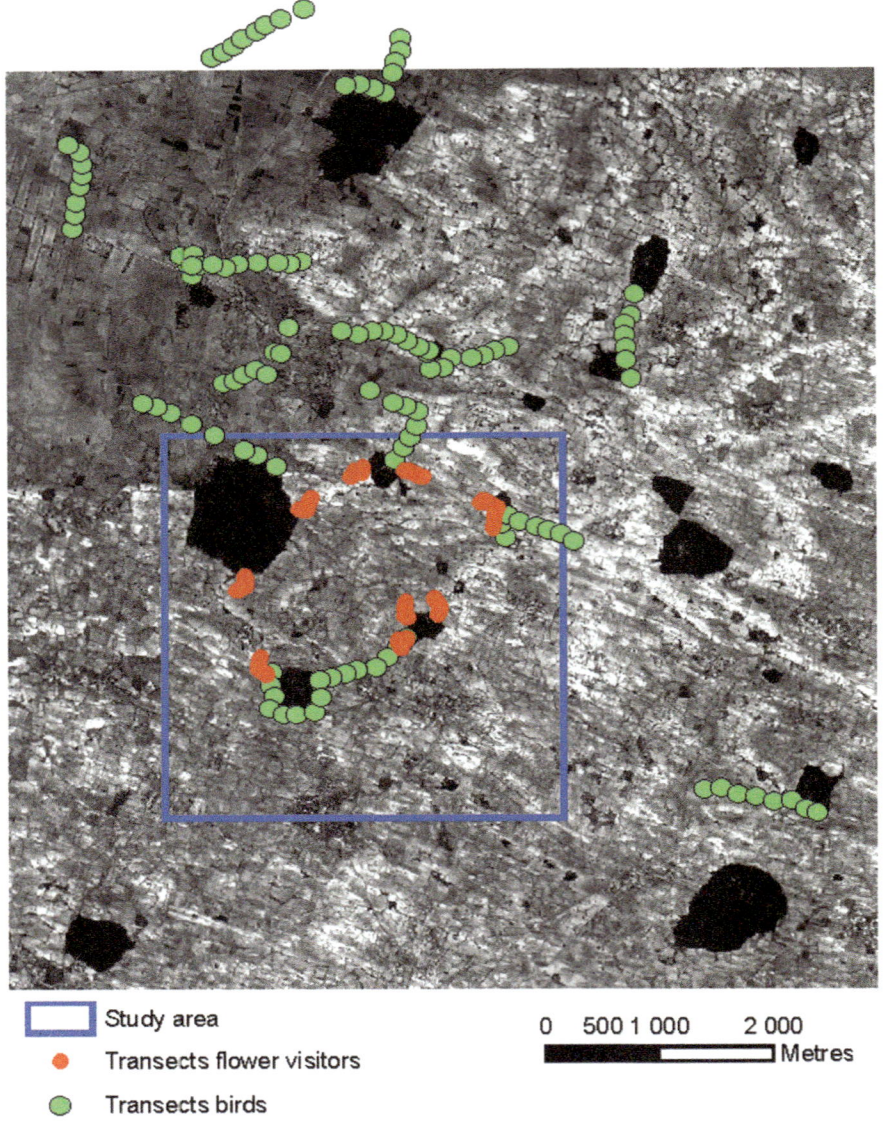

Study area

Transects flower visitors

Transects birds

0   500 1 000   2 000
Metres

Fig. 6.2 Study site as illustrated with an IKONOS satellite image

Geospatial Imaging, USA), which uses a slightly modified algorithm in comparison
to the standard Normalized Difference Vegetation Index (NDVI) (Lillesand et al.
2003).

Based on this analysis, we delineate forest patches of different sizes and larger
Opuntia clusters. Even though single trees and bushes are detected, the linearly

shaped hedges (generally 0.5–2.5 m wide) do not appear. However, the linear features appear clearly in the panchromatic band, which is therefore chosen for further analysis. We use the software ERDAS Imagine to generate a number of filtered images using different vertical, horizontal, and diagonal filtering algorithms applied on the original panchromatic image. The resulting images are thereafter compiled into one single image, which in turn is manually cleaned up in Adobe Photoshop, removing noise and speckle. Forests and *Opuntia* stands are also removed since they are already detected through the tNDVI index. Finally the image was auto-vectorized in ArcInfo (ESRI), creating an *Opuntia* network map of the area. However, in many places openings or gaps occur between the different hedges (and other types of vegetation), for which we assume that animals can cross these stretches of open ground.

6.2.2.2   GIS Analysis

We use two different GIS approaches to test whether the *Opuntia* hedges act as corridors and link forest patches and therefore function as least cost routes. The first approach is a vector-based network model, and the second is a raster-based cost-weighted distance approach.

For the network model, we assume that a model animal (representing several taxa) could cross open ground up to a distance of 20 m. This value is chosen to test the influence of cacti hedges on the connectivity of the studied landscape. In reality, distances will vary greatly among species and due to various contextual conditions, e.g. time of day, wind, resource availability, or season. To find a possible connection between hedges, we apply a 10 m buffer on all features. This creates the area of movement that animals have around each *Opuntia* hedge, stand, and forest patch, thereby closing 20 m gaps in all directions. This gives every object a width of at least 20 m.

The buffer zones are converted to raster format, merged together with the forest *Opuntia* stand layer, and then thinned to a line width of one pixel in ArcInfo. Finally they are reconverted back to vector format through auto-vectorizing. Using the Network Analyst extension network, GIS layers are established containing line segments and junctions. All gaps up to 20 m are closed via this method, but also all parallel hedges along roads are merged into one single network segment. For comparison, the shortest distances between the forest edges are measured (as the crow flies), with the same start and end points being also used to calculate the shortest network routes. The forests are named in clockwise order A–F (Sect. 6.3.1).

Subsequently we apply a raster-based least cost path analysis approach (Eastman 1989) on the entire study area to generate a cost surface and find surface connectivity between two or more points (Rothley 2005; Theobald 2005). Different features within the landscape are classified according to the difficulty (or the cost) for animal movement. With increasing friction, the cost per pixel becomes higher. In this study we test different levels of background costs. We set the cost per pixel to move along a hedge, while values for the "open landscape" pixel values are set to 2.5 and

10, respectively. Background costs for movement in forests larger than 1 ha is set to zero, which in practice means that it is cost-free for an animal to move within forest, since forest can be regarded as habitat. We set the centre of each forest as the start and end points for the calculation of the cost surfaces. The final distances for the different alternative routes are, however, in all cases calculated edge to edge between forests.

In order to estimate the coverage and spatial distribution of the *Opuntia* hedges, ArcInfo (ESRI) raster random creator is used to randomly select 50 plots of 1 ha each, which do not intersect with forests. From the total hedge length in each plot, an average width of the hedge was estimated.

6.2.2.3 Ecological Surveys

The development of the models outlined in Sects. 6.2.2.1 and 6.2.2.2 is based on data from two ecological surveys conducted on site, which assessed species movements along hedges. The first field survey focused on birds and was conducted during 1 month (November–December 2006), before and during the rainy season. The second survey focused on floral visitors and was conducted along *Opuntia* hedges in December 2007.

For the bird survey, we use point count observations ($n = 8$) in each of the 14 line transects (Fig. 6.2). The positional locations were determined with a GPS and set to be over 100 m apart. The counting of bird species and individuals was undertaken during 5 plus 5 min in a radius of 50 m. Notes were made of how they were connected to landscape elements, i.e. whether individuals were inside a forest patch, sitting on/foraging from *Opuntia* spp., connected to other wild woody plants (e.g. trees and shrubs scattered in the fields), standing/walking on ground, or just flying by. The transects were mainly set to end or start at a forest patch in order to investigate whether the abundance of birds declined further away from the forest patches.

The floral visitor survey was made along ten transects of 200 m in length each, including 20 points with 10 m intervals located along ten *Opuntia* hedges. Observations were made in a 5 m radius around each point. Three observers noted the species, the abundances of flower visitors on *Opuntia* flowers, and recorded the abundance of *Opuntia* flowers. At each point, observers spent roughly 5 min searching, counting, and recording visitors of *Opuntia* flowers. The location of each point was recorded using a handheld GPS devise.

6.3   Results

6.3.1   *Landscape Fragmentation and Connectivity*

The study area (900 ha) consists of 111 ha of forests (11%), with the remaining area covered with agricultural land that contains *Opuntia* hedges (see below for total length). Hedge length in the studied plots ranges between 0 and 550 m/ha, with a normal distribution and a mean average length of 254 m/ha (Fig. 6.3). The size of *Opuntia* hedges can vary greatly, but a mature hedge is generally 3–4 m in height and at least 2 m in width. Assuming a mean width of 2 m, the *Opuntia* hedges cover as much as 4% of the entire study area.

The studied landscape appears highly fragmented with low connectivity between the forest patches when observed at the scale of a Landsat satellite image (15 m resolution in the panchromatic band) (Fig. 6.4a). However, when utilizing high-resolution satellite imagery (IKONOS), an interesting landscape pattern emerges displaying a dense network of cactus hedges that seem to connect "isolated" forest patches (Fig. 6.4b and c).

The output of the network model suggests that the total length of *Opuntia* and forest stands is 220 km, compared to a length of 200 km extracted from the satellite image. The length increased even though some of the double *Opuntia* hedges along roads were reduced to a single line during the buffering/rasterizing and thinning procedure. Many of the gaps between hedges are closed when using 20 m as bridging distance. These distances (compared to as the crow flies) are in general 37% longer, which suggests a large number of alternative routes between the forests (see Fig. 6.5, Table 6.1).

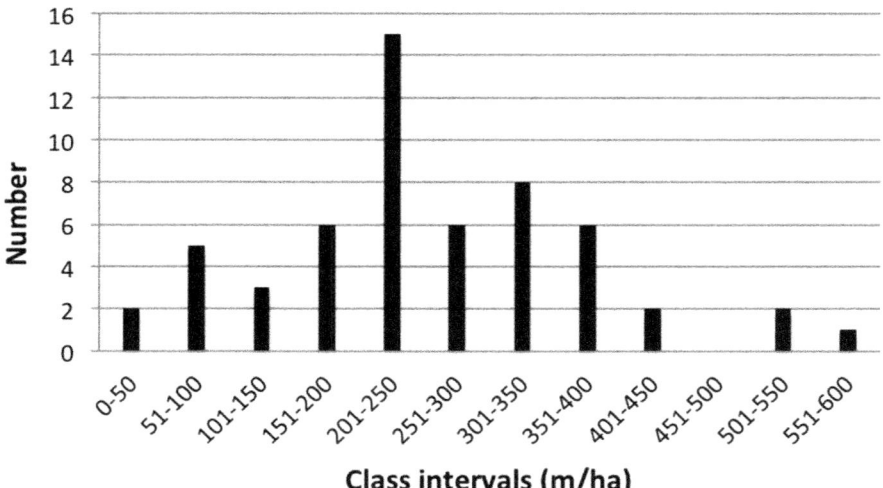

Fig. 6.3 Hedge sizes across the 50 random sample plots

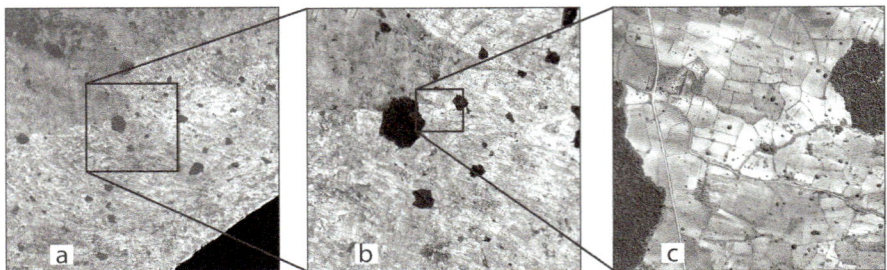

Fig. 6.4 Landscape fragmentation in the study site
Note: The used satellite images are (**a**) subset of a Landsat ETM + satellite image (20 km across), (**b**) subset of an IKONOS satellite image (6 km across), and (**c**) subset of IKONOS satellite image (1 km across)

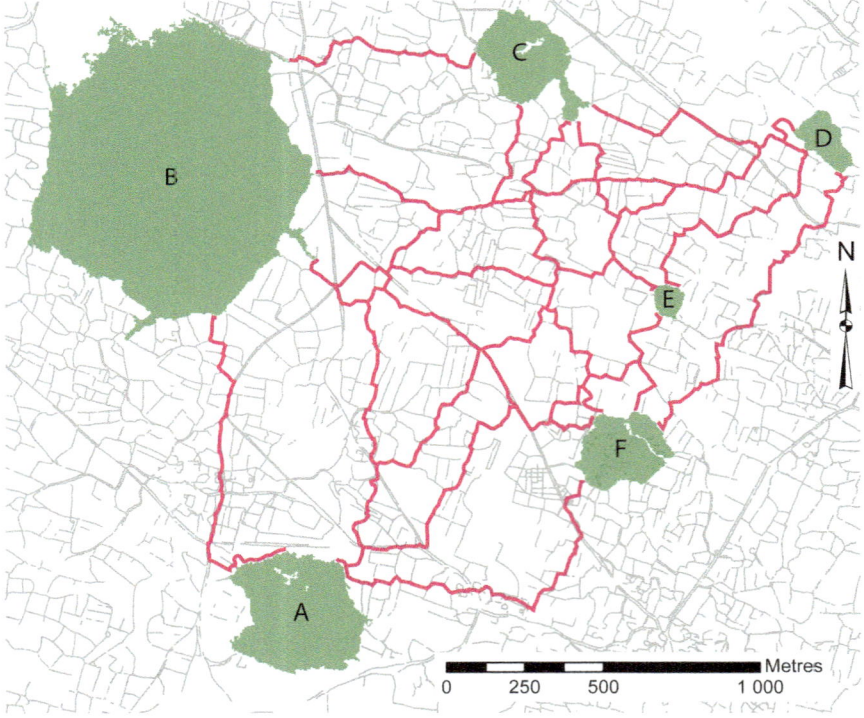

Fig. 6.5 Modelled shortest network routes between forest patches
Note: The study area is buffered to fill gaps of 50 m between hedges. If the maximum movement for an animal is 50 m from a hedge, 16% of the area is unreachable (light grey and black) when using cactus hedges as corridors of movement. If a movement limitation is set to 100 m, only 3% of the remaining area is not appropriate for animal movement (black)

Table 6.1 Distances between forest fragments

Route	Distance as crow flies (m)	Distance through network (m)	% increase in distance
A–B	888	1363	53
A–C	1793	2333	30
A–D	2232	3018	35
A–F	893	1431	60
B–F	1143	1464	28
C–F	1078	1529	42
B–C	646	819	27
C–D	719	1012	41
D–E	676	997	47
E–F	363	491	35
B–D	1682	2165	29
C–E	673	843	25
A–E	1444	2004	39
B–E	1199	1585	32
D–F	1123	1502	34

The hedges are also buffered, closing gaps up to 50 m and 100 m. However, it is not meaningful to create further networks from these layers, as the possible movement area is almost entirely filled. The results are instead presented as "leftover" areas that do not fall inside the buffering (Fig. 6.6). For 50 m gasps the overall area falling outside the connectivity area is 145 ha (i.e. 16% of the entire 900 ha study area). When using 100 m, the remaining area is only 28 ha (3%) of the entire study area.

The raster-based method using cost-weighted distances shows that with increasing background cost (i.e. when friction grows for crossing open land), the route distances will increase. Figure 6.6 shows that by utilizing this technique, the different forests can act as stepping-stones. When using a background cost of 2, the route distances increase by 15% (mean value) compared to the shortest alternative (i.e. as the crow flies). The increase in route length for a background value of 5 is 25% (mean value), and for a background value of 10, it is 35% (mean value). It is important to note that the total distance for different routes is calculated only between forest edges. This suggests that the width of forest patches has a negative impact on the route length since it is excluded from the final route length.

Both tested models demonstrate that landscape connectivity dramatically increases when including *Opuntia* hedges. The hedges can thus possibly provide a large variety of different potential alternative routes for the dispersal, foraging, and migration of different species. The outputs of the network model suggest that with a 10 m buffer around the *Opuntia* hedges, all of the forests patches included in study essentially have a full structural connectivity. If a buffer is set to 50 m or 100 m, only 16% and 3%, respectively, of the study area remains unconnected.

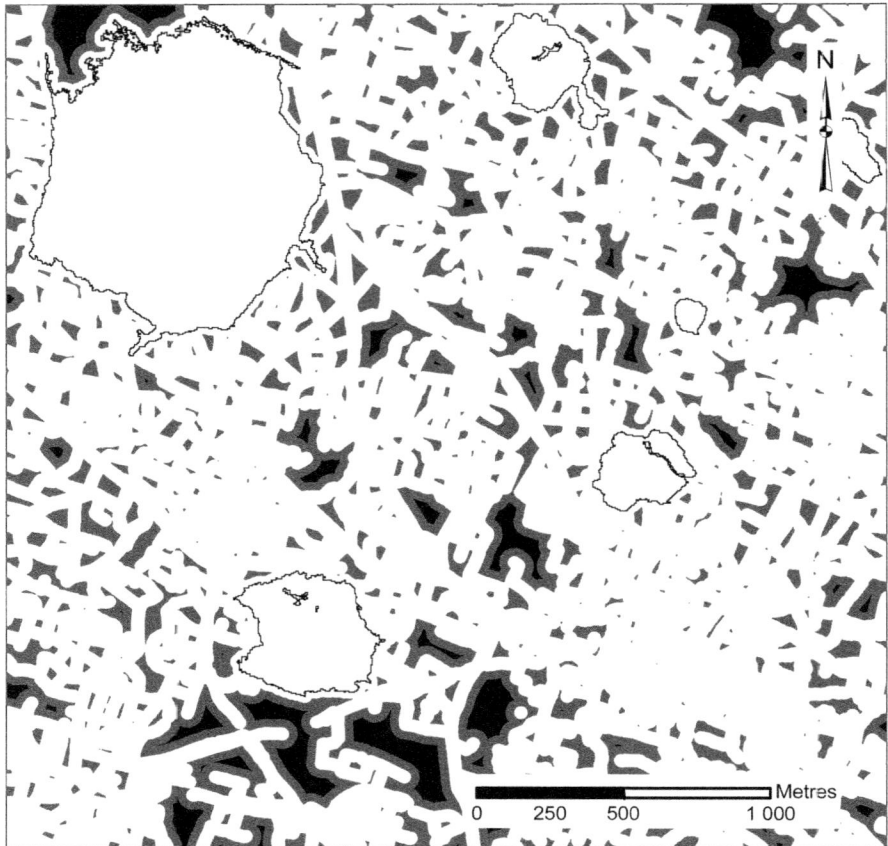

Fig. 6.6 Modelled routes from forest A to forest D
Note: The network route is created in the vector-based network model, and the three least cost paths
are derived through the raster-based modelling. The starting point in forest A is in the centre of the
forest for all three raster-based routes

6.3.2 Observation of Birds and Flower-Visiting Species at Opuntia Hedges

The results from the bird survey show that the abundance and diversity of birds does
not depend on the *Opuntia* hedges but that the density of other wild, woody plants in
the fields has had a positive effect. The average recorded number of bird individuals
was 1130 individuals per km^2, divided among 33 species (excluding the cores of
restricted forest patches). When linking the observed species to different landscape
elements, the average number of individuals per observation is 0.6 on *Opuntia*, 4 on
other woody plants growing on the fields, 0.8 on the ground, and 1.7 passing by. Six
bird species were recorded in connection to *Opuntia* hedges, of which the

Flower visitor abundance

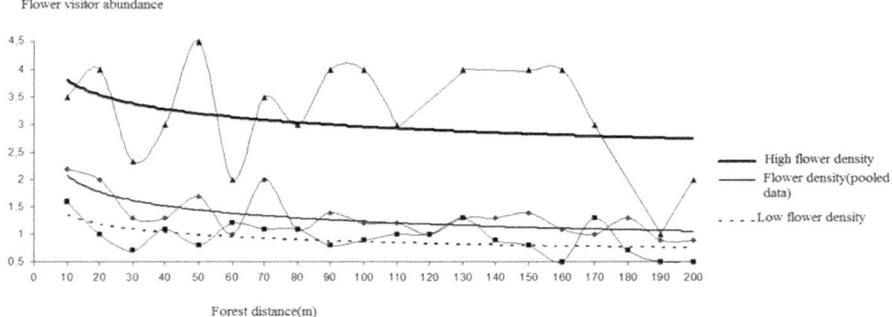

Forest distance(m)

Fig. 6.7 Forest distance in relation to the abundance of flower visitors to *Opuntia* hedges

insectivorous Madagascar cisticola (*Cisticola cherina*) was the most frequently observed species foraging and perching in and around the cacti hedges. The second most frequently occurring bird in *Opuntia* hedges was the common myna (*Acridotheres tristis*).

Insect richness along the 200 m hedges varied between 4 and 19 insect species between all 10 transects. In total, 19 insect species were observed visiting *Opuntia* flowers, with bees (3 species) and beetles (8 species) being the most abundant (45 and 31 individuals were recorded respectively). Flies (5 species) and wasps (1 species) were rarely seen, with only 9 and 2 individuals recorded, respectively. Ants (three species) were also observed on *Opuntia* flowers. Visitor species richness increased significantly with floral density (P = 0.01; R = 0.90). At high flower densities, flower visitors were abundant also at distances greater to 200 m from the forests (Fig. 6.7).

6.4   Discussion

6.4.1   *Multidimensional Benefits of* Opuntia *Hedges*

The geospatial analysis identified different patterns of landscape connectivity due to *Opuntia* hedges (Sect. 6.3.1). In particular, the study site (900 ha) contained *Opuntia* hedges with a total length of 220 km and an average density of 254 m/ha. It can be argued that the cacti network provides complete structural connectivity between the six forest patches in the study sites, as well as a whole range of redundant routes (Sect. 6.3.1, Fig. 6.5). Based on the high-resolution satellite imagery, our models suggest that if *Opuntia* hedges are included in connectivity models, then the forest patches exhibit a full structural connectivity, given that animals could cross open ground up to a distance of 20 m. These results are comparable to previous coarser spatial models of pollinating bees and seed-dispersing lemurs using Landsat 7 ETM images with a resolution of 30 m (e.g. Bodin et al. 2006). This study sets a movement range of 400–1400 m for bees and 500–2000 m for lemurs, with the results suggesting that the landscape was strongly associated with pollination and seed dispersal services. These services are still yet vulnerable for further loss as discussed below.

The forest patches studied in this chapter are arguably crucial habitats for various endemic species of plants (Tengö et al. 2007) and animals such as lemurs (Bodin et al. 2006). The long-term survival of many of these species, however, depends on maintaining viable sub-populations within these fragmented landscapes. Many species may depend highly on migration routes to reach other forest habitats for genetic exchange between sub-populations. For example, due to landscape fragmentation, the populations of endangered brown lemurs (*Eulemur collaris*) are subjected to inbreeding and homogenization in the eastern littoral forest (Bertoncini et al. 2017). Another example of a species relying on such corridors is the grey mouse lemur (*Microcebus murinus*) that uses corridors of exotic trees to move between littoral forest fragments (Andriamandimbiarisoa et al. 2015). The extensive networks of *Opuntia* hedges can possibly provide these necessary linkages between forest areas/patches and thus facilitate to the creation of a "genetic network" (Hanski and Gilpin 1991). Areas outside these connected zones (Fig. 6.5) should be prioritized for restoration efforts seeking to further restore landscape connectivity.

Our empirical findings support to some extent the assumption that *Opuntia* hedges attract some forest species into the agricultural matrix. In particular we observed that some bird species move throughout the agricultural landscape by using *Opuntia* hedges and other types of vegetation (e.g. trees, shrubs) as stepping-stones (Sect. 6.3.1). In particular, the cisticola (*Cisticola cherina*) is frequently spotted in *Opuntia* hedge (Sect. 6.3.2), with this bird species recognized by the local communities as performing pest regulation services by preying on crickets and other pest insects. The common myna (*Acridotheres tristis*), which is the second most frequently observed bird in *Opuntia* hedges, was originally introduced from India in the late nineteenth century to control the proliferating pest locusts (Langrand 1990).

Floral visitor abundance correlates with flower density along the cacti hedges. At high flower densities, flower visitors are abundant in distances above 200 m from the forest fragments (Sect. 6.3.2). Hedges can help flower visitors such as bees to move further into the agricultural landscape. This more extensive range may increase pollination services and can have positive effects to the yields of locally important crops such as beans. However, for flower-visiting insects, the contribution of *Opuntia* cacti in landscape connectivity could be strongly influenced by hedge management practices. For example, farmers cut and burn some portions of the *Opuntia* hedges for cattle feeding, before replanting them. Young hedges often have low flower density, compared to mature hedges, which might affect the attraction of flower-visiting insects and thus the overall contribution of the hedges on landscape connectivity.

The fruits, nectar, and cladodes of the *Opuntia* are also an important food source for several species such as lemurs, tortoises, birds (both ground-living and flying), and insects, especially during the dry periods when other food sources are limited

(Lingard et al. 2003). One such example is the endangered radiated tortoise (*Geochelone radiata*),[2] which is one of the four endemic tortoises in Madagascar. This tortoise species has been protected in Madagascar through a national law since 1961, is listed on CITES Appendix I[3] since 1975, and is included in the IUCN Red List as a critically endangered species. Its population has been experiencing a rapid decline (O'Brien et al. 2003) and is now considered to be one of the most threatened tortoise species in the world. The *Opuntia* cacti can provide a nesting ground and food source to these tortoises (O'Brien et al. 2003), with Leuteritz and Ravolanaivo (2005) finding that 36.4% of the egg-laying nests were associated to *Opuntia*. Feeding observations have also shown that the *Opuntia* was the second most important food source after grass (Leuteritz, 2003), as the cacti produce an abundant supply of fruits when other food sources are scarce (e.g. during dry periods or droughts, Sect. 6.2.1). During these periods the radiated tortoise is reported to almost exclusively feed on *Opuntia* (Lingard et al. 2003). It is also likely that *Opuntia* hedges provide shade, as the females of the species are known to be vulnerable to overheating when nesting (Leuteritz and Ravolanaivo 2005). Furthermore, the hedges could also possibly serve as corridors for the dispersal and/or migration of the tortoises, but this needs to be further substantiated.

Endangered lemurs such as the ring-tailed lemurs (*Lemur catta*) are another example of a species using *Opuntia* cacti as a food source (Simmen et al. 2003). Similar to other species, it is likely that the *Opuntia* hedges provide a durable food resource during critical periods of migration and especially during droughts. During these periods lemurs need to leave their usual habitats to find alternative foraging resources in other areas, hence relying to *Opuntia* fruits for sustenance.

Furthermore the *Opuntia ficus-indica* has a network of roots that spreads close to the ground surface. This root network stores water in the fleshy above-ground plant bodies and in the soil held by the thickened mass of roots. Following rainfall the rainwater held below- and above-ground can cause the sudden flowering of the cactus plant and the other plants around it (Kaufmann 2004). Moreover, Monjauze and Le Houerou (1965:102) noted that the cactus "increases the aeration of the ground, its permeability, the infiltration of rainwater, and the availability of water..." (cited from Kaufmann 2004). The prickly pear fences can also limit wind damage to plants and the soil, with this type of hedging locally called *valandraketa* (i.e. cactus fencing or cactus enclosures).

The incorporation of the *Opuntia* in the pastoral economy has allowed local communities in southern Madagascar to become increasingly sedentary and less dependent on pasturing herds over large distances. This is largely because the cacti can be used as fodder during the dry season when other livestock feed is insufficient

---

[2]This species is endemic to the spiny forests of southern Madagascar and lives in dry regions of brush, thorn bushes, and woodlands, feeding on grass and succulent plants (O'Brien et al. 2003; Leuteritz et al. 2005).

[3]The Appendix I of the Convention on International Trade in Endangered Species of Wild Fauna and Flora (CITES) lists the CITES-listed animal and plant species that are the most endangered and threatened with extinction.

Table 6.2 Potential ecosystem services provision from *Opuntia* hedges

Type	Ecosystem service
Provisioning	- Food for humans. Fruits are edible, are rich in vitamins, and are often used as an emergency food source during dry spells
	- Fodder and water resource for cattle
Regulating	- Fencing to protect agricultural fields from cattle grazing
	- Fencing to protect food crops from sand drift
	- Enhance soil moisture
	- Erosion regulation
	- Fire break
Cultural	- Demarcation of sacred forests
Supporting	- Habitat for flora and fauna (e.g. offers shelter, protection from predation, nesting ground, shade)
	- Landscape connectivity (e.g. enhances gene flow by facilitating migration and species dispersal)
	- Food source for native animal species (e.g. fruits for lemurs and radiated tortoise and nectar/pollen for endemic insects and nectarivorous birds)

(Kaufmann 2004). *Opuntia* can also provide other livelihood opportunities in this marginal environment. For example, by diversifying and cultivating several varieties of *Opuntia*, it is possible to extend the harvest period and store fruits for several months (and sometimes up to a year). Occasionally, during periods of food scarcity, local communities also consume *Opuntia* cladodes (of *raketasonjo* and *raketambazaha*) that can be eaten raw, grilled, or cooked. In this respect *Opuntia* can enhance food security in a region that generally receives adequate rainfall for food crop agriculture (Kaufmann 2004). Some scholars have also discussed the potential of *Opuntia* seed oil for livelihood improvement in semi-arid Madagascar (Hanke et al. 2018). Considering the above examples, it could be argued that introducing the *Opuntia* has increased the resilience of the local communities, as it contributes in multiple ways to their sustenance in a part of Madagascar that is constantly facing water scarcity (Kaufmann 2004).

Table 6.2 summarizes some of the ecosystem services possibly provided by the *Opuntia*, as indicated above. These ecosystem services possibly include various provisioning (i.e. natural products), regulating (i.e. benefits obtained from the regulation of ecosystem processes), cultural services (i.e. nonmaterial benefits obtained from ecosystems), and supporting services (i.e. services necessary for the production of all other ecosystem services) (Millennium Ecosystem Assessment 2005).

6.4.2   Limitations and Future Research

Opuntia hedges may provide similar functions to that of other linear landscape elements such as hedgerows (Fitzgibbon 1997; Beier and Noss 1998; Baudry et al.

2000; Graham et al. 2018). Studies have suggested that hedgerows can provide surrogate habitat to different native species in human-modified environments, where the natural habitats occupy only a small fraction of the landscape (Dover and Sparks 2000; Hannon and Sisk 2009; Graham et al. 2018). They can also provide safe cover for both local and larger-scale movement and may facilitate access to resources or habitat, which might otherwise be too risky or too remote (Hinsley and Bellamy 2000; Merckx et al. 2010; Graham et al. 2018).

Most of the studies that have explored the biodiversity conservation benefits of linear landscape elements have been undertaken in developed countries, mainly in Europe and North America (see above). However, the possible positive biodiversity outcomes of such landscape elements have been under-researched in the developing world despite its potential for biodiversity conservation outside of protected areas (including in SSA) (Kull et al. 2013).

This study essentially provided a preliminary landscape analysis to highlight some of the potential benefits of such landscape elements in SSA. However, more extensive studies would be needed to fully understand the different benefits and the potential trade-offs of such landscape elements in regard to biodiversity conservation. First, future studies should consider the movement capabilities of the different species in the specific study context. To enhance the accuracy of the estimates, it would be important to consider the habitat requirements and dispersal abilities of the focal species. This could help establish properly the assumptions regarding meaningful patch sizes and the length, width, and gaps in corridors. Furthermore more extensive studies would be needed to assess the actual provision of ecosystem services from such landscape elements. This should include, among others, surveys with local communities to understand the type, extent, and timing of the obtained ecosystem services. This can go a long way towards identifying whether the benefits of *Opuntia* hedges can enhance the resilience of local communities during difficult times of the year (e.g. dry season) or even during climatic shocks (e.g. droughts).

Finally, it should be noted that the raster-based least cost analysis cannot determine gap lengths that are bridged. On the other hand, the network approach offers more control over the distances that are bridged, making it possible to find alternative routes if the first modelled route is broken. Such points should be considered properly in future studies that explore the landscape connectivity benefits of such landscape elements.

6.4.3   Policy Implications and Recommendations

It has often been pointed out that biological diversity cannot be preserved through conservation alone (e.g. Folke 2006). A major incentive for biodiversity conservation in SSA is to secure, restore, and develop the capacity of ecosystems to generate ecosystem services (Millennium Ecosystem Assessment 2005; IPBES 2018). In this sense there is an increasing need for landscape management approaches to evolve, embrace a more holistic sustainability science mindset, and move away from rigid solutions, accepting that change is the rule rather than the exception (Folke 2006).

At the same time, reducing the high poverty levels in SSA countries such as Madagascar without jeopardizing their unique natural capital is a major sustainability challenge (Chaps. 1, 7 Vol. 1). This is echoed in the Madagascar National Development Plan, which has been formulated in line with the Sustainable Development Goals (SDGs), and aims to alleviate poverty but at the same time to protect, conserve, and valorise natural capital in the country (Ministère de l'economie et de planification 2015). Key elements of this plan seek to reduce land use change and landscape fragmentation and manage to the extent possible invasive alien species, integrating a directive to promote regulatory mechanisms for the effective management of invasive species (Ministry of Environment and Forest Madagascar, United Nation Environment Program 2016).

However, as outlined throughout this chapter, some alien species such as the *Opuntia* can indeed provide valuable ecological functions in fragmented landscapes and at the same time provide valuable ecosystem services to local communities. Management interventions based on the management of *Opuntia* hedges can, in theory, not only contribute to key conservation goals related to SDG15 (Life on Land) but also create valuable synergies with SDGs related to local livelihoods such as SDG2 (Zero Hunger) and SDG1 (No Poverty) and adaptation to climate change: SDG13 (Climate Action).

Deepening our knowledge of not only the negative impacts of alien species but also of their possible contributions can be an important means for developing useful case-by-case management plans for such species. These plans should encourage further research on alien and invasive species, considering that their management must go hand in hand with socioeconomic development realities. In this respect, the focus of this chapter is in line with both the Madagascar National Biodiversity Action Plan and the National Development Plan and offers a starting point for developing novel ecosystem management plans using *Opuntia* in the Androy region, as a means of addressing multiple sustainability challenges.

6.5   Conclusions

This chapter outlined the results of a preliminary landscape analysis regarding the landscape connectivity effects of *Opuntia* cacti hedges in southern Madagascar. We used a combination of literature review, geospatial analysis, and ecological surveys to identify the potential benefits of these hedges for landscape connectivity and ecosystem services provision. Our preliminary analysis points that these hedges can (under some circumstances) enhance landscape connectivity by creating corridors between forest patches. As such they can facilitate species movement (i.e. dispersal and migration) in an analogous way to other linear landscape structures described elsewhere in the academic literature. Furthermore the hedges can provide diverse provisioning, regulating, and possibly cultural services, contributing manifold to the wellbeing of local communities in southern Madagascar.

Unlike most other alien species in other parts of the world, the *Opuntia* can provide critical ecosystem functions and services in a heavily human-dominated landscape. In this respect it can possibly enhance biodiversity conservation outside protected areas. Thus our study underlines that landscape management interventions based on introduced species can have some potential. Despite its preliminary nature, this study can offer some important insights for developing sustainable landscape management policies in Madagascar and possibly other parts of SSA. However it is important to assess more properly the landscape connectivity benefits and the provision of ecosystem services through further studies. These possible benefits need to be considered on a case-by-case basis, taking into account the underlying environmental and socioeconomic contexts.

Acknowledgements We express our gratitude to Johan Colding and Erik Andersson for valuable comments on the manuscript, to Ray Evans for proof reading, Markus Franzen and Nivo Raharison for knowledge and field data, Marcin Jarzebski for drawing Fig. 6.1, and to all the people receiving us in field in the Androy region, in particular all the inhabitants of Amboanaivo. This work was financed by Ph.D. grants from the Stockholm University and supported by grants from Sida/Sarec and the Swedish Research Council.

References

Andriamandimbiarisoa L, Blanthorn TS, Ernest R, Ramanamanjato JB, Randriatafika F, Ganzhorn JU, Donati G (2015) Habitat corridor utilization by the gray mouse lemur, Microcebus murinus, in the littoral forest fragments of southeastern Madagascar. Madagascar Conservation Development 10(S3):p144–p150

ANGAP, FOFIFA, WWF (2001) Annexe 2. Proposition d'un programme de gestion de raketa mena et de valorisation des espèces dans le sud

Baudry J, Bunce RGH, Burel F (2000) Hedgerows: an international perspective on their origin, function and management. J Environ Manag 60:7–22

Beier P, Noss RF (1998) Do habitat corridors provide connectivity? Conserv Biol 12:1241–1252

Bertoncini S, D'Ercole J, Brisighelli F, Ramanamanjato, Capelli C, Tofanelli S, Donati G (2017) Stuck in fragments: population genetics of the endangered collared brown lemur Eulemur collaris in the Malagasy littoral forest. American Journal of Anthropology 163:1–11. https://doi.org/10.1002/ajpa.23230

Binggeli P (2003) Opuntia spp. prickly pear, raiketa, rakaita, raketa. In: Goodman SM, Benstead JP (eds) The natural history of Madagascar. University of Chicago Press, London

Bodin Ö, Tengö M, Norman A, Lundberg J, Elmqvist T (2006) The value of small size: loss of forest patches and threshold effects on ecosystem services in southern Madagascar. Ecol Appl 16(2):440–451

Boy G, Witt A (2013) Invasive alien plants and their management in Africa. CABI Africa, Nairobi

CITES, The Convention on International Trade in Endangered Species of Wild Fauna and Flora http://www.cites.org/eng/app/index.shtml

Cushman SA (2006) Effects of habitat loss and fragmentation on amphibians: a review and prospectus. Biol Conserv 128:231–240

Dammhahn M, Randriamoria MT, Goodman SM (2017) Broad and flexible stable isotope niches in invasive non-native Rattus spp.in anthropogenic and natural habitats of central eastern Madagascar. BMC Ecol 17:16. https://doi.org/10.1186/s12898-017-0125-0

Decary R (1930) L'Androy (extrême sud de Madagascar): essai de monographie régionale. Vol. I. Géographie physique et humaine. Société d'Editions Géographiques, Maritimes et Coloniales, Paris.

Dover J, Sparks T (2000) A review of the ecology of butterflies in British hedgerows. J Environ Manag 60:51–63

Eastman J (1989) Pushbroom algorithms for calculating distances in raster grids. Proceedings of Autocarto 9, Autocarto: 288–297

Elmqvist T, Pyykönen M, Tengö M, Rakotondrasoa F, Rabakonandrianina E, Radimilahy C (2007) Patterns of loss and regeneration of tropical dry forest in Madagascar: the social institutional contewt. PLoS One 2(5):e402. https://doi.org/10.1371/journal.pone.0000402

EPP PADR (2006) Programme Regional de developpement rural PRDR Androy http://www.mpae. gov.mg/wp-content/uploads/pdf/GTDR/PRDR%20ANDROY.pdf

Fitzgibbon CD (1997) Small mammals in farm woodlands: the effects of habitat, isolation and surrounding land-use patterns. J Appl Ecol 34:530–539

Folke C (2006) The economic perspective: conservation against development versus conservation for development. Conserv Biol 20:686–688

Ganzhorn JU, Lowry PP, Schatz GE, Sommer S (2001) The biodiversity of Madagascar: one of the world's hottest hotspots on its way out. Oryx 35:346–348

Goodman SJ, Patterson B (1997) Natural change and human impact in Madagascar. Smithsonian Institution Press, Washington DC

Graham L, Gaulton R, Gerard F, Staley JT (2018) The influence of hedgerow structural condition on wildlife habitat provision in farmed landscapes. Biol Conserv 220:122–131

Groombridge B (1992) Global biodiversity: status of the Earth's living resources. Chapman and Hall, London

Gutiérrez RJ, Harrison S (1996) Applying Metapopulation theory to spotted owl management: a history and critique. In: McCollough DR (ed) Metapopulations and wildlife conservation. Island Press, Washington DC, pp 167–185

Haddad NM, Brudvig LA, Clobert J, Davies FK, Gonzalez A, Holt RD, Lovejoy TE, Sexton JO, Austin MP, Collins CD (2015) Habitat fragmentation and its lasting impact on Earth's ecosystems. Sciences Advances 2015:1. https://doi.org/10.1126/sciadv.1500052

Hänke H, Barkmann J, Müller C, Marggraf R (2018) Potential of Opuntia seed oil for livelihood improvement in semi-arid Madagascar. 13(1)

Hannon LE, Sisk TD (2009) Hedgerows in an Agri-natural landscape: potential habitat value for native bees. Biol Conserv 142:2140–2154

Hanski I, Gilpin M (1991) Metapopulation dynamics: brief history and conceptual domain. Biol J Linn Soc 42:3–16

Harrison PA, Berry PM, Simpson G, Haslett JR, Blicharska M, Bucur M, Dunford R, Egoh B, Garcia-Llorente M, Geamănă N, Geertsema W, Lommelen E, Meiresonne L, Turkelboom F (2014) Linkages between biodiversity attributes and ecosystem services: a systematic review. Ecosyst Serv 9:191–203

Hinsley SA, Bellamy PE (2000) The influence of hedge structure, management and landscape context on the value of hedgerows to birds: a review. J Environ Manag 60:33–49

IPBES (2018) In: Archer E, Dziba L, Mulongoy KJ, Maoela MA, Walters M (eds) The IPBES regional assessment report on biodiversity and ecosystem services for Africa. Secretariat of the Intergovernmental Science-Policy Platform on Biodiversity and Ecosystem Services, Bonn

IPBES (2019) The IPBES global assessment report on biodiversity and ecosystem services. Secretariat of the intergovernmental science-policy platform on biodiversity and ecosystem services. Bonn. Germany

Jones JPG, Rasamy JR, Harvey A, Toon A, Oidtmann B, Randrianarison MH,Raminosoa N, Ravoahangimalala OR (2008) The perfect invader: a parthenogenic crayfish poses a new threat to Madagascar's freshwater biodiversity. Biol Invasions. https://doi.org/10.1007/s10530-008-9334-y

Kaufmann JC (2004) Prickly pear cactus and pastoralism in Southwest Madagascar. Ethnology 43:345–361

Keller I, Largiader CR (2003) Recent habitat fragmentation caused by major roads leads to reduction of gene flow and loss of genetic variability in ground beetles. Proceedings of the Royal Society of London Series B-Biological Sciences 270:417–423

Koechlin J (1972) Flora and vegetation of Madagascar. In: Battistini R, Richard-Vindard G (eds) Biogeography and ecology in Madagascar. W Junk B.V, Hague

Koechlin J, Guillaumet JL, Morat P (1974) Flore et végétation de Madagascar. J. Cramer Verlag, Vaduz

Kremen C (2005) Managing ecosystem services: what do we need to know about their ecology? Ecol Lett 8:468–479

Kremen C, Ostfeld RS (2005) A call to ecologists: measuring, analyzing, and managing ecosystem services. Front Ecol Environ 3:540–548

Kruess A, Tscharntke T (1994) Habitat fragmentation, species loss, and biological-control. Science 264:1581–1584

Kull CA, Tassin J, Moreau S, Ramiarantsoa HR, Blanc-Pamard C, Carriere SM (2012) The introduced flora of Madagascar. Biol Invasions 14:875–888

Kull CA, Carriere SM, Moreau S, Ramiarantsoa HR, Blanc-Pamard C, Tassin J (2013) Melting pots of biodiversity: tropical smallholder farm landscapes as guarantors of sustainability. Environment 55:6–16

Langrand O (1990) Guide to the birds of Madagascar. Yale university press, New haven

Laurance WF, Campbell MJ, Alamgir M, Mahmoud MI (2017) Road expansion and the fate of Africa's tropical forests. Front Ecol Evol. https://doi.org/10.3389/fevo.2017.00075

Leuteritz TEJ (2003) Observations on the diet and drinking behaviour of radiated tortoises (Geochelone radiata) in Southwest Madagascar. Afr J Herpetol 52:127–130

Leuteritz TEJ, Ravolanaivo R (2005) Reproductive ecology and egg production of the radiated tortoise (Geochelone radiata) in southern Madagascar. Afr Zool 40:233–242

Leuteritz TEJ, Lamb T, Limberaza JC (2005) Distribution, status, and conservation of radiated tortoises (Geochelone radiata) in Madagascar. Biol Conserv 124:451–461

Lillesand TM, Kiefer RW, Chipman JW (2003) Remote sensing and image interpretation. Wiley, New York

Lingard M, Raharison N, Rabakonandrianina E, Rakotoarisoa JA, Elmqvist T (2003) The role of local taboos in conservation and management of species: the radiated tortoise in southern Madagascar. Conserv Soc 1:223–246

Lundberg J, Moberg F (2003) Mobile link organisms and ecosystem functioning: implications for ecosystem resilience and management. Ecosystems 6:87–98

Mack RN, Simberloff D, Lonsdale WM, Evans H, Clout M, Bazzaz FA (2000) Biotic invasions: causes, epidemiology, global consequences, and control. Ecol Appl 10:689–710

Marshall MB, Casewell RN, Vences M, Glaw F, Andreone F, Rakotoarison A, Giulia Zancolli G, Woog F, Wüster W (2018) Widespread vulnerability of Malagasy predators to the toxins of an introduced toad. Curr Biol 28:R635–R655

Merckx T, Feber RE, Mclaughlan C, Bourn NAD, Parsons MS, Townsend MC, Riordan P, Macdonald DW (2010) Shelter benefits less mobile moth species: the field-scale effect of hedgerow trees. Agric Ecosyst Environ 138:147–151

Middleton K (2002) Oportunities and risks: a cactus pear in Madagascar. Acta Hortic 581

Millennium Ecosystem Assessment (2005) Ecosystems and human Well-being: synthesis. Island Press, Washington, DC

Ministère de l'economie et de planification (2015) Plan National de Developpement 2015–2019.109p

Ministry of Environment and Forest Madagascar, United Nation Environment Program (2014) Status and trend of biodiversity. In: Fifth National report to the convention of biological diversity Madagascar. MEF and UNEP (eds). Antananarivo.pp10–38

Ministry of Environment and Forest Madagascar, United Nation Environment Program (2016) National biodiversity Strategy and action plans

Monjauze A, Le Houerou HN (1965) Le role des *Opuntia* dans l'economie agricole Nord Africaine Extrait du bulletin de l'Ecole Nationale Superieure d'Agriculture de Tunis 8–9:85–164

Mooney HA, Cleland EE (2001) The evolutionary impact of invasive species. Proc Natl Acad Sci U S A 98:5446–5451

Mooney HA, Hobbs RJ (2000) Invasive species in a changing world. Island Press, Washington DC

Myers N, Mittermeier RA, Mittermeier CG, da Fonseca GAB, Kent J (2000) Biodiversity hotspots for conservation priorities. Nature 403:853–858

O'Brien S, Emahalala ER, Beard V, Rakotondrainy RM, Reid A, Raharisoa V, Coulson T (2003) Decline of the Madagascar radiated tortoise Geochelone radiata due to overexploitation. Oryx 37:338–343

Perrings C, Halkos G (2018) Agriculture and the threat to biodiversity in sub-Saharan Africa. Environ Res Let 10:095015

Pratt CE, Constantine KL, Murphy ST (2017) Economic impacts of invasive alien species on African smallholder livelihoods. Glob Food Sec 14:31–37

Reh W, Seitz A (1990) The influence of land-use on the genetic-structure of populations of the common frog Rana-Temporaria. Biol Conserv 54:239–249

Rothley K (2005) Finding and filling the "cracks" in resistance surfaces for least-cost modeling. Ecol Soc 10(1):4

Russell AC, Meyer JY, Holmes ND, Pagad S (2017) Invasive alien species on islands: impacts, distribution, interactions and management. Environ Conserv 44(4):359–370

Sakai AK, Allendorf FW, Holt JS, Lodge DM, Molofsky J, With KA, Baughman S, Cabin RJ, Cohen JE, Ellstrand NC, McCauley DE, O'Neil P, Parker IM, Thompson JN, Weller SG (2001) The population biology of invasive species. Annu Rev Ecol Syst 32:305–332

Schroter D, Cramer W, Leemans R, Prentice IC, Araujo MB, Arnell NW, Bondeau A, Bugmann H, Carter TR, Gracia CA, de la Vega-Leinert AC, Erhard M, Ewert F, Glendining M, House JI, Kankaanpaa S, Klein RJT, Lavorel S, Lindner M, Metzger MJ, Meyer J, Mitchell TD, Reginster I, Rounsevell M, Sabate S, Sitch S, Smith B, Smith J, Smith P, Sykes MT, Thonicke K, Thuiller W, Tuck G, Zaehle S, Zierl B (2005) Ecosystem service supply and vulnerability to global change in Europe. Science 310:1333–1337

Simmen B, Hladik A, Ramasiarisoa P (2003) Food intake and dietary overlap in native Lemur catta and Propithecus verreauxi and introduced Eulemur fulvus at Berenty, southern Madagascar. Int J Primatol 24:948–967

Tengö M, Johansson K, Rakotondrasoa F, Lundberg J, Andriamaherilala JA, Rakotoarisoa JA, Elmqvist T (2007) Taboos and forest governance: informal protection of hot spot dry forest in southern Madagascar. Ambio 36:8

Theobald DM (2005) A note on creating robust resistance surfaces for computing functional landscape connectivity. Ecol Soc 10(1)

Tsutsui ND, Suarez AV, Holway DA, Case TJ (2000) Reduced genetic variation and the success of an invasive species. Proc Natl Acad Sci U S A 97:5948–5953

Vitousek PM, Dantonio CM, Loope LL, Rejmanek M, Westbrooks R (1997) Introduced species: a significant component of human-caused global change. N Z J Ecol 21:1–16

Chapter 7
The Legacy of Mine Closure in Kabwe, Zambia: What Can Resilience Thinking Offer to the Mining Sustainability Discourse?

Orleans Mfune, Chibuye Florence Kunda-Wamuwi, Tamara Chansa-Kabali, Moses Ngongo Chisola, and James Manchisi

7.1   Introduction

Mining is an integral component of many national economies in sub-Saharan Africa (SSA). The mining sector extracts over 60 metals and minerals throughout the continent, contributing significantly to local employment, foreign exchange earnings and national gross domestic products (GDP) (Hilson 2002) (Chap. 1 Vol. 1). Zambia, for example, derives nearly 80% of her foreign exchange earnings from the mine sector (Simutanyi 2008). If well managed, the extraction of lucrative minerals such as copper, diamond, gold and tin can provide poor countries with large revenue streams that can enhance economic development and contribute to poverty alleviation (Pegg 2006; Gajigo et al. 2012) (Chap. 1 Vol. 1).

However, there have been substantial concerns over the sustainability of mineral resources and mining in SSA. For example, some of the dominant perspectives in the mining literature relate to mineral resources availability and the implications of their exploitation on national development and the environment. This type of research on mineral resources availability or exhaustibility tends to focus on how they can be used sustainably to ensure their benefits for future generations (Prior et al. 2012, 2013; Guirco and Cooper 2012; Lebre and Corder 2015). Even though such studies have focused on the non-renewability of mineral resources, the research on resource exploitation implications has brought to the fore the many negative environmental and social outcomes associated with mining in SSA (Kitula 2006; Glaister and Mudd 2010).

O. Mfune (✉) · C. F. Kunda-Wamuwi · T. Chansa-Kabali · M. N. Chisola · J. Manchisi
University of Zambia, Lusaka, Zambia
e-mail: Orleans.mfune@unza.zm; tamara.kabali@unza.zm

© Springer Nature Singapore Pte Ltd. 2020
A. Gasparatos et al. (eds.), *Sustainability Challenges in Sub-Saharan Africa II*,
Science for Sustainable Societies, https://doi.org/10.1007/978-981-15-5358-5_7

Studies on the negative implications of mining often frame impacts in the context of the resource curse and Dutch disease hypotheses (Arellano-Yanguas 2011; John 2011; Koitsiwe and Adachi 2015). For example, studies drawing on the resource curse hypothesis argue that mining has multiple negative social and economic outcomes at the national level such as corruption, armed conflict, weakening of national institutions, deceleration of economic growth and underinvestment in human capital (Simutanyi 2008; Prior et al. 2013; Marais et al. 2017). Conversely, the Dutch disease thesis postulates that mining inhibits economic diversification and discourages exports not related to mining, which tend to improve trade balances and foreign exchange earnings (Marais et al. 2017).

Studies on the sustainability impacts of mining in SSA have focused on countries as diverse as Gambia, Ghana, South Africa and Zambia (Kitula 2006; Glaister and Mudd 2010; Simutanyi 2008; Tembo et al. 2005; Appiah and Osman 2014; Kakonge 2006). At the local level, the sustainability impacts mainly revolve around the multiple socio-economic and environmental complications from mining operations. For example, some of the environmental impacts that are rather well-documented in the literature include chronic soil degradation, chemical contamination and pollution (Tembo et al. 2005; Hilson 2002; Kitula 2006). Some of the main socio-economic impacts of mining include public health risks from pollution and the disruption of local livelihoods due to mine development (Wasylycia et al. 2014; Yabe et al. 2015).

The concept of 'sustainable mining' has become increasingly popular in the literature and has sought to capture such sustainability concerns in the minerals and mining sector (Azapagic 2004; Giurco and Cooper 2012; Laurence 2011; Limpitlaw 2004; World Bank 2002). However, unsurprisingly, there are radically different views on what sustainable mining really means. For example, some scholars have defined sustainable mining through the perspective of the continued availability or the exhaustibility of mining resources as depleting assets (Laurence 2011; Lebre and Corder 2015), while others have defined it through the lens of the Brundtland definition of sustainable development (Hilson and Murk 2000; Azapagic 2004). Azapagic (2004: 640), for example, notes that 'following the Brundtland definition of sustainable development, the main challenge for the sector is to clearly demonstrate that it contributes to the welfare and well-wellbeing of the current generation, without compromising the potential of future generations for a better quality of life. Achieving this goal requires a systems approach that enables balancing of economic, environmental and social concerns'.

According to Azapagic's (2004) definition, sustainable mining must embrace the triple bottom line of sustainable development, balancing the social, economic and environmental considerations during mining operations. In this definition, the 'social' often reflects issues pertaining to local communities in areas of mineral extraction (Solomon et al. 2008). However, the temporal dimensions of mining pose a major challenge because by its nature, mineral development is a process that inevitably comes to an end in a given area. This raises the question of whether (and the extent to which) mining can be considered 'sustainable', particularly when considering resource availability and the economic aspects enshrined in the concept of sustainable mining. As Lebre and Corder (2015) point out, due to the non-renewability of its feedstock (i.e. minerals), it is physically impossible to prolong the lifetime of a mine forever.

The notion of 'sustainable mining' becomes even more complicated when considered from the perspective of local community outcomes. If the decline and closure of a mine is inevitable, how can communities whose livelihoods are often tied to a mine remain sustainable well beyond the mine's lifetime? It is here, perhaps, that we urgently need to conceptualize sustainable mining in ways that embrace this aspect of mine decline and closure, which in many ways represent a major disturbance in the lifetime of a mine and its dependent community.

This is a major gap in the academic literature, as most studies adopting global, regional and national perspectives on mining sustainability tend to disregard the local scale and thus de-emphasize local community experiences with mining (Limpitlaw 2004). Furthermore, most studies from developing regions such as SSA rarely address the final stage of the mining life cycle, i.e. mine decline and closure. Indeed most of the literature on mine decline and closure is based on experiences from the global North (Laurence 2002, 2006a, b; Peck and Sinding 2009) and, to some extent, China (Andrews-Speed et al. 2005). Literature from SSA is largely restricted to South Africa (Binns and Nel 2003; Marais 2013; Marais et al. 2005, 2017), despite the obvious implications of mine closure on mine-dependent communities in other SSA countries (see above). Indeed, there seems to be a general lack of interest in the last stages of the mine life cycle in most SSA studies on sustainable mining (Marais et al. 2017), including Zambia (Simuntanyi 2008).

The above suggests that when engaging with the concept of sustainable mining, it is important to consider how local communities might be affected by (and cope with) shocks related to mine decline or closure. From a local community perspective, sustainable mining involves a commitment not only towards environmental sustainability but also to the socio-economic development of mining communities both during the life cycle of the mine and well beyond its closure.

Ensuring the sustainability of mining activities across its different phases and criteria and its effective contribution to sustainable development in the continent is a major challenge in most parts of SSA. Acknowledging the limitations of the current mining literature and practices, this chapter aims to contribute to the current discourse on mining and sustainability, by focusing on the societal impacts of mine decline and closure on local communities. We focus on Kabwe, one of Zambia's oldest mining towns, as an illustrative example of the negative sustainability outcomes of mine closure, particularly on mine-dependent communities. The focus is explicitly on the socio-economic changes induced by mine closure in Kabwe, using a framework anchored on resilience thinking (Sect. 7.2.1) that accounts for (and systematizes the evidence of) changes following mine closure. Resilience thinking is one of the few approaches that can integrate elements related to ecological, social and economic sustainability over various temporal and spatial scales while accounting for shocks and stresses (Davies et al. 2012). In stretching the notion of sustainability to embrace resilience, we are aware that this chapter is paddling in contested conceptual waters (Redman 2014; Marchese et al. 2018; Elmqvist et al. 2019).

However, from the point of view of this chapter, the two concepts are strongly intertwined and often difficult to separate (Saunders and Becker 2015).

Section 7.2 outlines the key concepts related to resilience thinking in the context of mining and the sustainability challenges manifesting in Kabwe following mine collapse. Section 7.3 systematizes the information obtained through a historical analysis and expert interviews, to present the evolution of the Kabwe mine across the different stages of the adaptive cycle. Section 7.4 discusses the main implications and highlights how resilience thinking can guide the development of policies and interventions aimed at enhancing the ability of mine-dependent communities to adjust to the changes brought by mine decline and closure.

7.2 Methodology

7.2.1 Research Approach

Similar to the concept of sustainability, the concept of resilience has been defined in different ways across different contexts (see Chap. 6 Vol. 1). For example, ecological resilience refers to the ability of a natural ecosystem to accommodate a disturbance (or recover from it) and maintain its functions (Lankford and Beale 2007; Suding et al. 2004). Resilience has become very popular for the study of coupled social-ecological systems, whose ecological and socio-economic components are strongly interlinked (Mats 2012; Plummer and Armitage 2007; Adjer et al. 2005; Hirst 2017).

For a local community, this would entail bouncing back to a pre-disturbed structure and function, following a disturbance (e.g. natural disaster) (Chap. 6 Vol. 1). This definition assumes that once a disturbance derails a system from its trajectory, then self-correcting forces and adjustments eventually return it to that initial trajectory (Simmie and Martin 2010). However, local communities live and operate in social-ecological systems that hardly change in a predictable and linear manner (Pisano 2012) (Chap. 6 Vol. 1).

Thus the adaptive capacity of individuals and communities within larger social-ecological systems (Saunders and Becker 2015) is of particular importance in resilience thinking. In a sense, resilience reflects the ability of a system to adapt its structure and functions successfully in response to a disturbance (Simmie and Martin 2010). The key features of this notion of resilience are change and adaptability. For example, a community is resilient if it can respond or adapt to change in such a way that maintains its development path (or indeed improves it). This type of resilience emphasizes on the ability of a local community to adapt to the challenges and changes encountered during (and after) a disturbance by drawing on its individual, institutional and collective resources (Saunders and Becker 2015) (Chap. 6 Vol. 1). While the local community may not bounce back to its pre-disturbance state, it may evolve to another state when dealing with the outcomes of the original disturbance (Saunders and Becker 2015; Simmie and Martin 2010).

Table 7.1 Adaptive cycles in socio-economic systems

1. *Rapid growth or exploitation stage (r-phase)*
    The system is characterized by an abundance of resources
    Period of rapid growth as actors exploit resources and opportunities
3. *Release stage/creative destruction stage (omega phase or Ω-phase)*
    Disturbance phase that occurs suddenly (e.g. market shock, technology failure, bankruptcy)
    Structural decline and loss of growth momentum likely to follow the disturbance
    Decline of system resilience to potential shocks is proportional to the length of the conservation stage
    Few actors remain operational to restart the next stage
	2. *Conservation stage (k-phase)*
	    Period of stability and high connectedness between various actors within the system
	    Few actors/organizations come to dominate the system
	    The growth patterns of the system become more rigid
	4. *Reorganization stage (alpha phase or α-phase)*
	    System begins reorganization and renewal
	    System is open to new possibilities and new activities
	    Experimentation and innovation as new actors emerge in the system

Source: Simmie and Martin (2010), Pisano (2012) and Resilience Alliance (2017)

Fig. 7.1 Schematic representation of the adaptive cycle. (Source: Angeler et al. (2015) based on Gunderson and Holling (2002))

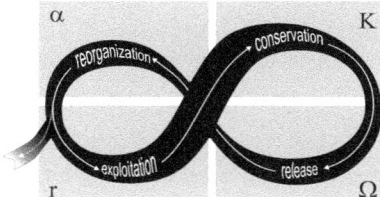

Another central theme in resilience thinking is how social-ecological systems change over time following disturbances. The concept of 'adaptive cycles' aptly illustrates this process between different stages of change. According to resilience scholars, when facing disturbances, social-ecological systems can go through four main phases, namely, growth, conservation, release and reorganization (Pisano 2012; Resilience Alliance 2017). Table 7.1 and Fig. 7.1 illustrate these distinctive stages for a socio-economic system.

The transition from the exploitation or growth phase to the conservation stage tends to be slow and reflects a period of incremental growth, development and fairly predictable system behaviour (Pisano 2012). The release stage reflects the disturbance stage leading to rapid and dramatic changes and increased uncertainty within the system (Resilience Alliance 2017). It is the release stage that leads to a period of reorganization as new possibilities open up.

The four stages of the adaptive cycle can be conceptualized as having two transitions or opposing nodes: a forward and a backward loop (Fig. 7.1). While the forward loop describes the combined stages of exploitation and conservation, the backward loop combines the stages of release and reorganization, illustrating the behaviour of a system, before and after a disturbance (Fig. 7.1).

Numerous studies have employed resilience thinking to analyse mining issues. Xu and Hassink (2017), for example, draw on resilience thinking to examine industrial dynamics in two coal-mining regions of China. In particular, they examine how old and mine-dependent economic paths interact with emerging industrial paths in the two towns. Wasylicia-leis et al. (2014) use resilience thinking to describe the various components of the socio-ecological system of a mine-dependent community and the challenges brought about by the dominance of mining in the local economy. Similarly, Gibson and Klink (2005) use a resilience lens to analyse the impact of mining on aboriginal communities and especially explain how mine closure impacts are distributed, experienced and mediated within these communities.

In this chapter, we employ resilience thinking as an analytical lens that allows for the examination of issues related to mining sustainability. In particular, we use the concept of adaptive cycles in two ways. First, we use these cycles as a frame for categorizing the main changes that have occurred in a mining town (Kabwe, Zambia; see Sect. 7.2.2), using mine closure as the manifestation of disturbance. This acknowledges the fact that in order to propose appropriate interventions to enhance mining sustainability, it is critical to appreciate at which stage the system is within the cycle and what the past trends have been (Sects. 7.3, 7.4.1, 7.4.2, 7.4.3 and 7.4.4). Second, we use the adaptive cycle to illustrate how resilience thinking might be used to develop policies and interventions to advance sustainable mining communities (Sect. 7.4.5).

7.2.2 Study Site

The study site is Kabwe town in central Zambia. The town was founded in 1904 under the name Broken Hill. Today, Kabwe is the capital of the Central Province of Zambia. Figure 7.2 shows the location of Kabwe and the mine, which is situated in the southern part of the town. The mine occupies an area of 2.5 km^2 (Kabwe Municipal Council 2016).

Kabwe grew from a small settlement to a town with a population of more than 200,000 people due to the lead and zinc mine operated for nearly a century until its closure in 1994. The mine (and its operator) was the central institution in Kabwe's socio-economic activity, contributing more than anything else to the town's expansion. Today, despite some small-scale mining activities and plans to reopen the old mine, the town typically exemplifies what may be termed a *bust* mining town with others referring to it as a ghost town (Chewe 2016; Kabwe Municipal Council 2016).

The mine closure (and the subsequent decline of mining) brought to the fore two main sustainability challenges discussed in more length below. The first concerns the environmental and health legacy of a century of largely uncontrolled and unregulated mining for lead, which is one of the most potent toxic minerals. This earned Kabwe the label of the world's most toxic or poisonous town (Carrington

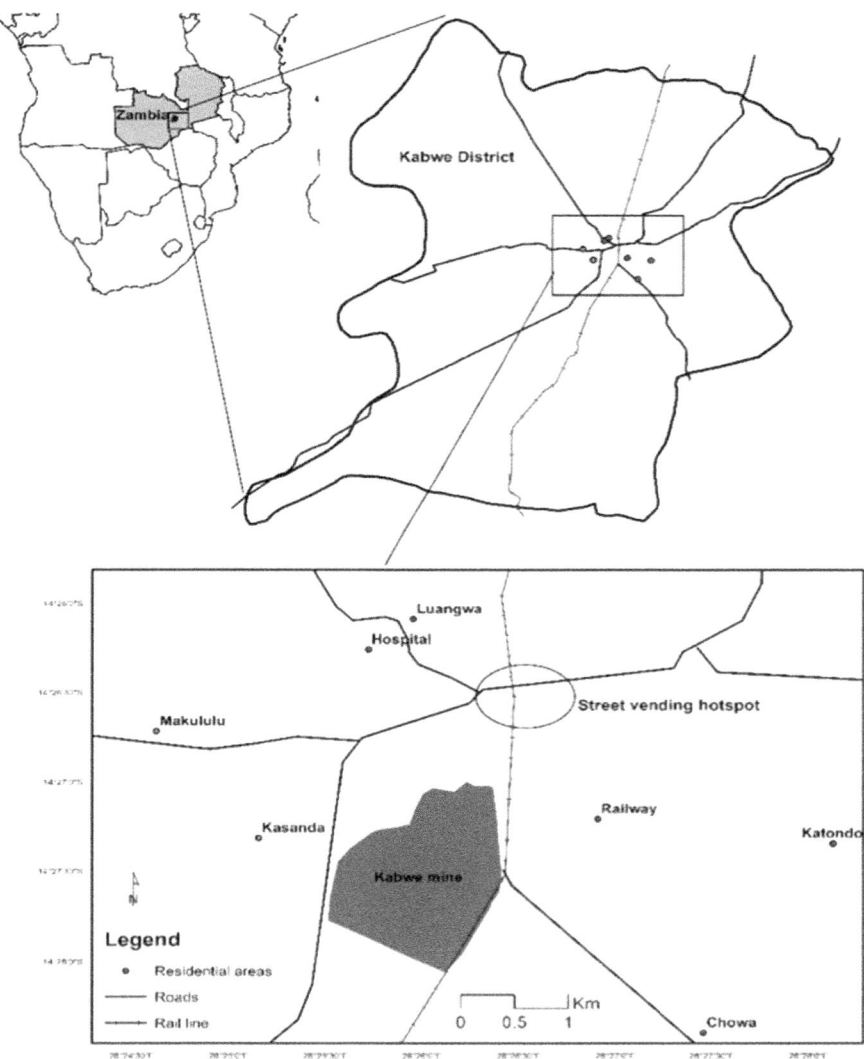

Fig. 7.2 Location of the study site

2017). Several studies have demonstrated the environmental and health risks that the local residents experience following the closure of the mine, including (a) the distribution of lead, cadmium and zinc concentrations in soils around Kabwe (Tembo et al. 2005), (b) impact of lead pollution on wildlife (Nakayama et al. 2011) and (c) effects of lead poisoning on children living in the vicinity of the old mine (Yabe et al. 2015). All these studies paint the bleak picture of the legacy of the mine, both from an environmental and a public health perspective. For example,

Yabe et al. (2015) demonstrate the alarming levels of lead poisoning in children in the townships surrounding the mine, even two decades after the closure of the mine. These studies have attracted the attention of the national government, environmental NGOs and other actors to the health plight of the town following a century of lead mining operations.

The second major sustainability challenge relates to the socio-economic wellbeing of the local communities. The Zambia Consolidated Copper Mines (ZCCM), a state company, operated the mine. ZCCM was often perceived as an extension of the state, as it developed public infrastructure and provided social services (Simutanyi 2008). At its peak, the Kabwe mine employed directly over 5000 people and indirectly even more in downstream industries. Furthermore, it owned and maintained housing, roads and water and sanitation infrastructure in all mining townships. In addition, it provided sports facilities and other recreational services that benefited the entire population of Kabwe. It is thus not surprising that through these diverse activities the mine contributed manifold to the welfare of the local community and that its collapse precipitated various hardships (Mankapi 2001). In a similar way that Kabwe was branded as a poisonous town (see above), it was also dubbed a ghost town due to the ensuing economic slump, as some firms relocated while others closed (Chewe 2016).

7.2.3   Data Collection and Analysis

In order to understand how the various actors and sectors have been affected by mine closure, we use a combination of historical analysis, institutional analysis and qualitative analysis of expert interviews. The main secondary data sources include demographic records from Zambia's Central Statistics Office (CSO), historical material and policy and planning documents. The historical and secondary data is complemented with empirical information obtained through 21 in-depth interviews conducted with former miners, local residents, local businesspersons and representatives of both the central and the local government (Table 7.2). These in-depth

Table 7.2 Characteristics of the expert interview respondents

Stakeholder	Number of respondents
District administration	1
Municipal council	1
Vendors Association of Zambia	1
Former senior government official	1
Kabwe District heads of government departments	2
Local business persons	2
Ex-miners	5
Long-term residents	8

interviews included mainly open-ended questions, which allowed the respondents to elaborate freely on their answers.

In order to be included in the study, participants had to be either long-term residents that were present during mine closure or representatives of institutions that are key stakeholders in Kabwe. The former can help elucidate the long-term changes experienced in the town through their experiences, while the latter can provide unique insights about the drivers and impacts of mine closure due to their institutional knowledge. The long-term residents and ex-miners were recruited mainly through snowball sampling from local guides that helped identify appropriate respondents that were still living and operating in the study area. The participants identified in this way were then also asked to identify other ex-miners or other long-term residents that could participate in the study as key informants.

The information collected from all these sources was systematized and used to populate the four stages of the adaptive cycles outlined in Sect. 7.2.1. The main findings are presented in Sect. 7.3 and are then critically discussed (Sects. 7.4.1, 7.4.2, 7.4.3 and 7.4.4) to identify policy- and practice-relevant implications and recommendations (Sect. 7.4.5).

7.3 Results

It can be argued that the Kabwe community has gone through all four stages of the adaptive cycle and is now undergoing the early phases of the reorganization stage, with a new emerging socio-economic regime (Table 7.3). Below we discuss in more detail the different stages that the Kabwe community has experienced from the perspective of the mine closure, through the synthesis of historical information and qualitative information from the interviews (Sect. 7.2.3).

The beginning of the rapid growth stage (r-phase) of Kabwe can be traced in the first decade of the twentieth century, when lead and zinc deposits were discovered in the area (Table 7.3). Following this discovery, Kabwe was initially developed as a small settlement operated by the Rhodesia Broken Hill Development Company. The mining company is essentially the pioneer actor that seized an opportunity for investment and laid the foundation for the development of the town. The connection to the rail line in 1906 and the development of a hydropower plant to service the mines and settlement boosted the growth potential of the settlement. This was the first hydropower plant in the region. Between 1920 and 1926, the labour force working in the mines increased from 500 to 3500 men (Mufunda 2015). In 1930, the settlement was declared a township and later in 1957 a municipality (Kabwe Municipal Council 2016). By 1963, the settlement grew into one of the most vibrant and important towns in Zambia, as the production of lead and zinc increased dramatically, reaching 250,000 tons per year (Kabwe Municipal Council representative: personal communication, 4th June 2017).

Kabwe entered the conservation stage (k-phase) during the postcolonial era (Table 7.3). During this period, the mine community transformed into a major

Table 7.3 Adaptive cycle of Kabwe's socio-economic system

1. *Rapid growth or exploitation stage (r-phase) (1902–1966)*	2. *Conservation stage (k-phase) (1967–1980s)*
1902: The discovery of lead and zinc leads to the rapid development of a new mining community (Kabwe)	1967: The railway company made Kabwe its headquarters, becoming the second largest employer in the area
1904: Kabwe town is established as an organized mine settlement operated by the Rhodesia Broken Hill Development Company Limited	1967: The Kabwe Industrial Fabrics was established in Kabwe
1906: The rail line reached Broken Hill town (now Kabwe)	1969: Nationalization of mines in the country
1920s: In the early 1920s, the mine employed 500 people, increasing to over 3000 by 1926	1975: General Pharmaceuticals Limited established
1964: Independence of Zambia	1981: Mulungushi Textiles Plant set up
1966: Kabwe gets its current name, which means 'ore'	4. *Reorganization stage (alpha phase or α-phase) (2000 to present)*
3. *Release stage/creative destruction stage (omega phase or Ω-phase) (1990s–2000)*	Emergence of new actors, such as informal sector businesses, public service and the educational sector
Mine decline and eventual closure due to the combination of technological factors and macro-economic conditions	Emergence of Kabwe as a market town
Breaking of the linkages between the mines and the railway company, milling companies and the social sector	Survival of some of the previous main actors such as KIFCO and milling companies. These actors have become part of the reorganization stage
Unemployment, decaying infrastructure and poor social services. Major companies close (e.g. Kafue Textiles), while others are restructured or privatized (e.g. Zambia Railways)	Emergence of new investors to take advantage of opportunities in the service and agricultural sector

town and became the provincial capital, attracting skilled professionals in a number of major economic sectors. In particular, the mine expanded and directly employed over 5000 people while also operating several subsidiaries such as milling companies. The mine owned over a third of all properties in the municipality of Kabwe, including nearly 2000 housing units (former government official: personal communication, 3rd June 2017). Furthermore, during this period the national rail company established its headquarters in Kabwe, which was followed by the development of textile, leather and pharmaceutical industries, consolidating the town's vibrant economy (Table 7.3). At that time, the economic prospects of Kabwe looked bright, and most respondents did not perceive that a bust was imminent (former government official: personal communication, 3rd June 2017). Although the mine company experienced some challenges during this period (e.g. a slump in metal prices and loss of business when mineral exports to South Africa were halted), these were all viewed as temporary challenges. The conservation stage lasted up to the late 1980s, with the bust coming in June 1994 when the closure of the mine was announced.

The main factor behind this bust was the antiquated technology used in the mine, which led to higher operational costs that consequently made it unprofitable to continue production (ex-miner: personal communication, 4th June 2017). Although the mine still contained between 1 and 1.6 Mt of lead and zinc silicate ore in situ, some of the technology dated back to the colonial period and required replacement and modernization in order for the mine operations to remain profitable. Unfortunately, when the mine was closing, the declining economic situation prevented the state-owned mine company from investing in new technology, leading to a premature closure (District Administration representative: personal communication, 3rd June 2017).

The release stage (Ω-phase) was sudden and unanticipated by the local residents and the main stakeholders (Table 7.3). Local residents were the most affected, including over a thousand employees that were released into the labour market. Most of the miners were relatively young, but with very limited non-mining skills to enable them to find other employment following mine closure (long-term resident: personal communication, 5th June 2017). Besides the loss of direct jobs and business decline, the impacts were also felt by other segments of the local community such as the suppliers of goods and services to the mining company. Furthermore, the sudden loss of mining jobs of the miners (who were some of the highest paid residents) greatly reduced the demand for certain goods and services. For example, businesses in the recreation and hospitality sector were among the most affected. As one resident put it, 'when miners got paid, we all knew it, there was life in town and there was business for the pubs, but when the mines closed business slumped' (businessperson: personal communication, 5th June 2017).

Furthermore, mine closure dashed the hopes of local residents and migrants that were still searching for jobs in the mines. These were mostly concentrated in Katondo and Makululu (see Fig. 7.2), two unplanned settlements around the two mine townships of Kabwe (Kasanda and Chowa). These unplanned areas also received many of the miners who lost their jobs and could not keep their accommodation in the mine townships.

Apart from the local communities, mine closure affected other actors such as the national government. Besides the obvious loss of revenue through taxes and dividends, the government also assumed key responsibilities of the mining company such as operating the Kabwe Mine Hospital and maintaining the road infrastructure in mine townships. However, the government struggles to carry out these responsibilities effectively, as indicated by the fact that nearly all of the township streets that were well-paved during the operation of the mine were in a state of disrepair during the field visits. Furthermore, the sporting and recreational facilities that were taken over by private companies and charity organizations following mine closure have either changed use or lack maintenance.

The Kabwe Municipal Council is another important actor that experienced the negative impacts of mine closure. During mine operation, Kabwe was among the best-performing municipalities in the country due to two main reasons (representative of Kabwe Municipal Council, personal communication, 4th June 2017). First, as the mine company owned a significant proportion of properties across the municipality, the council benefited from municipal taxes paid by the mines. For example, the mining company owned approximately 1000 and 600 housing units in the Kasanda and Chowa townships, respectively. This was in addition to office premises, sports facilities and various other properties. Second, the council (much like the national government) benefited from the fact that the mining company had assumed some of the services that should otherwise have been provided by the council itself. In this sense, mine closure not only considerably reduced the revenues of the local authority but also added the extra burden of taking over services such as water reticulation and waste management that were previously provided by the mine company. An indicator of the failure of the municipal council to offer these services efficiently is the intermittent water supply in the former mine townships (long-term resident: personal communication, 5th June 2017).

Mine closure exposed a number of weaknesses in the ability of the national and the local government to respond effectively to the disturbance caused by mine closure. Clearly, the response of the national government was slow and ineffective to match the scale of its impacts on the local community as acknowledged by representatives of both the national and the local government (former government official: personal communication, 3rd June 2017; Kabwe Municipal Council representative, 4th June 2017). These respondents suggested that it took the national government a few years before reconciling itself to the realities of mine closure and to start enacting measures to return the Kabwe town to a development trajectory. In this regard, the reorganization stage started in the 2000s and possibly continues to this date (Table 7.3).

One of the notable government responses to mitigate the effects of mine closure has been to transform Kabwe into a town specializing in the provision of education and public services. Over the past decade, two public universities were created from smaller pre-existing colleges. In addition, Kabwe hosts military training centres, trades training colleges, private colleges and a nursing school. While this appears to have curbed, to some extent, the negative outcomes of mine closure, these changes have not benefited several of the social groups affected by mine closure, including informal settlement dwellers, former miners and other residents with low skills to enter the civil service. Thus, despite employment generation in the public service and educational sectors, the absence of large labour-intensive industries has condemned the vast majority of the local residents to the informal sector (Vendors Association of Zambia representative: personal communication, 6th June 2017). In fact, Kabwe is now home to the second largest informal settlement in southern Africa and hosts the headquarters of the Vendors Association of Zambia, an organization that represents the interests of informal sector traders.

Having examined the history behind the closure of the Kabwe mine and its effects on the environment, the local community and other stakeholders, Section 7.4 discusses how resilience thinking can provide a powerful lens to inform the development of sustainable mining communities beyond the life cycle of a mine. It must be noted that due to specific contextual issues, the findings from Kabwe cannot be generalized to all mining communities in SSA. However, it is illustrative of how the adaptive cycle framework can be used to identify policies and interventions to mitigate mine impacts and develop sustainable and/or resilient mining communities.

The post-closure phase at Kabwe took place without any visible planning to mitigate the impacts of mine closure. For such mines, resilience thinking needs to be applied following a different approach. Sections 7.4.1, 7.4.2, 7.4.3 and 7.4.4 identify ways to develop policy interventions in each of the stages of the adaptive cycles, as a means of contributing to sustainable mining communities.

7.4 Discussion

7.4.1 Exploitation Stage

In mining contexts, the exploration and early stages of mine development usually reflect the exploitation stage of the adaptive cycle (Sects. 7.2.1 and 7.3). This stage also includes the mine planning phase and is crucial not only for the success of the mining operations but also for building a resilient and sustainable mine community. It is critical that at this stage, community perspectives are considered and possibly integrated in the mine plans. Currently, environmental impact assessment (EIA) processes are usually the main mechanism to take into consideration community voices during mining development in most SSA countries (Kakonge 2006; Appiah and Osman 2014).

However, the EIA has obvious limitations, as the assessment of mining impacts focuses on expected impacts on existing communities pre-mining (Solomon et al. 2008). Thus, most mining EIAs in SSA tend to emphasize the resettlement and compensation of the local communities expected to be affected by mining operations (Appiah and Osman 2014). Moreover, there are serious concerns about the monitoring or evaluation of actual impacts during mine operations or after mine closure, following EIA completion (Solomon et al. 2008). Indeed, community dynamics may change drastically long after the start of mining operations or following mine closure, in ways that have not been anticipated during the original EIAs.

Considering the above, it becomes crucial to encompass community perspectives, values and other issues that are not addressed by planning tools such as EIAs. Such approaches should go beyond simply prescribing employment, housing and/or social services to miners and affected communities (as was the case for Kabwe) or suggesting shallow mitigation measures for environmental impacts. Instead they should adopt longer-term perspective that considers community wellbeing well beyond the life of the mine. In this vein, Laurence (2006a, b) notes that if mines are to contribute to sustainable development, then mine closure objectives and impacts should be considered from project inception. This should require the development of closure plans that define a vision of the end point of the mining process and ensure that the full benefits of mining projects (e.g. revenues, expertise) are used to develop local communities or broader regions in such ways that will remain vibrant after mine closure (Laurence 2006a, b).

However, mine operators cannot achieve such long-term visions and plans alone. Designing effectively such visions and plans would require extensive cooperation (and possibly leadership in some aspects) between the government (both local and national), the private sector and other actors. Ideally such plans should be incorporated in regional or city plans designed by the relevant regional governments or municipalities. In keeping with resilience thinking, it would be important for such plans to not only take into consideration the predicted effects of mining but also build in elements of surprise or unpredictability. From a mine closure perspective, this is crucial as only a small fraction of mine closures have been planned in advance, with conversely most closures being premature and sudden (Lebre and Corder 2015).

It is worth noting that in Kabwe, community perspectives were not considered, nor was a closure plan developed. Furthermore, EIAs were not mandatory during the development of the Kabwe mine. Thus, without a prior understanding of the possibility of mine closure (or its possible effects) and a closure plan, many community members were simply unprepared for the socio-economic changes that occurred with mine closure. Similarly, the municipal government was unprepared to assume the provision of some public services (e.g. water, sanitation, waste management, health) that were previously provided by the mine company.

7.4.2 Conservation Stage

As explained in Sect. 7.2.1, the conservation stage reflects the phase of stability within the system but also the period in which the system is the least resilient. In a mining context, this tends to coincide with the operational period of the mine when it is well established and strongly connected to other sectors. This also tends to be a period of affluence for mining communities, and as exemplified in Kabwe's case, it is not uncommon to be perceived as a business-as-usual stage that will not change in the foreseeable future. Indeed, there may be a tendency to optimize operations at this stage to gain immediate benefits (Pisano 2012). For example, in Kabwe's case the economic benefits generated through mining operations were invested in social infrastructure. Even though such investments are undoubtedly beneficial to local communities, it is also important to invest in assets and processes that will ensure the long-term resilience of the local communities to unexpected changes posed by mine closure.

In our view, this entails cultivating the mindset that the conservation stage will inevitably end and will have important ramifications to local communities and other relevant stakeholders. The economic output of the mine and the affluence of the broader mining community during this stage may offer a crucial foundation for developing coping mechanisms for the release stage (see below). For example, mining companies (and other ancillary companies) could invest some of the profits generated during the conservation stage to build capacity and new skills that could enable miners (and the broader communities) to enjoy a decent living during the release stage. Furthermore, savings and/or insurance schemes can be designed to enable mine employees to have robust financial capital to allow coping with any unexpected changes, including mine collapse. Similarly, municipalities could use the revenues generated through mining-related taxes and higher property values to develop asset management funds to allow the maintenance of infrastructure and social services well after mine closure. Perhaps, even more crucial at this stage would be efforts led by municipal and central governments to use the generated revenue to spur the generation of new employment opportunities and diversify local economies through sectors unrelated to mining. For example, Marais et al. (2017) note that diverse and vibrant local economies can help mitigate the negative economic outcomes of mining decline or closure.

None of the above was done in Kabwe, which meant that nearly all economic sectors had no other feasible alternative when the mine closed. For example, the lack of prior efforts to develop (let alone implement) employment diversification plans left Kabwe with few employment opportunities unrelated to mining that could absorb the local residents after mine closure. Furthermore, the challenges facing local residents depending on the mine for their livelihoods were compounded by the lack of any operational insurance or saving scheme to mitigate the effects of employment loss.

7.4.3   Release Stage

Drawing on resilience thinking, disturbances in social-ecological systems tend to have both short-term immediate impacts and longer-term impacts that may lead to permanent changes (Pisano 2012; Simmie and Martin 2010). In mining contexts, mine closure is essentially such a disturbance that can have far-reaching consequences. For example, mine closure in Papua New Guinea reduced service demand and employment opportunities, affecting other members of the local community seeking mine employment (Jackson 2002). Similarly, the loss of business opportunities and the lack of employment (even several years following mine closure) are among the most important consequences of mine closure in South Africa (Marais et al. 2005; Binns and Nel 2003).

In order to cope effectively with the immediate impacts of mine closure, there is a need to develop emergency plans for mine-dependent communities. To deal with longer-term impacts, it is critical to develop long-term recovery plans (see Sect. 7.4.4 on the reorganization stage). When developing emergency plans, it is important to understand how local communities are affected by mine closure, both in terms of the affected groups and the severity of the impacts (Xavier et al. 2015; Laurence 2002). This is necessary to determine what types of relief programmes can be implemented to help local communities cope with the immediate changes, with examples including retraining former mine workers, developing new businesses and offering economic stimuli packages to save/assist businesses affected by mine closure. Critical for the implementation of such plans would be mobilizing assistance and funds from various sources, including central governments, local governments, NGOs and aid agencies (see Chap. 5 Vol. 1).

In the case of Kabwe, there seem to have been no emergency plans in place following mine closure. The local community was not expecting mine closure and was certainly not psychologically prepared for it. It also lacked the basic information and guidance to implement actions that could have mitigated the immediate effects of mine closure. Indeed, some of the respondents that experienced the closure confirm that this period remains one of the most painful episodes in the history of the town (ex-miner: personal communication, 6 June 2017). In fact, most respondents described the mine closure as a sudden event that was followed by a chaotic period. Several respondents also attested that there was not any preparedness or emergency plan in place to mitigate the effects of the sudden closure. This resulted in the failure of the mine-dependent local residents and stakeholders to quickly adjust to the changes brought by the closure.

7.4.4 Reorganization Stage

The outcomes of mine closure generally tend to be viewed as mostly undesirable. However, in resilience thinking, disturbances can also provide opportunities for learning (Pisano 2012). In this regard, the reorganization stage needs to revolve around recovery plans that take into consideration the lessons learnt from the specific mine closure (as well as from other similar contexts in other parts of the world). Depending on the production and trade of a single mining commodity can increase the vulnerability to associated external shocks, for example, changes in international trading regulations or commodity prices (University of Colorado 2010) (see also Chap. 3 Vol. 1). Thus, it would be vital to critically appraise resource endowments and competitive advantages and develop appropriate recovery plans that steer areas highly dependent on mining towards a more diversified economy (Marais et al. 2017). To increase their legitimacy, such recovery plans must have a long-term perspective and reflect the input and efforts of various actors (World Bank 2002; Laurence 2002, 2006a, b). While it is highly likely that local and national governments would lead such efforts to increase legitimacy, the vision of recovery plans must draw on (and ideally integrate) the views and resources of other stakeholders from the private sector, civil society and local communities.

In Kabwe, there have been some strides towards revitalizing the local economy (and the city more broadly), capitalizing on the competitive advantages offered by the geographical location, social structures and history of the town. For example, Kabwe is located between the two most urbanized regions of Zambia (i.e. Lusaka and the Copperbelt region), allowing its emergence as an important transit town that provides key services related to transportation, hotel accommodation and other amenities (Sect. 7.3). Furthermore, the town has been historically associated with some of the most prominent education and training centres in Zambia, influencing the government to upgrade some of the existing educational centres in the city to universities as a means of diversifying and boosting economic activity.

Although these developments suggest that Kabwe is still in the early phases of the reorganization stage, more substantial efforts would be needed to ensure decent employment for the entire local community. Some of these options could include further enhancing Kabwe's role as a market town, taking advantage of the opportunities offered by its surrounding areas and location along major rail and road transport networks. Besides offering market-related services to the surrounding agricultural areas, Kabwe also has a huge potential for developing agro-processing plants, which could stimulate industrial growth. Such opportunities could absorb a sizable portion of the labour force that is currently occupied in the large informal sector.

7.4.5 Policy and Practice Implications and Recommendations

This chapter suggests that mine operations can have substantial economic benefits at the local, regional and national level, offering decent job opportunities for local communities but also driving regional and national economic development (Sects. 7.3, 7.4.1 and 7.4.2). In this respect mining can be instrumental in spurring poverty alleviation and the delivery of broader societal benefits, through, for example, the development and operation of important infrastructure. However, many of these benefits are precarious and can be lost following the end of mining operations, especially if this happens suddenly and in an unplanned fashion (Sects. 7.3, 7.4.3 and 7.4.4). At the same time, poorly regulated mine operations have been linked to negative environmental and public health outcomes, which need to be considered properly when seeking to promote the sustainability of the mining sector (Sects. 7.1 and 7.3). The above suggests that mining sits at the interface of many SDGs such as SDG1 (No poverty), SDG3 (Good health and wellbeing), SDG8 (Descent work and economic growth), SDG9 (Industry, innovation and infrastructure) and SDG12 (Responsible consumption and production), to name a few (WEF 2016; Monteiro et al. 2019).

Throughout this chapter we discuss how resilience thinking can offer important insights for ensuring mining sustainability during the planning and operation of mining activities, especially related to disturbances posed by mine closure. Unfortunately, most of the large-scale mines operating currently in Zambia (and other parts of SSA) were designed without resilience considerations. In fact, many of these mines were designed at a time when environmental and social impact assessments were not mandatory. The failure to address the socio-economic challenges posed by mine closure (especially at the level of the local community) reflects the weak current policy and regulatory frameworks related to mining in Zambia. For example, at the national level, the Zambia Mines and Minerals Development Act (2015) prescribes that mining licences should be issued on the condition that prospective operators develop plans for generating local employment during mining operations.

Although this measure aims at developing local communities, it does not prescribe how to sustain the same communities once mining operations cease. The only obligations that operators have to comply with, following mine closure, relate to environmental rehabilitation and the removal of hazardous substances. Sadly, this is not unique to Zambia as, for example, the notions of mine closure or downscaling are not embedded in national mining policies in South Africa, despite the growing experience in mine downscaling in the country (Marais et al. 2017).

We believe that resilience thinking can indeed inform the development of policy, practice and planning approaches aiming at mitigating the possible short-term and long-term negative socio-economic outcomes of mine closure. These approaches can span different levels, ranging from the national level to the municipal and company level.

At the national level, one of the current shortcomings of national and regulatory frameworks on mining and minerals is the lack of an explicit focus on the socio-economic dimensions of mine closure. While existing frameworks emphasize on the environmental measures that mine operators need to put in place after closure, there is no explicit guidance on what should be done to ensure the financial and economic viability of mine-dependent communities and individual households in the eventuality of unexpected mine closure or downscaling. It is recommended, in this regard, that national mining policies should include clear provisions that focus on how to assure the financial and economic stability of mine-dependent communities after closure. In the case of Zambia, this could include policies and programmes aimed at local economic diversification, job creation and enhancing the employability of ex-miners.

At the municipal level, it would be critical for mining municipalities to start planning from the exploitation phase (i.e. when the mine is still in its infancy) for changes that may materialize following mine closure or downscaling (Sects. 7.3 and 7.4.1). This is especially critical for towns where mining is the dominant economic activity. In such cases, municipalities must develop long-term sustainability plans that are anticipatory of a post-mining economy. Due to its intuitive nature, the concept of adaptive cycles can possibly guide the development of such plans, for example, by conceptualizing how possible changes in mining operations might emerge and manifest and identifying mitigation measures for the impacts of such changes. Furthermore, mining municipalities should put in place emergency or preparedness plans, which could allow for the effective response to the rapid socio-economic outcomes of mine closure.

At the company level, mine operators must be motivated (and possibly incentivized) to invest in the development of mechanisms and schemes that could empower their employees to cope with the changes catalysed by mine downscaling or closure. Some of the mechanisms already discussed in this chapter include capacity building to develop non-mining skills, as well as the creation of savings and insurance schemes that could act as a livelihood buffer in the eventuality of mining closure.

Finally, it is important to recognize that different mines may be at various stages of development at a given point in time. While some mines may be at the exploratory stage or design stage, others may be at the closure stage. Mines in their design stage have the opportunity to integrate resilience principles into mine planning at the onset of mine development. This could possibly allow them to develop more robust systems for enhancing the resilience of mine-dependent communities against disturbances.

7.5   Conclusions

This chapter focused on Kabwe, a community in Zambia that has been affected severely by mine closure. The chapter utilized a framework based on resilience thinking to synthesize information from historical sources and expert interviews with different stakeholders and local residents. The results highlight the significant socio-economic changes that the local mining community experienced during the four phases of the adaptive cycle.

We demonstrate that the notion of the adaptive cycle is a powerful analytical frame that could be applied to explain the evolution of mining operations and identify interventions in the context of mineral resources and mining. In particular, resilience thinking can inform the development of policies and interventions seeking to mitigate the short-term and long-term socio-economic outcomes of mine closure. Even though it is ideal to embed resilience thinking in planning during the earliest stages of the mine life cycle (i.e., when developing mine plans), the reality is that most of the mines currently operating in SSA were developed when there was little awareness of mining sustainability and particularly how mine closure might affect it. This is especially true in Zambia for most of the large mines operating in the Copperbelt region. However, resilience thinking can still be applied at various stages of the mine life cycle to aid governments, local communities and companies in their response to change precipitated from mine closure.

A key outcome of this chapter is that sustainability assessments in the mining sector must pay more attention to the final stages of mining. Unexpected mine closure can have significant and dire implications for mine-dependent communities, as exemplified in the case of Kabwe. In this regard, we suggest that the notion of sustainable mining should not be restricted simply to sustainability impacts during the lifetime of the mine. It should also expand to consider how to build resilient mining communities that are able to cope with the outcomes of mine decline and closure.

Acknowledgements This chapter is based on preliminary results from a much larger research project funded by the Japanese Trust Fund and African Development Bank that seeks to understand how local communities negotiate boom-and-bust mining cycles in Zambia. The Project was undertaken under the auspices of the Education for Sustainable Development in Africa (ESDA) Next Generation Researchers Programme, hosted by the United Nations University Institute for Advanced Study of Sustainability (UNU-IAS).

References

Adjer WN, Hughes TP, Folke C, Carpenter SR, Rockstrom J (2005) Social-ecological resilience to coastal disasters. Science 30:1036–1039

Andrews-Speed P, Ma G, Shao B, Liao C (2005) Economic responses to the closure of small-scale coal mines in Chongqing, *China*. Resour Policy 30:39–54

Angeler DG, Allen CR, Garmestani AS, Gunderson LH, Hjerne O, Winder M (2015) Quantifying the adaptive cycle. PLOS ONE 10(12):e0146053. https://doi.org/10.1371/journal.pone.0146053

Appiah DO, Osmar B (2014) Environmental impact assessment: insights from mining communities in Ghana. J Environ Policy Manage 16(4):1–20

Arellano-Yanguas J (2011) Aggravating the resource curse: decentralisation, mining and conflict in Peru. J Dev Stud 47(4):617–638

Azapagic A (2004) Developing a framework for sustainable development indicators for the mining and minerals industry. J Clean Prod 12:639–662

Binns T, Nel E (2003) The village in the Game Park: local response to the demise of coal mining in Kwazulu-Natal, South Africa. Econ Geogr 79:41–66

Carrington D (2017) The world's most toxic town: the terrible legacy of Zambia's lead mines. https://www.theguardian.com/environment/2017/may/28/the-worlds-most-toxic-town-the-terrible-legacy-of-zambias-lead-mines. Retrieved on 30th June 2017

Chewe F (2016) The rebirth of Kabwe town. Zambia Daily Mail, August 2, 2016

Davies J, Maru Y, May T (2012) Enduring community value from mining: conceptual framework. Working paper CW007. Ninti One Limited, Alice Springs

Elmqvist T, Andersson E, Frantzeskaki N, McPhearson T, Olsson P, Gaffney O, Takeuchi K, Folke C (2019) Sustainability and resilience for transformation in the urban century. Nat Sustain 2 (4):267–273

Gajigo O, Mutambatsere E, Ndiaye G (2012) Gold mining in Africa: maximising economic returns for countries. Africa Development Bank, Tunis

Gibson G, Klink J (2005) Canada's resilient north: the impact of mining on aboriginal communities. Pamatisiwin: A J Aborig Indigenous Communities 3(1):126–139

Giurco D, Cooper C (2012) Mining and sustainability: asking the right questions. Miner Eng 29:3–12

Glaister BJ, Mudd GM (2010) The environmental costs of platinum–PGM mining and sustainability: is the glass half-full or half-empty. Miner Eng 23:438–450

Gunderson LH, Holling CS (2002) Panarchy: understanding transformations in human and natural systems. Island Press, Washington, DC

Hilson G (2002) Small-scale mining in Africa: tackling pressing environmental problems with improved strategy. J Environ Dev 11(2):149–174

Hilson G, Murk B (2000) Sustainable development in the mining industry: clarifying the corporate perspective. Resour Policy 26:227–238

Hirst E (2017) The unit of resilience: Unbeckoned degrowth and the politics of (post) development in Peru and the Maldives. J Polit Ecol 24:463–475

Jackson TR (2002) Capacity building in Papua New Guinea for community maintenance during and after mine closure. International Institute of Environment and Development, London

John D (2011) Is there really a resource curse? A critical survey of theory and evidence. Glob Gov 17:167–184

Kabwe Municipal Council (KMC) (2016) Transformation of Kabwe into a City. KMC, Kabwe

Kakonge J (2006) Environmental impact assessment in sub-Saharan Africa: the Gambian experience. Impact Assess Apprais 24(1):57–64

Kitula AGN (2006) The environmental and socio-economic impacts of mining on local livelihoods in Tanzania: a study of Geita District. J Clean Prod 14:405–414

Koitsiwe K, Adachi T (2015) Australia mine boom and Dutch disease: analysis using VAR method. Procedia Econ Finan 30:401–408

Lankford B, Beale T (2007) Equilibrium and non-equilibrium theories of sustainable water resources management: Dynamic River basin and irrigation behavior in Tanzania. Glob Environ Chang 17(2):168–180

Laurence DC (2002) Optimising mine closure outcomes for the community-lessons learnt. Mineral Energy-Raw Mater Rep 17(1):27–38

Laurence D (2006a) Optimising mine closure outcomes for the community- lessons learnt. Mineral Energy-Raw Mater Rep 17(1):27–38

Laurence D (2006b) Optimisation of the mine closure process. J Clean Technol 14:285–298

Laurence D (2011) Establishing sustainable mining operations: an overview. J Clean Prod 19:278–284

Lebre E, Corder G (2015) Integrating industrial ecology thinking into the management of mining waste. Resources 4:765–786

Limpitlaw D (2004) Mine closure as a framework for sustainable development. Conference on sustainable development practices on mine sites. University of Witwatersrand, 8–10th March 2004

Mankapi L (2001) The closure of the Kabwe mine and its impact on the socio-economic transformation of Kabwe urban. Unpublished masters dissertation, University of Zambia

Marais L (2013) Resources policy and mine closure in South Africa: the case of the Free State Goldfields. Resour Policy 38:363–372

Marais L, Pelser A, Botes L, Redelinghuys N, Benseler A (2005) Public finances, service delivery and mine closure in Koffiefontein (Free State, South Africa): from stepping stone to stumbling block. Town Reg Plan 48:5–16

Marais L, Rooyen D, Nel E, Lenka M (2017) Responses to mine decline downscaling: evidence from secondary cities in the south African goldfields. Extract Ind Soc 4(1):163–171

Marchese D, Reynolds E, Bates ME, Morgan H, Clark SS, Linkov I (2018) Resilience and sustainability: similarities and differences in environmental management. Sci Total Environ 613-614:1275–1283

Mats W (2012) Resilience thinking versus political ecology: understanding the dynamics of small scale labour-intensive landscapes. In: Plieninger T, Bieling C (eds) Resilience and the cultural landscapes: understanding and managing change in human landscapes. Cambridge University Press, Cambridge, pp 95–11

Monteiro NBR, da Silva EP, Neto MM (2019) Sustainable development goals in mining. J Clean Prod 228:509–520

Mufunda M (2015) History of mining in Broken Hill: 1902 to 1929. Unpublished masters thesis, University of the Free State

Nakayama SMM, Ikenaka Y, Hamada K, Muzandu K, Choongo K, Teraoka H, Mizuno N, Mayumi I (2011) Metal and metalloid contamination in roadside soil and wild rats around a Pb-Zn mine in Kabwe, Zambia. Environ Pollut 159:175–181

Peck P, Sinding K (2009) Financial assurance and mine closure: stakeholder expectations and effects on mine operating decisions. Resour Policy 34:227–233

Pegg S (2006) Mining and poverty reduction: transforming rhetoric in reality. J Clean Prod 14:376–387

Pisano U (2012) Resilience and sustainable development: theory of resilience, systems thinking and adaptive governance. ESDN quarterly report no. 26

Plummer R, Armitage D (2007) A resilience based framework for evaluating adaptive co-management: linking ecology, economics and society in a complex world. Ecol Econ 61:62–74

Prior T, Giurco D, Mudd G, Mason J, Behrisch J (2012) Resource depletion, peak minerals and the implications for sustainable resource management. Glob Environ Chang 22:577–587

Prior T, Daly J, Mason L, Giurco D (2013) Resourcing the future: using foresight in resource governance. Geoforum 44:316–328

Redman CL (2014) Should sustainability and resilience be combined or remain distinct pursuits? Ecol Soc 19(2):37

Resilience Alliance (2017) Adaptive cycle. www.resalliance.org/adaptivecycle. Retrieved 20 May 2017
Saunders WSA, Becker JS (2015) A discussion of resilience and sustainability: land use planning recovery from the Canterbury earthquake sequence, New Zealand. Int J Disaster Risk Reduct 14:73–81
Simmie J, Martin R (2010) The economic resilience of regions: towards an evolutionary approach. Econ Soc 3:27–43
Simutanyi N (2008) Copper mining in Zambia: The development legacy of privatisation. ISS Paper 2008(165):16
Solomon F, Katz E, Lovel R (2008) Social dimensions of mining: research, policy and practice challenges for the minerals industry in Australia. Resour Policy 33:142–149
Suding KN, Gross KL, Houseman GR (2004) Alternative states and positive feedback in restoration ecology. Trends Ecol Evol 19(1):46–53
Tembo DB, Sichilongo K, Cernak J (2005) Distribution of copper, lead, cadmium and zinc concentrations in soils around Kabwe town in Zambia. Chemosphere 63:497–501
University of Colorado Center of the American West (2010) Boom and Bust in the American West. www.centerwest.org/wp-content/uploads/2010/12/boombust,pdf. Retrieved on 12th May 2017
Wasylycia-Leis J, Fitzpatrick P, Fonseca A (2014) Mining communities from a resilience perspective: managing disturbance and vulnerability in Itabira, Brazil. Environ Manag 53:481–495
World Bank (2002) It's not over when it's over: mine closure around the world. World Bank, Washington, DC
World Economic Forum (2016) Mapping mining to sustainable development goals: an atlas. World Economic Forum, Geneva
Xavier A, Veiga MM, van Zyl D (2015) Introduction and assessment of a socio-economic mine closure framework. J Manage Sustain 5(5):38–49
Xu X, Hassink R (2017) Exploring adaptation in uneven economic resilience: a tale of two mining regions. J Reg Econ Soc:1–27
Yabe J, Nakayama SMM, Ikenaka Y, Yohannes YB, Bortey-Sam N, Oroszlany, Muzandu K, Choongo K, Kabalo AN, Ntapisha J, Mweene A, Umemura K, Ishizuka M (2015) Lead poisoning in children from townships in the vicinity of a lead-zinc mine in Kabwe, Zambia. Chemosphere 119:941–947

Chapter 8
Knowledge Co-production in Sub-Saharan African Cities: Building Capacity for the Urban Age

Zarina Patel, Ntombini Marrengane, Warren Smit, and Pippin M. L. Anderson

8.1 Introduction

Transdisciplinary approaches for the co-production of urban knowledge are gaining traction globally, as a means of creating better science-policy connections for enhancing societal benefits and achieving urban sustainability (International Expert Panel on Science and the Future of Cities 2018) (Chap. 9 vol. 2). The assumption that underpins this normative acceptance is that it is necessary to integrate multiple sources of knowledge in order to address complex sustainability challenges in cities (Polk 2015a, b) (Chap. 9 Vol. 2). This chapter focuses on the application of this assumption in the context of cities in sub-Saharan Africa (SSA), as it provides fertile grounds for research and knowledge generation considering that the continent enters its urban age (UN-Habitat 2008). Unlike countries of the global North, many SSA countries are undergoing an "urban revolution" and are projected to reach the 50% urbanisation threshold around 2030 (UN DESA 2011, cited in Pieterse and Parnell 2014: 1) (Chap. 1 Vol. 1). This dynamism and unpredictability regarding the scale of demographic and spatial transitions that are currently unfolding in SSA cities (and which are projected into the future) adds to the complexity of decision-making and policy development. Growth and development in a global context is constrained by limits to growth, and the desire to achieve low-carbon futures lends urgency to understanding the potential of alternate knowledge configurations to address such complexities in urban contexts (Chap. 9 Vol. 2).

Z. Patel (✉) · P. M. L. Anderson
Environmental and Geographical Science, University of Cape Town, Cape Town, South Africa
e-mail: zarina.patel@uct.ac.za; pippin.anderson@uct.ac.za

N. Marrengane · W. Smit
African Centre for Cities, University of Cape Town, Cape Town, South Africa
e-mail: ntombini.marrengane@uct.ac.za; warren.smit@uct.ac.za

As primary and secondary cities across SSA expand and transform at an unprecedented rate, they are no longer viewed exclusively as political capitals or colonial monuments (Njoh 2009). In fact, since the colonial period, cities in the continent have come to be perceived both as spaces of deprivation, informality and poverty (Pieterse and Parnell 2014; Pieterse 2011) and engines of economic growth and technological excellence (South African Cities 2004). Leveraging the potential for economic growth and the role of technology must therefore be done in ways that address the multiple dimensions of sustainable development, with a specific focus on achieving social and environmental justice (Patel 2009; Cole 2015) (Chap. 8 Vol. 1).

However, SSA cities experience multiple sustainability challenges. For example, many of the environmental challenges faced by SSA cities are immediate and are closely linked to the lack of access to clean and safe basic services, such as drinking water and sanitation services among others (Chap. 1 Vol. 1; Chap. 4 Vol. 2). Traditional brown agenda challenges, associated with developing contexts, are rife in SSA cities, including overcrowding, indoor air pollution, water pollution, and fires (Hardoy et al. 2013) (Chap. 1 Vol. 1). The vulnerability of the urban poor on natural hazards is due to diverse socio-economic and environmental factors and severely undermines their adaptive capacity and ability to thrive alongside the escalating risks posed by climate change (Chap. 1 Vol. 1; Chap. 2 Vol. 2). Related to the above, the provision of urban social services such as healthcare, education, housing and sanitation has become a key urban sustainability challenge in many SSA cities in recent decades (Parnell and Walanege 2014) (Chaps. 1, 5 Vol. 1; Chap. 4 Vol. 2). The compounding effects of exceeding ecological limits and failing to deliver on the basic rights of the social protection floor[1] render many SSA cities both unsustainable and unjust (Cole 2015).

As mentioned above, SSA cities are evolving during the Anthropocene era, shaped to large degree by the global imperative of fostering low-carbon smart cities, with technology-based solutions to mitigate environmental impacts and address resource scarcity. Translating these imperatives into SSA contexts, considering the prevailing scale and severity of deprivation, as well as inequality and lack of access to adequate basic services for the majority of urban dwellers living in informal slums, poses significant societal and policy challenges (Pieterse and Parnell 2014) (Chap. 1 Vol. 1). Improving access to urban services is unsurprisingly difficult given the predominance of informal systems in these areas (Breda van and Swilling 2019). Spatial and demographic urban transition in SSA must be harnessed by introducing new ways of accessing data on both access to resources and services but also on the multiple and complex ways in which urban residents live and interact in cities.

Traditionally, government officials have assumed the responsibility of tackling such sustainability challenges whilst at the same time effectively managing limited

---

[1]The 'social protection floor' is defined as a *global and coherent social policy concept that promotes nationalty defined strategies that protect a minimum level of access to essential services and income security for all.... A national Social Protection Floor is a basic set of rights and transfers that enables and empowers all members of a society to access a minimum of goods and services and that should be defended by any decent society at any time* (ILO and WHO 2009: 1).

resources (UN-Habitat 2008: 2014) (Chap. 5 Vol. 1). However, the assumption that governments can effectively drive such processes locally is based on the premise of strong governments. The reality in SSA cities is that the local government is often weak, with limited capacity to counter the invisible hand of power exerted by neoliberal agents shaping technological choices (Pieterse and Parnell 2014). Furthermore, the sheer scale of the technological innovation that would be required to address the imperatives of resource decoupling, decarbonisation and biodiversity restoration in SSA cities points to the need for reconfiguring research alliances to forge innovation (Chap. 8 Vol. 1).

It is within this context, where urban practitioners from the public and private sector have to increasingly find ways to collaborate and address constructively the current urban sustainability challenges in SSA, including the demand for (and access to) basic services and growing inequality. Several scholars have pointed out that the ability of SSA cities in meeting social service delivery (and generally enhancing urban sustainability) should no longer be the exclusive domain of local governments and should be perceived as a major governance challenge (Andersen et al. 2009).

Knowledge partnership between academic researchers, local government officials, civil society and local communities is a possible way of bringing together urban stakeholders to explore and design solutions to pressing urban sustainability challenges (Polk and Kain 2015). This need for knowledge partnerships gains further significance in the context of the Sustainable Development Goal 11 (SDG 11) that seeks to make cities and human settlements inclusive, safe, resilient and sustainable.

Indeed, despite the compelling case for transdisciplinary approaches to urban knowledge production, there is little evidence about the mechanisms through which these partnerships can occur and be successful. Patel et al. (2017) and Simon et al. (Simon et al. 2016) show that the success in meeting SDG 11 depends on the availability and accessibility of robust datasets, as well as the reconfiguration of governance systems that can catalyse urban transformation (Chaps. 1, 5 Vol. 1). The significance of reliable data and the centrality of urban knowledge for SDG 11 are obvious when honing in on the indicator frames. The UN Secretary General's Independent Advisory Group called for a data revolution, in which statistical systems must be strengthened at local, national and international levels and new means of collecting data of high quality and coverage should be promoted (IEAG 2014). Given the patchy and inconsistent datasets related to urban processes and services in SSA, specifically in informal contexts, establishing and leveraging informal and formal collaborations could enhance stakeholder dialogue and catalyse the development of additional capacities and innovations (Anderson et al. 2013; Patel et al. 2015).

However, forging such research collaborations and improving data collection, analysis, storage and sharing capabilities is a major sustainability challenge for many SSA countries, especially in urban contexts (Chaps. 1, 5 Vol. 1). This is possibly due to the combined effects of lack of capacity and resources for (and experience in) undertaking such processes. However, forging stronger transdisciplinary processes in urban contexts does not only relate to SDG11 but also to SDG17, which seeks to strengthen SDG implementation and revitalise partnerships for sustainable development.

The aim of this chapter is to elucidate some critical characteristics of knowledge co-production approaches when engaging with urban sustainability challenges in SSA, especially (but not only) from experience in South Africa. We do not attempt to undertake a comprehensive review of the debates on knowledge co-production but to rather focus on demonstrating how the mechanics of two types of knowledge configurations can assist in understanding and solving urban sustainability challenges. The specific focus is on (a) knowledge partnerships between universities and city authorities and (b) knowledge co-production with civil society. We draw on three applied and transdisciplinary urban research projects conducted at the African Centre for Cities (ACC)[2] at the University of Cape Town. Throughout this chapter, we trace how salience, legitimacy and credibility emerged in these projects and how it catalysed the formation of effective knowledge partnerships (Cash et al. 2002a). This analysis serves to enrich our understanding of how SSA cities operate, how local government stakeholders engage with other urban stakeholders invested in the city's performance and how processes in SSA cities can be analysed and understood. Considering that most of the current research on assessing the impact of urban knowledge co-production processes comes from a northern perspective (Breda van and Swilling 2019), this chapter contributes to this emerging literature from an African perspective.

Section 8.2 provides an overview of current debates on (urban) knowledge co-production, with a focus on their application in SSA cities. Section 8.3 introduces the three case study projects and outlines the adopted methodological approaches based on the concepts of salience, legitimacy and credibility. Section 8.4 outlines how these three key elements of knowledge co-production emerged in the three case studies. Section 8.5 identifies the main lessons learnt from the three projects and some of the main policy implications for tackling urban sustainability challenges in SSA in the context of SDG11 and 17.

8.2 Urban Knowledge Co-production

8.2.1 Urban Knowledge Co-production: Conventional Wisdom

Knowledge co-production has increasingly gained attention in the interdisciplinary field of urban studies (Swilling 2014; Andersen et al. 2009; Polk and Knutsson 2008;

---

[2]The ACC is part of the University of Cape Town (UCT). It is an interdisciplinary research and teaching programme focusing on the dynamics of unsustainable urbanisation processes in Africa. The ACC has embedded knowledge co-production in its research philosophy and undertakes research projects related to sustainable urban transitions and urban management.

Cash et al. 2002a; Gibbons et al. 1994). This new paradigm is pushing knowledge production beyond the traditional view of the university as a ground for exploring the frontiers of science and technology, with a one-way flow of knowledge outwards to user groups (Pieterse 2013; May and Perry 2006). This shift towards "requesting" universities to work with other knowledge brokers in a bidirectional process of knowledge creation through transdisciplinarity and knowledge co-production receives increasing traction in order to increase relevance (Oldfield and Patel 2016). Below, we outline the main debates shaping the assumptions and accepted wisdom about urban knowledge co-production, namely, (a) what constitutes legitimate knowledge, (b) what is knowledge co-production, (c) what is the role of science and knowledge in co-production processes, (d) what are the implications of a shift to transdisciplinarity and (e) the question of research methods.

Regarding (a), there is a long debate about what constitutes knowledge (Nonaka 1994) and from where it gets its legitimacy. Choo (1996) proposes a broad definition of knowledge which places value on tacit knowledge, explicit knowledge and experiential knowledge. Tacit knowledge is defined as highly personalised and foregrounded in repeated exposure and experience of a particular environment (Nonaka 1991). Whilst tacit knowledge can include technical knowledge, it is not based on scientific principles but on experiential learning. Such an example in the context of urban development is the lived experiences of urban residents, whether in high- or low-income areas, and the ways in which they interact with city services (Polk 2015a, b). By contrast, explicit knowledge is defined as *formal and systematic* (Nonaka 1991: 98) and is based on the scientific method and hard evidence.

Regarding (b), knowledge co-production is based on the assumption that both tacit and explicit knowledge must co-exist, as policy must be based on an understanding of material conditions, consequences and robust norms, with the success of implementation being determined by an understanding of factors affecting and motivating action. Although the term knowledge co-production has enjoyed increasing popularity in the academic literature across various disciplines, there is a lack of a unified body of literature on the concept (Jasanoff 2004; Watson 2014). The initial terminology of knowledge co-production, which emerged in the 1980s, was used to describe the participation of citizens in the provision of urban public services outside of the built environment. Rather than simply being a single transaction where the state responds to citizen service delivery demands (Whitaker 1980; Brudney and England 1983), a partnership is developed between distinct stakeholder groups to enhance and move forward ideas and methods on service delivery. This creates a new outcome that each group may have been unable to achieve on its own.

Various terms have been used to describe this type of interaction, including knowledge co-production, interactive knowledge production, knowledge integration and knowledge co-creation, but there are some basic tenants across the competing classifications (Ostrom 1996; Jasanoff 2004; Hessels and Van Lente 2008; Pohl et al. 2010; Edelenbos et al. 2011). Nonaka (1991) explains that knowledge creation is moving beyond the 'processing of objective information' and instead requires the open review and assessment of the insights (and intuitions) of stakeholders, which are then translated and developed into new knowledge. Understanding that there is no universally accepted definition of knowledge co-production, for the purpose of this chapter, we use the definition of Edelenbos et al. (2011) that draws on experience in the water sector. The process of knowledge co-production can be described as the outcome of bringing together scientific knowledge, bureaucratic knowledge and stakeholder knowledge (Edelenbos et al. 2011). This framing of knowledge co-production builds on the idea that whilst knowledge itself is useful, it needs to be relevant to broader constituencies (i.e. stakeholders) for it to be transformative and produce value. It must also be rigorous in order to meet the standards of the scientific community, as well convey (and generate) useful information that can shape and inform policy development and implementation.

Regarding (c), a common thread in the knowledge co-production literature is that science can no longer be viewed as a closed and hallowed domain. Jasanoff (2004) takes this a step further and highlights the role of knowledge co-production in addressing the hierarchical relationship between knowledge and power that gives science hegemony. Through the democratisation of knowledge (Delanty 2001), the pursuit of science is no longer the exclusive province of academics and scientists. Oldfield and Patel (2016) argue that this democratisation of knowledge does not eradicate power, but knowledge co-production relationships (which are reconfigured to be mutually accountable) certainly keep this power in check. This is particularly the case in the field of urban studies as *urban knowledge cannot be isolated from the conditions of its production and concepts must be related to specific circumstances in order to make sense of them* (Andersen et al. 2009: 9). That specificity comes from the perspective of those who live the urban experience and know intimately the constraints and opportunities at the city, neighbourhood and street scale (Andersen et al. 2009). Assembling the data and creating the narratives to explore urban conditions requires the input and voice of communities that occupy those geographies. Pohl et al. (2010) stress that all stakeholders (whether inside or outside academia) have relevant knowledge based on their unique perspectives and proximity to the issue being addressed. Andersen et al. (2009: 9) posit that urban knowledge co-production is necessarily an *action-oriented, multidisciplinary and contextually defined knowledge generation process that has the express purpose of enhancing the operations of cities and the quality of life of its residents through practical action.* Knowledge co-production has value in that it presents a possible way to bridge and integrate scientific knowledge and technocratic expertise, with the understandings and experiences of citizens outside the walls of power (Coburn 2007).

Regarding (d), transdisciplinarity is central to urban knowledge co-production, as it is necessary to work across academic and policy networks (Hanson and Polk 2018). Expertise from various disciplines is a starting point towards the assessment and solution of the complex and multidimensional aspects of urban development (Lemos and Morehouse 2005). Gibbons et al. (1994) have explained the shift towards knowledge co-production in recent years by comparing and contrasting the conventional academic scholarship (which depends on disciplinary expertise), with current transdisciplinary methodologies, theories and practices. They argue that traditional science and scholarship is generated and disseminated with an academic audience in mind. However, transdisciplinarity does not merely entail knowledge generation using insights from more than one discipline but requires deliberate examination of theoretical and applied knowledge across disciplines (Pohl and Hirsch Hadorn 2007). Transdisciplinary approaches must recognise the (at times) competing perspectives that inform the types of problems to be solved. This process provides a space for assessing and valuing practical knowledge and scientific work. Thus, stretching the spaces of knowledge production beyond science and academia into the domains of the lived experience, policy and practice provides the potential for more effective policy formulation and implementation (Polk 2015a, b).

Regarding (e), methodology is a critical and highly contested aspect of knowledge co-production and transdisciplinary processes. Relevant concepts such as reflection and reflexivity have started becoming more popular outside the humanities (Gibbons et al. 1994; May and Perry 2010). What distinguishes transdisciplinarity from other approaches that seek to integrate divergent bodies of knowledge is that the research is conducted jointly between different stakeholders in a process of collective problem-solving that does not perceive them as objects of enquiry (Swilling 2014). In a sense, the relationship between subject and object in such research processes is no longer separate but is rather co-constituted (Oldfield and Patel 2016). In the transdisciplinary shift towards application in social contexts (Gibbons et al. 1994), academics should be able to reflect and acknowledge how their research (and its presentation) can influence audiences outside academia, transforming its purpose from knowledge generation to advocacy (Swilling 2014).

Considering the above, the transition from traditional scientific and disciplinary approaches to new methods for knowledge generation requires "contextualised" research (Gibbons et al. 1994). Research and experimentation processes are becoming increasingly more integrated in the applied and the social sciences. The distinction between Mode 1 and Mode 2 research approaches can illustrate this shift, with Mode 1 approaches classified as disciplinary and Mode 2 as transdisciplinary (Andersen et al. 2009; Gibbons et al. 1994). Hessels and Van Lente (2008) have illustrated succinctly the main points of departure between traditional scientific methods and the more fluid approach of knowledge co-production (Table 8.1).

However, despite the compelling arguments for the value of Mode 2 engagements, there is still the need to build multiple practices for effective urban knowledge co-production that are responsive to different contexts. Patel et al. (2015) examine

Table 8.1 Characteristics of Mode 1 and Mode 2 approaches to research

	Mode 1	Mode 2
Problem-solving	Academic context	Context of application
Knowledge base	Disciplinary	Transdisciplinary
Extent of organisational unity/ diversity	Homogeneous	Heterogeneous
Process of knowledge production	Autonomy	Reflexivity/social accountability
Quality assurance of knowledge	Traditional quality (peer review)	Novel quality control

Source: Hessels and Van Lente (2008)

how urban knowledge co-production may challenge notions of 'best practice' in formulating responses to urban sustainability challenges in SSA cities. This work contributes to an in-depth understanding of the possibilities for identifying alternative solutions at the city level, which promote both experimentation and flexibility towards urban sustainability. However, there remains room to explore what the critical dimensions that are found in urban knowledge co-production are in resource-constrained environments where a departure from standard practice is necessary.

8.2.2   Applicability of Urban Knowledge Co-production Processes in African Cities

Urban scholars in SSA are increasingly engaged in knowledge co-production partnerships that aim to better understand the relationship between rapid urbanisation and how to manage/solve the sustainability challenges posed by population growth and increasing demand for basic services (Sect. 8.1) (Chap. 1 Vol. 1). As SSA cities grow, more emphasis is required on urban planning and management. However, growing demand does not automatically translate into greater capacity, particularly in the SSA context, where the scale and scope of the backlog in service delivery adds complexity to future and anticipatory planning and delivery (Chap. 1 Vol. 1; Chap. 4 Vol. 2).

The literature on knowledge co-production suggests that capacity to produce knowledge can be enhanced through partnerships (Sects. 8.1 and 8.2.1). However, in tandem with the growth in knowledge co-production partnerships, there is also a growing need to prove the value of these partnerships and provide evidence to support the effectiveness of transdisciplinary research processes. However, providing such evidence is not straightforward. For example, Hanson and Polk (2018) suggest that the relationship between collaboration and impacts/outcomes is not always direct due to challenges related to attribution and the time needed for policy change.

Cash et al. (2002a) identify three criteria for effective sustainability research, namely, salience (or relevance), credibility and legitimacy. According to Cash et al. (2003: 8086), *credibility involves the scientific adequacy of the technical evidence and arguments. Salience deals with the relevance of the assessment to the needs of decision-makers. Legitimacy reflects the perceptions that the production of information and technology has been respectful of stakeholders' divergent values and beliefs, unbiased in its conduct, and fair in its treatment of views and interest.* These criteria are assumed to both help stakeholders in filtering knowledge and influence the decision-making process by assigning value and significance to this knowledge (Jones 1999; Cash et al. 2002b; Lemos and Morehouse 2005).

However, it is not always true that the knowledge processes and products that are relevant, credible and legitimate will be implemented and taken up in policy (Cash et al. 2002a). However, some scholars caution that a lack of readily visible impact should not be interpreted as meaning the process and products are 'useless'. Cash et al. (2002a) instead suggest that the impact lag/gap can be better understood in terms of 'conditions, context and efforts of the involved institutions' (Godin and Dore 2005; cited in Polk 2018: 2).

Towards this end, Cash et al. (2002a) introduce the element of boundaries to explore the contextual and institutional factors related to the effective application of knowledge in decision-making. For example, in the context of SSA cities, the institutional boundaries of each stakeholder largely inform what is considered salient, credible and legitimate. In some instances, the creation of (or location of) a boundary organisation can enable stakeholders from different groups to start working collectively beyond their institutional boundaries. Such boundary organisations often "inhabit" the space between stakeholders groups, such as academia, policy-/decision-makers, urban practitioners and civil society. According to Hellström and Jacob (2003: 235), such boundary organisations "occupy the space between science, policy and business concerns" and enable effective communication across institutional limits. Such entities are crucial elements of an enabling environment for knowledge co-production in urban SSA contexts.

The efforts of involved institutions on knowledge co-production processes can be assessed by focussing on the actual commitment of the key stakeholders. Smit et al. (2015) argue that this is particularly important in SSA, as research driven by international donors is one of the leading sources of urban sustainability knowledge. In such contexts, the inputs of external consultants and technical advisors have traditionally been valued over the institutional knowledge and experience of officials in local authorities and local communities (Smit et al. 2015) (Chap. 4 Vol. 2). Thus, the commitment of key urban stakeholders to this knowledge (and the process of deriving it) cannot be underestimated and can pose a barrier to effective knowledge co-production. Ensuring stakeholder commitment to the knowledge co-production process can create a space or an agora (Gibbons et al. 1994; Cornell et al. 2013) for critical dialogue, where all relevant voices (including those often marginalised or not heard in urban management) are given a platform to share knowledge and experiences. Furthermore, the commitment of key decision-makers, in addition to local communities, academics and civil society, can also help steer donor-funded processes towards strengthening locally driven policy development and implementation, over and above the elicitation of information and data.

It is also worth noting that the conditions under which urban knowledge co-production processes occur affect the extent to which each involved stakeholder can contribute knowledge and expertise to these processes (Polk 2015a, b). For example, close interactions between stakeholders can catalyse new learnings and understandings of urban management approaches at city and project level. Furthermore, the perspectives and experiences of the stakeholders involved in co-production processes can be shared with other stakeholders to further enhance the effectiveness of these processes. Polk and Knutsson (2008) explore a concept of mutual learning which focuses on the informal exchanges that can occur in knowledge co-production and suggest that such mutual learning can contribute to the production of legitimate and credible knowledge. Such efforts may not only validate experiential knowledge but more importantly build and strengthen the knowledge of researchers tasked with articulating the full scope of urban sustainability challenges and other stakeholders that are well positioned to address it.

8.3 Methodology

8.3.1 Research Approach

This chapter illustrates some of the critical aspects associated with knowledge co-production in partnerships between universities, government and civil society in SSA. As discussed in Sect. 8.2, there is no single model of knowledge co-production. In fact, different urban contexts 'demand' different approaches towards knowledge co-production (Patel et al. 2015). In this chapter, we focus on the relationship between the quality of knowledge collaborations and its influence on outcomes. To understand this relationship, we undertake an empirical investigation of three urban knowledge co-production partnership programmes in SSA (Sect. 8.3.2). By drawing on the literature summarised in Sect. 8.2, we develop an analytical framework focusing on the criteria of salience, legitimacy and credibility (Cash et al. 2002a, 2003).

In our study, salience refers to information which is deemed important for decision-making by each actor. Credibility refers to the scientific merit and technical robustness of information. Legitimacy refers to the perceived fairness and consideration of the perspectives of all relevant stakeholders. Even though our analysis is divided into these categories, it must be noted that these categories are difficult to separate completely, as they are interrelated and mutually constituted. Our analysis is based on project documents, reports from workshops and reports to funders.

8.3.2 Description of the Studied Knowledge Co-production Partnerships

The three programmes used in this chapter to highlight urban knowledge co-production practices include (a) the CityLab Programme, (b) the Mistra Urban Futures Knowledge Transfer Programme (KTP) and (c) the State of Cities in Africa (SOCA) programme. The key features that are used to describe these programmes are duration, funding, stakeholders and aims (Table 8.2). In all these programmes, the African Centre for Cities (ACC) served as the intermediary between academia and the relevant stakeholders engaged in knowledge partnerships to address societal and sustainability challenges.

Table 8.2 Key features of the studied knowledge co-production partnerships

	CityLabs	Knowledge Transfer Programme (KTP)	State of Cities in Africa (SOCA)
Aims	Investigated urban dynamics in the Cape Town city-region through multi-partner research processes	Sought to co-produce knowledge to support sustainable urban transitions in Cape Town	Documented urban conditions on a national basis Established a benchmark against which to measure the success of urban policies
Scale	Neighbourhood to city scale	City of Cape Town	National
Duration	Phase 1: 2008–2013 Phase 2: 2013–ongoing	Phase 1: 2012–2016 Phase 2: 2016–2019	2009–2013
Approach	Organised across different research themes: - Central City (2008–2013) - Philippi (2008–2015) - Climate Change (2009–2012) - Urban Flooding (2008–2012) - Healthy Cities (2009–) - Urban Ecology (2010–2013) - Public Culture (2012–) - Sustainable Human Settlements (2012–2019) - Urban Violence (2012–2015)	Embedded researchers in local government agencies with different foci, such as energy governance, climate change adaptation, green economy, space economy, transit-oriented development and SDGs Facilitated the City Officials Exchange Programme Held a joint programme on governance systems	Created a knowledge network for urban practitioners and policy-makers across the continent to promote information dissemination and knowledge exchange and enhance opportunities for peer learning
Mode of engagement	- Seminar series - Co-authoring processes - Collaborative research processes - Public engagements	- Embedded researchers supported policy processes - City Officials Exchange Programme was used to co-author publications	- Workshops - Virtual engagements (e.g. through Skype) - Webportal (refer to www.urbanafrica.net)
Partners	- University of Cape Town - City of Cape Town - Western Cape provincial departments - Civil society	- University of Cape Town - City of Cape Town	- University of Cape Town - National government departments in partner cities - Cities Alliance
Funders	Funding for each CityLab was provided according to salience of the focus area. Funders included (a) Western Cape provincial government, (b) City of Cape Town, (c) Eskom and Vodacom, (d) Mistra	- Mistra Urban Futures - City of Cape Town	- Cities Alliance - World Bank Institute - German Society for International Cooperation (GIZ)

(continued)

Table 8.2 (continued)

	CityLabs	Knowledge Transfer Programme (KTP)	State of Cities in Africa (SOCA)
	Urban Futures, (e) German Society for International Cooperation (GIZ) and (f) South African lotteries board		
Outputs	- Special journal issue in *Ecology and Society* (Anderson and Elmqvist 2012) - Three edited books (Brown-Luthango 2012; Cirolia et al. 2016) - Various journal publications - Postgraduate degrees (many theses and dissertations were linked to the individual CityLabs)	- Various journal publications through the City Officials Exchange Programme (over 20) - Journal papers by embedded researchers and programme managers - Policy briefs - Strategy documents - Conference presentations	- State of Cities reports for the partnering countries

Note: Embedded research 'describes a mutually beneficial relationship between academics and their host organizations whether they are public, private or third sector. The relationship typically provides the researcher with greater access to the host organization with benefits for collecting data and research funding. For the host organization the relationship provides a bridge to academia and academic knowledge, networks and critical approaches to developing organizational policies and practices' (McGinity and Salokangas 2014: 3)

8.3.2.1 CityLab Programme

The CityLab Programme aimed at establishing closer connections between academia, local government and broader society. Each CityLab focused on a specific theme as a means of understanding and responding to urban sustainability challenges in Cape Town (Anderson et al. 2013) (Table 8.2). The CityLabs were set up to catalyse transdisciplinary engagement, both across academic disciplines and between academia and the broader society. By using Cape Town as a 'laboratory,' the CityLabs sought to create working partnerships across the domains of research, governance, management and lived experience, thus creating policy-relevant knowledge.

The first generation of CityLabs operated between 2008 and 2013 and included four thematic labs undertaken at the scale of the entire city of Cape Town, namely, the Urban Ecology CityLab, the Healthy Cities CityLab, the Flooding CityLab and Climate Change CityLab/Think-Tank. Two additional CityLabs, the Central and the Philippi CityLabs, were geographically bounded and limited to specific regions of the city. The Central CityLab focused on central Cape Town and explored issues related to densification. The Philippi CityLab focused on the area of Philippi and explored the multiple urban issues related to informal settlements, government spending and infrastructure.

Each CityLab encouraged the adoption of an original research approach. Whilst bounded by the broad imperative of responding to sustainability challenges relevant to the South African urban condition and to work towards knowledge dissemination, each CityLab leader was given full responsibility to run the CityLab in a configuration they saw most appropriate. This provided some degree of authority to the CityLab leader, to inform the scope and process of the lab through their own personal strengths and working knowledge of the field. This lack of a top-down, institutional approach served to validate each CityLab as an authoritative entity in its own field.

8.3.2.2 Knowledge Transfer Programme

The Knowledge Transfer Programme (KTP) is a partnership between the City of Cape Town and the ACC, under the umbrella of Mistra Urban Futures, a global network focused on knowledge co-production for urban sustainability. The KTP seeks to co-produce knowledge for supporting sustainable urban transitions in Cape Town by ensuring more defensible and legible urban policies to support sustainable development. This aim was understood and approached as both a policy/practice goal and an academic challenge. There was an implicit acknowledgement that if urban policy is to become more robust, then broader knowledge must contribute fruitfully to policy development and decision-making processes within the city. This partnership entailed the inclusion of academic methods and research as a means of generating evidence-based knowledge jointly with the practice-based knowledge typically informing policy processes in the city. The KTP was also committed towards increasing the legibility of policy processes, in ways that challenge and shape academic discourses about cities and urban transitions.

During the first phase of the KTP, four PhD researchers worked closely with city counterparts embedded in local government departments related to four policy areas: climate change adaptation and mitigation, energy governance, green economy and space economy. In the second phase, four researchers were embedded in local government departments tasked with transit-oriented development, cultural planning and the implementation of the SDGs. These embedded knowledge engagements provided opportunities for the development of alternate, robust and relevant policy responses whilst generating new insights into the internal operations of the local government.

The first phase of the programme had a reciprocal knowledge exchange where over two dozen city officials were granted a 2-month writing sabbatical, during which they were paired up with academics to co-author journal articles and an edited book on policy-relevant issues (Scott et al. 2019). These publications resulted in increasing the legibility of policy endeavours, by documenting and making accessible previously opaque policy processes, thereby situating local policy experiences and innovations in a much broader context and set of debates.

8.3.2.3 State of Cities in Africa Programme

The SOCA programme sought to catalyse urban research at a continental scale and facilitate the preparation of national 'State of Cities' reports. These reports sought to document for the first time contemporary urban conditions in baseline reports of demographic, economic and environmental patterns at the city scale across SSA. The SOCA Project supported networks of urban stakeholders in different countries, anchored by local learning institutions to engage in baseline studies of national urban systems. Subsequently, the project supported local actors to mobilise resources for the collection, analysis and monitoring of urban data in order to provide empirical evidence for urban policy-making. The selected countries included Ethiopia, Botswana, Malawi, Ghana and Tanzania.

The ACC served as a technical advisor and secretariat for the SOCA programme, providing resource mobilisation and management support. Eventually, it aspired to nurture and catalyse an emergent urban research centre of excellence, in order to take up the cities' research agenda in different national contexts. The SOCA programme was concluded in 2013, but the principles of consolidating research expertise and knowledge institutions in SSA have been adopted by the African Urban Research Initiative (AURI). AURI promotes and fosters interdisciplinary applied research through partnerships with urban research centres and think-tanks. Currently, the AURI network has 20 members in 16 SSA countries, each with solid research capacity and expert knowledge of their urban systems.

8.4 Results

8.4.1 Salience

All CityLabs achieved a high degree of salience in that they were created to either reflect a pressing sustainability challenge within Cape Town or were a direct response to interests expressed by the provincial or local government. The CityLab themes spanned geographical boundaries and issues of global (e.g. urban responses to climate change, urban ecology) and local importance. Despite the breadth of the CityLabs foci, all of them were bounded in (and aimed to tackle) urban sustainability challenges pertinent to Cape Town.

The Knowledge Transfer Programme (KTP) capitalised on the track record of the CityLab Programme on urban knowledge co-production. The KTP was launched in Cape Town in 2012 as a knowledge partnership between the City of Cape Town (CCT) and the ACC. The established relationships between the ACC and the city through the CityLab Programme made possible the identification of policy areas that could benefit from engagement with embedded researchers. Strong relationships between researchers and city officials were already established, facilitating an agreement on knowledge collaborations.

In most cases, the suggested priorities were mutually agreed, with only one major exception. In one case, the university believed that there would be strong benefits from embedding a well-respected academic with a background in architecture and urban design. However, this was not deemed a desirable match by the local government, as it had already identified spatial development frameworks and environmental management as two priority areas that required capacity.

In 2009, the ACC embarked on the State of Cities in Africa (SOCA) programme. The purpose of the SOCA programme was to respond to the need of city managers, mayors and established urban scholars, to measure the sustainability impacts of urbanisation in SSA in a tangible manner. Rather than using anecdotal information to gauge the rate of urban growth (and more precisely the growth of unplanned and unserviced settlements), these decision-makers urgently needed robust and well-organised research on urban systems across the continent. The SOCA programme sought to meet that demand, so simply designing a demand-driven programme ensured salience. In this process, the ACC operated as an intermediary, bringing together urban stakeholders in the different national contexts to articulate their information needs. It then used these as the defining principles upon which the respective State of Cities projects would be developed. This required working with central and local government agencies responsible for the delivery of urban services, civil society organisations working for the urban poor and academic institutions with built environment training programmes. The collective and upfront agreement of participating stakeholders ensured that those issues requiring the most urgent attention remained at the top of the research priority list.

8.4.2 Credibility

Each CityLab was headed by an academic researcher that directed the type of engagement and the outputs throughout the duration of the CityLab. Each leader had training and expertise pertinent to the theme of their CityLab, and each CityLab was positioned in relation to relevant academic debates. The CityLabs generally aimed at delivering academic outputs in the form of peer-reviewed special issues or book publications (see Table 8.2). This served to give each CityLab academic credibility, which was anchored to some extent in a related disciplinary culture. For example, the Urban Ecology CityLab decided early on to produce an entirely academic output, namely, a special issue in the international peer-reviewed journal *Ecology and Society* (Anderson and Elmqvist 2012). This particular CityLab, under the direction of its ecologically trained leader, sought credibility through the traditionally recognised route in the discipline of ecology, i.e. peer-reviewed journal publications. On the contrary, the past work experience of the Philippi CityLab leader in the civil society sector (community development) was reflected in the greater effort of this CityLab to establish credibility with the local communities, through meetings and delivery of community services. In this respect, whilst each CityLab strived to achieve credibility, there was always the question of 'credibility by whose measure'. This raises questions around how success is measured or monitored (Petts et al. 2008) (see Sect. 8.5).

The credibility gained through the CityLabs and previous knowledge engagements between the City of Cape Town (CCT) and UCT were key in building trust between the two institutions over a long period of time, thus creating the conditions for deeper engagement. The credibility gained through the successful track record of knowledge co-production in the CityLabs was also recognised by Mistra Urban Futures (MUF), who partnered with ACC to support the KTP. The embedded researchers provided evidence-based and theoretically grounded inputs in the policy process to increase the defensibility of the outcomes. At the same time the embedded researchers used their experiences and understandings of the local government from within, to sharpen their own research questions and methods, thus ensuring relevance through engaged scholarship. The City Officials Exchange Programme entailed local government officials working with academics in pairs as writing partners to publish a journal article on the policy-relevant work they had been involved in. Through this process of academic writing and publishing, the work of the CCT became more legible and visible, without changing the authorial voice of officials involved in the policy processes. The credibility offered by these research publications gave greater traction to policy positions in decision-making processes within the local government and provided participants with opportunities for upward mobility within local government structures.

Unlike the two previous co-production processes, the SOCA programme occupied an uncharted space where academics, government officials and civil society organisations agreed to work together in an ambitious exercise of collecting and analysing data on SSA cities. Given its limited resources, the programme did not entail primary data collection but instead relied heavily on previously published data from government agencies that had not been analysed at the city level. In addition, the SOCA programme planned to use data collected by local authorities at the city level to assess their performance in key areas such as basic service provision to city residents, municipal income and expenditure and employment. As the data was generated using existing local and national government sources (augmented through new research based on data gaps), the credibility and authenticity of the new data was not contested. In this way, the SOCA programme also sought to generate data on key urban processes in contexts that previously lacked adequate data and engage with state agencies in producing and communicating this data to audiences outside of government.

8.4.3   Legitimacy

In the CityLabs, the involvement of various stakeholders in agenda setting was critical to ensure legitimacy. However, reaching consensus on terminology and shared goals was not always easy. For example, the establishment of the Healthy Cities and Flooding CityLabs was protracted, as time was required to collectively agree on the frameworks, terminology and methods. Finding and agreeing on the actual methods was critical to ensure the legitimacy of these particular CityLabs, as method selection can often complicate trans- and interdisciplinary research (Ramadier 2004). Debates over methods sometimes slowed the research process and, in some instances, even caused the departure of certain actors. There were also evident shifts in perceptions around issues such as climate change, particularly from within local government, as was witnessed in the Climate Change Think-Tank. The ACC often played the role of 'broker' (Godfrey et al. 2010) to help systematically navigate differences between knowledge partners in joint workshops, further providing legitimacy to the process.

Later on, practical issues became critical, such as where to hold meetings. For example, this was a constant debate in the Philippi CityLab, where various participating civic groups voiced discomfort over meeting at the university. As a result, it was necessary to find new venues within the community to ensure broader stakeholder attendance and participation. Similarly, when the Urban Ecology CityLab approached a landscape planner to present a recent design informed by biodiversity, the invitation was only readily accepted (after the initial decline) when the proposed venue and field trip allowed for the presentation outside the confines of a conventional academic setting. These are only some of the examples of how CityLabs attempted to enhance legitimacy by understanding and incorporating the voices of different stakeholders.

Within the KTP, the local government counterparts highly valued aspects related to policy defensibility, credibility of academic arguments and the rigour of academic methods employed by the embedded researchers. This process was considered important in equipping local government officials with tools, evidence and reasoning to argue their case during decision-making processes. The fact that the embedded researchers were engaged in academic study gave them credibility within the CCT, compared to knowledge engagements with consultants. Whilst each of the embedded researchers had specific tasks within the local government (and were, to varying degrees, engaged in mainstream institutional work), they were primarily acting as 'researchers' during their participation in the programme (Patel et al. 2015). Thus, research was a legitimate part of their daily work, which is a luxury seldom afforded to local government officials. The spaces occupied by the embedded researchers allowed them unprecedented access to data and an intimate knowledge of local government processes. However, they also had to balance the multiple roles they simultaneously held. In this process, time proved to be one of the biggest challenges in ensuring legitimacy, as balancing these multiple roles left embedded researchers 'time-stressed'.

By being simultaneously involved in policy development and research, the embedded researchers had to navigate the very different timeframes within which the local governments and universities operate. For example, local government must often deliver interventions irrespective of the credibility of the facts at hand, as it cannot afford waiting to get facts perfectly right before acting on urgent matters. On the other hand, academic research is a slow process that requires numerous iterations and refinements to ensure validity and robustness. Thus, it was critical to partner the embedded researchers with the most appropriate local government officials to help them navigate these complexities and institutional cultural differences to achieve mutual benefit. Finding the right fit between research focus and local government priorities was not only important for salience (Sect. 8.4.1) but also critical for ensuring legitimacy. Building trust and support to sustain these partnerships over a 3-year period was largely achieved by finding the right fit between researchers and local government counterparts. This was in turn instrumental for building legitimacy and mutual respect between them.

The secondary aim of the City Officials Exchange Programme was to address questions of policy defensibility through research. Local government officials highly valued the credibility of producing a journal article or book chapter. Furthermore, engaging with academic literature and discourses on the policy topics that local government officials had been working on for extended periods was also highly valued. It was perceived that this engagement added value to the policy work of the involved local government officials whilst also validating their practice-based skill set. Furthermore, their day-to-day work experience was useful for engaging with (but also contributing to and challenging) academic literature and discourse. This added further legitimacy to the production of academic articles that was seldom realised under traditional knowledge partnership arrangements.

One of the main contributions of the SOCA programme was the overarching guiding principle that it should be demand driven. Unlike other donor-facilitated projects, this demand had to be articulated explicitly by (and include voices from) local structures involved in the development and decision-making processes. As part of the preparatory phase in each country, the local universities seeking to host this project were required to engage with government actors and representatives from civil society that worked in specific urban sectors such as housing, water or employment. This was not always easy to be translated into practice, especially in contexts where the government was particularly dominant and/or civil society was weak and fragmented. However, the inclusion of this demand-driven approach in project design meant that the research agenda had to not only consider the views of decision-makers within government but also at least try to include the perspectives of representatives of the urban poor (e.g. civil society organisations). This effort to ensure broader participation sometimes led to tensions over who had the final say in determining the scope of the project, both in terms of selecting study cities and entry points for evaluating urban trends. For example, in the case of Ghana, the timing of the national census provided a window of opportunity to collect urban data systematically, hence providing legitimacy and support for SOCA activities in that country.

8.5   Discussion

8.5.1   *Reflections on Urban Knowledge Co-production*

The three research programmes discussed in Sects. 8.3 and 8.4 highlight the diversity of knowledge co-production approaches, as well as the differing scales at which the principles of co-production can be embedded in urban research in SSA. Between them, they illustrate the varied methodologies and the expanding boundaries of knowledge co-production approaches in SSA and respond to the call for addressing knowledge and capacity gaps in cities in the region (and in articulating the concrete sustainability challenges in doing so) (Pieterse 2010: 2011).

The connecting thread across the three programmes is a recognition that no single actor can fully understand or address the diverse sustainability challenges posed by urbanisation in SSA, ranging from the rapidly changing urban form to poverty, violence and biodiversity conservation in increasingly populated urban areas (Chap. 1 Vol. 1). This created a fertile ground for developing knowledge co-production partnerships based on shared interests and not on contractual relationships. It also created an opportunity for engaging constructively with the private sector (e.g. urban practitioners), urban residents, civil society and officials from other local government agencies.

Table 8.3 summarises how the criteria of credibility, salience and legitimacy emerged in each of the three case studies on urban knowledge co-production. Whilst there are some similarities, there are also significant differences between case studies. This suggests that context matters and that there is no single model for knowledge co-production that fits all cases, especially in highly diverse regions such as SSA.

All three research programmes show that the success of knowledge co-production processes depends on various 'soft' factors, including the alignment of interests, commitment to outcome and process and the right fit between research focus and policy priorities. The model then is 'no model' (Patel et al. 2015), as the factors that influence credibility, salience and legitimacy have to be navigated on a case-by-case basis and in a broader context of openness and willingness to experiment between involved actors (e.g. local government and university in the case of the KTP).

When assessing the effectiveness of the research programmes, history and context matter. In some cases, assessing effectiveness is far from straightforward. For example, although the depth and reach of the KTP can be measured through publications, events (e.g. conference, workshops, seminar presentations), op-eds, news items, blogs, graduated PhD students and policy development and outcomes (Table 8.2), the full impact of the programme cannot be easily captured in the short term. This poses important questions, and indeed challenges, on how to think about (and assess) the impact of knowledge co-production processes. Thus, the policy and practice impacts of such processes will have to be tracked longitudinally. In this sense, given the *longue duree* of building relationships and realising impact,

Table 8.3 Comparison of research programmes across the criteria of credibility, salience and legitimacy

Programme	Credibility (technical adequacy)	Salience (relevance to decision-making)	Legitimacy (fairness)
City lab (sub-city or city scale)	Develop a memorandum of understanding (MoU) between the local government, the province and the university	Identify joint focus areas among the university and the local government authorities. Main selection criteria were based on policy needs and researcher fit	Conduct research in partnership with community-based organisations and leaders. The research reflected different perspectives including that of academic facilitators, community participants and local government representatives
Knowledge Transfer Programme (city scale)	Develop formal agreement between the local government and the university to establish the City Officials Exchange Programme. Develop academic knowledge outputs including books and journal articles	Enable bidirectional knowledge transfer to better inform and document policy options and decision-making processes	Establish the joint governance of the programme through equal participation in decision-making, co-funding and in-kind arrangements
State of Cities in Africa programme (regional scale)	Establish a coalition of willing participants, reflecting the authority of national and local government, voices of the urban poor and expertise of university researchers	Adopt an iterative process of data gathering on urban indicators agreed between different stakeholders at the city, regional and national levels	Utilise data and information from various sources in order to present a realistic picture of urban conditions

long-term funding could be critical to achieve objectives (effectiveness) and equal power sharing in producing policy and scholarly outputs (legitimacy).

It is worth noting that credibility, salience and legitimacy were achieved incrementally and on negotiated terms that took into account the perspectives, priorities and resources of the broad group of involved stakeholders. The salience of each co-production process went beyond the commitment of each stakeholder to engage in knowledge co-production. Salience was also visible in the commitment of each partner prior to the launch of the collaborative research activity. Thus, collaborating with a wider group of actors with different skills, but motivated by the same issues, reinforced the centrality of the research questions and approach. Legitimacy came from the creation of a space within each project to engage with (and even question when needed) not only the academics but also the influence and priorities of policy- and decision-makers. In other fora, these priorities might have eclipsed the voices and experience of community members or local government officials without access to the same platforms and resources.

Finally, the three case studies suggest that urban knowledge co-production processes in resource-constrained environments, such as those encountered in most SSA cities, might require more than the combination of salience, credibility and legitimacy. In such contexts, greater reflexivity, acknowledgement and commitment to the learning process are equally necessary features of urban knowledge co-production (Patel et al. 2017; Roux et al. 2017).

8.5.2 Policy Implications and Recommendations

This chapter has highlighted how different types of knowledge co-production processes can provide credible, salient and legitimate information to guide decision-making in SSA cities. Fostering such knowledge partnerships can catalyse (through different channels) progress in implementing multiple SDGs, especially SDG11 and 17. Based on the lessons learnt from these processes, we discuss below some of the major policy implications.

The experience accumulated through the different CityLabs suggests that the exposure of local government officials to the viewpoints of the different stakeholders involved in knowledge co-creation processes can indeed catalyse real change in views and practices. In order to incentivise officials to participate and remain committed, it would be important to ensure that the issues being addressed are of shared importance. Roux et al. (2017) reiterate this when they argue that what to partner about is as important as who to partner with.

It is also important to be clear about the main output of these processes as well as the timing. For example, some of the CityLabs specifically focussed on co-producing new policies,[3] which required long-term collaboration with government departments and other stakeholders (in one case, the co-production process took 5 years). These examples show that co-production processes can result in innovative policies that shift the thinking of government decision-makers. Furthermore, they also show that co-production can be useful for implementing these new policies. However, an important precondition is that the key government agencies involved need to be committed to the co-production process.

In some cases (e.g. the Urban Ecology CityLab), the process was ongoing and question driven within (and between) the original knowledge partners. In such cases where no immediate policy outcome is evident, it is important to retain the creative

---

[3]For example, the Sustainable Human Settlement CityLab, in addition to ongoing ad hoc policy support, facilitated a 4-year process to co-produce the Living Cape Framework. This new policy framework for the Western Cape provincial government sought to guide future investment in human settlements and create more functional and equitable cities and towns in the province. The new policy framework signifies an important shift in how the provincial government thinks about human settlements. There is currently an ongoing process underway to implement this new framework through 'testbeds of innovation', to pilot the proposed new approaches to intervene and learn through experimentation.

and generative relationships, as they might inform future policy development. Relationships can be maintained through seminar series, joint teaching or guest lectures and exchanges in knowledge products such as publications and policy briefs. Given that the success of partnerships has been shown to depend on long-term established relationships, these engagements serve to tether and efficiently build on knowledge and perspectives in the absence of windows for policy change. For example, a recent research project in green infrastructure mapping for the local government builds on the relationships and work established through the CityLab process.

In the KTP, the embedded researchers played different roles in respective policy processes during Phase 1.[4] Local government officials who were engaged in the writing exchanges and within academic debates were able to situate local policy innovations in a global context. Being able to draw on global debates in the literature and wider case study material provided credibility to the positions being put forward by officials to political decision-makers.

During Phase 2, embedded researchers were able to integrate academic insights and research into policies and projects. There was also some involvement in policy implementation, changing practices and establishing the implications of major new policies such as the transit-oriented development strategy and the adoption of the SDGs (Patel et al. 2017).

The outcomes of developing national State of Cities reports in partner countries were to (a) build urban knowledge and organisational capacity to respond effectively to the unique challenges and opportunities faced by SSA urban policy-makers, planners and development practitioners by creating access to integrated international best practices; (b) provide more detailed knowledge and information about their own national urban realities; and (c) develop effective means to rapidly increase their skills and capabilities in urban management.

Despite some common findings and lessons learned, it is important to note that numerous contextual factors affected each process and the stakeholder buy-in. Any similar projects following the approaches outlined in this chapter, whether in Cape Town or elsewhere in SSA, must therefore pay attention to goals, relationships and processes during knowledge co-production. A key lesson from the experiences outlined here is that there is no singular approach to partnering for knowledge co-production. Given the significance of context in shaping what is possible, policy development through knowledge partnerships cannot depend on 'best practice' but will of necessity be emergent (Patel et al. 2015).

---

[4]Interventions related to climate change adaptation included efforts to influence the institutional resourcing and functioning of this portfolio within the local government. In energy governance, the embedded researcher added capacity by aligning local policy directions with national policy imperatives. With regard to the space economy, new tools for decision-making and planning were co-developed between the researcher and his counterparts in the local government. In the green economy sector, engagement with academic debates led to an evidence-based entrenching of this policy direction.

8.6   Conclusions

This chapter has outlined the approach and lessons learned from three knowledge co-production research processes related to SSA cities. A common thread is that effective collaborative practices cannot be perfectly predetermined, as they are strongly shaped by context. The success of knowledge partnerships has been shown to be predicated on history, as well as past performance, which have in turn shown to influence both credibility and salience. Legitimacy was shown to increase in programmes that had deliberately built in opportunities that facilitated power sharing, including decisions on meeting venues, agenda setting and the forms of knowledge products emerging as joint outcomes. However, across the three programmes, the evidence of radical policy shifts is at best thin. Yet, the benefits of data and evidence generation based on rigorous processes and scientific outputs have left the partnering institutions in a stronger position to both navigate policy change and approach research with added confidence. Across the three programmes, it can therefore be concluded that knowledge co-production has resulted in building the capacity and commitment of the respective knowledge partners.

Acknowledgements   Funding from Mistra Urban Futures and the National Research Foundation, South Africa, is duly acknowledged.

References

Andersen HT, Nolmark H, Atkinson R, Muir T, Troeva V (2009) Urban knowledge arenas. Final report of cost action C20. European Cooperation in Science and Technology, COST, Bruxelles

Anderson P, Elmqvist T (2012) Urban ecological and social-ecological research in the City of Cape Town: insights emerging from an urban ecology CityLab. Ecol Soc 17(4)

Anderson PML, Brown-Luthango M, Cartwright A, Farouk I, Smit W (2013) Brokering communities of knowledge and practices: reflections on the African Centre for Cities CityLab programme. Cities 32:1–10

Breda van J, Swilling M (2019) The guiding logics and principles for designing emergent trans-disciplinary research processes: learning experiences and reflections from a transdisciplinary urbna case study in Enkanini informal settlement, South Africa. Sustain Sci 14:823–841

Brown-Luthango M (2012) Community-university engagement: the Philippi city lab in cape town and the challenge of collaboration across boundaries. High Educ 65:309–324

Brudney JL, England RE (1983) Toward a definition of the coproduction concept. Public Adm Rev 43(1):59–65

Cash D, Clark W, Alcock F, Dickson N, Eckley N, Jager J (2002a) Salience, credibility, legitimacy and boundaries: linking research, assessment and decision-making. John F. Kennedy School of Government, Harvard University, Faculty research working paper series RWP02–046, pp 1–24

Cash D et al (2002b) Salience, credibility, legitimacy and boundaries: linking research, assessment and decision making, KSG working papers series RWP02-046. SSRN: https://papers.ssrn.com/sol3/papers.cfm?abstract_id=372280

Cash D, Clark W, Alcock F, Dickson N, Eckley N, Guston D, Jager J, Mitchell R (2003) Knowledge systems for sustainable development. Proc Natl Acad Sci U S A 100(14):8086–8091

Choo CW (1996) The knowing organisation: how organisations use information to construct meaning, create knowledge and make decisions. Int J Inf Manag 16(5):329–340

Cirolia LR, Görgens T, van Donk M, Smit W, Drimie S (eds) (2016) Upgrading informal settlements in South Africa: A partnership-based approach. UCT Press, Cape Town

Coburn J (2007) Community knowledge in environmental health science: co-producing policy expertise. Environ Sci Pol 10:150–161

Cole M (2015) Is South Africa operating in a safe and just space? Using the doughnut model to explore environmental sustainability and social justice. Oxfam Research Reports. www.oxfam. org

Cornell S, Berkhout F, Tuinstra W, Tabara JB, Jager J, Chabay I, De Wit B, Langlais R, Mills D, Moll P, Otto IM, Petersen A, Pohl C, Van Kerkhoff L (2013) Opening up knowledge systems for better responses to global environmental change. Environ Sci Policy 28:60–70

Delanty G (2001) Challenging knowledge: the university in the knowledge society. The Society for Research into Higher Education & Open University Press, Buckingham

Edelenbos J, Van Buuren A, Van Schie N (2011) Co-producing knowledge: joint knowledge production between experts, bureaucrats and stakeholders in Dutch water management projects. Environ Sci Policy 14:675–684

Gibbons M, Limoges C, Nowonty H, Shwartzman S, Scott P, Trow M (1994) The new production of knowledge: the dynamics of science and research in contemporary societies. Sage, Stockholm

Godin B, Dore C (2005) Measuring the impacts of science: beyond the economic dimensions, INRS urbanisation, culture et societe, paper presented at the HIST lecture. Helsinki Institute for Science and Technology Studies, Helsinki Finland. http://www.csiic.ca/PDF/Godin_Dore_Impacts.pdf

Godfrey L, Funke N, Mbizvo C (2010) Bridging the science-policy interface: a new era for south African research and the role of knowledge brokering. S Afr J Sci 106(5–6):44–51. Retrieved from http://www.scielo.org.za/scielo.php?script=sci_arttext&pid=S0038-23532010000300013&lng=en&nrm=iso&tlng=es

Hanson S, Polk M (2018) Assessing the impact of transdisciplinary research: the usefulness of relevance, credibility, and legitimacy for understanding the link between process and impact. Res Eval 27(2):132–144

Hardoy JE, Mitlin D, Sattherthwaite D (2013) Environmental problems in an urbanising world. Finding solutions in cities in Africa, Asia and Latin America. Routledge, London

Hellström T, Jacob M (2003) Boundary organisations in science: from discourse to construction. Sci Public Policy 30(4):235–238

Hessels LK, Van Lente H (2008) Re-thinking new knowledge production: a literature review and research agenda. Res Policy 37:740–760

Independent Expert Advisory Group on a data revolution for sustainable development (2014) A world that counts: mobilising the data revolution for sustainable development. www. datarevolution.org

International Expert Panel on 'Science and the Future of Cities' (2018) Endorsed by nature sustainability

International Labour Organization (ILO) and World Health Organization (WHO) (2009). Social Protection Floor Initiative. The sixth initiative of the CEB on the global financial and economic crisis and its impact on the work of the UN system. Manual and strategic framework for joint UN country operations. Retrieved August 21, 2019, from https://www.social-protection.org/gimi/RessourcePDF.action?ressource.ressourceId=14484

Jasanoff S (2004) Ordering knowledge, ordering society. In: Jasanoff S (ed) States of knowledge: the co-production of science and the social order. Routledge, London, pp 13–45

Jones BD (1999) Bounded rationality. Annu Rev Polit Sci 2:297–321

Lemos MC, Morehouse BJ (2005) The co-production of science and policy in integrated climate assessments. Global Environmental Change-Human and Policy Dimensions 15(1):57–68

May T, Perry B (2006) Cities, knowledge and universities. Transformations in the image of the intangible. Soc Epistemol 20(3–4):259–282

May T, Perry B (2010) Social research and reflexivity. Sage, London

McGinity R, Salokangas M (2014) Introduction: 'embedded research' as an approach into academia for emerging researchers. Manag Educ 28(1):3–5

Njoh AJ (2009) Urban planning as a tool of power and social control in colonial Africa. Plan Perspect 24(3):301–317

Nonaka I (1991) The knowledge creating company. Harv Bus Rev 69:96–104

Nonaka I (1994) A dynamic theory of organizational knowledge creation. Organ Sci 5(1):14–37

Oldfield S, Patel Z (2016) Engaging geographies: questions of power and positionality in knowledge production. S Afr Geogr J 98(3):505–514

Ostrom E (1996) Crossing the great divide: co-production, synergy and development. World Dev 24:1073–1087

Parnell S, Walanege R (2014) Sub-Saharan African urbanisation and global environmental change. In: Parnell S, Pieterse E (eds) Africa's urban revolution. Zed Books, London, pp 35–59

Patel Z (2009) Environmental justice in South Africa: tools and trade-offs. Social Dynamics: A Journal of African Studies 35(1):94–110

Patel Z, Greyling S, Parnell S, Pirie G (2015) Co-producing urban knowledge: experimenting with alternatives to 'best practice' for Cape Town, South Africa. Int Dev Plan Rev 37(2). https://doi.org/10.3828/idpr.2015.15

Patel Z, Greyling S, Arfvidsson H, Moodley N, Primo N, Wright C (2017) Local responses to global agendas: the benefits of breaking with best practice in piloting the urban sustainable development goal in Cape Town. Sustain Sci 12:785–797. https://doi.org/10.1007/s11625-017-0500-y

Petts J, Owens S, Bulkeley H (2008) Crossing boundaries: interdisciplinarity in the context of urban environments. Geoforum 39(2):593–601

Pieterse E (2010) Cityness and African urban development. Urban Forum 21:205–219

Pieterse E (2011) Grasping the unknowable: coming to grips with African urbanism. Soc Dyn J Afr Stud 37(1):5–23

Pieterse E (2013) City/university interplays amidst complexity. Territorio 66:26–32

Pieterse E, Parnell S (2014) Africa's urban revolution in context. In: Parnell S, Pieterse E (eds) Africa's urban revolution. UCT Press, Cape Town, pp 1–17

Pohl C, Hirsch Hadorn G (2007) Principles for designing transdisciplinary research. Oekom, Munich

Pohl C, Rist S, Zimmerman A, Fry P, Gurung CS, Schnieder F, Speranza CI, Kiteme B, Boillat S, Serrano E, Hardorn GH, Wiesmann S (2010) Researchers' roles in knowledge co-production: experience from sustainability research in Kenya, Switzerland, Bolivia and Nepal. Sci Public Policy 37(4):267–281

Polk M (ed) (2015a) Co-producing knowledge for sustainable urban development: joining forces for change. Routledge, London

Polk M (2015b) Transdisciplinary co-production: designing and testing a transdisciplinary research framework for societal problem solving. Futures 65:110–122

Polk M, Kain J (2015) Co-producing knowledge for sustainable urban futures. In: Polk M (ed) Co-producing knowledge for sustainable urban development: joining forces for change. Routledge, Abingdon, pp 1–23

Polk M, Knutsson P (2008) Participation, value rationality and mutual learning in transdisciplinary knowledge production for sustainable development. Environ Educ Res 14(6):643–653

Ramadier T (2004) Transdisciplinarity and its challenges: the case of urban studies. Futures 36:423–439

Roux DJ, Nel JL, Cundill G et al (2017) Transdisciplinary research for systemic change: who to learn with, what to learn about and how to learn. Sustain Sci 12:711–726. https://doi.org/10.1007/s11625-017-0446-0

Scott D, Davies H, New M (eds) (2019) Mainstreaming climate change in urban development: lessons from Cape Town. UCT Press, Cape Town

Simon D, Arfvidsson H, Anand G, Bazaz A, Fenna G, Foster K, Jain G, Hansson S, Evans LM, Moodley N, Nyambuga C, Okolo M, Ombara DC, Patel Z, Perry B, Primo N, Revi A, van Niekerk B, Wharton A, Wright C (2016) Developing and testing the urban sustainable development goals, targets and indicators – a five-city study. Environ Urban 28(1):49–63

Smit W, Lawhon M, Patel Z (2015) Co-producing knowledge for whom, and to what end? reflections from the African centre for cities in cape town. In: Polk M (ed) Co-producing knowledge for sustainabile cities: joining forces for change. Routledge, Abingdon, UK, pp 47–69

South African Cities Network (2004) State of the cities report 2004. The South African Cities Network, Johannesburg

Swilling M (2014) Rethinking the science-policy interface in South Africa: experiments in knowledge co-production. S Afr J Sci 110(5/6):1–7

UN-Habitat (2008) The state of the African cities report 2008. Nairobi, UNHABITAT

Watson V (2014) Co-production and collaboration in planning – the difference. Plan Theory Pract 15:62–76

Whitaker GP (1980) Coproduction: citizen participation in service delivery. Public Adm Rev 40 (3):240–246

Part III
Synthesis

Chapter 9
Harnessing Science-Policy Interface Processes for Tackling Sustainability Challenges in Sub-Saharan Africa

Graham Paul von Maltitz

9.1 Introduction

In an ideal world, policy-makers and decision-makers should have access to easily understandable, up-to-date and unbiased scientific information to inform sound policy formulation to achieve a sustainable development. Such policies should result in the intended and desired social and environmental outcomes and should not lead to unintended consequences. As discussed throughout these two volumes, the policy considerations in the context of sustainable development in sub-Saharan Africa (SSA) tend to be complex and almost always involve some level of trade-offs.

In particular trade-offs between environmental goals and developmental needs are quite common in SSA (Chap. 1 Vol. 1). Trade-offs can also manifest in the timescale over which competing impacts are achieved. For example, environmentalists often make the case that short-term costs are justified if they allow the achievement of long-term environmental benefits, such as reduced climate change or biodiversity conservation. However, short-term social benefits with long-term negative environmental impacts may be expedient from a political perspective. Even when there are supposedly win-win solutions (often referred to as synergies), when digging deeper into potential consequences and impacts, it is likely to reveal some initially unforeseen negative outcomes.

Usually policy-makers require the best available advice against which to make complex decisions in the context of sustainable development. Considering the multifaceted nature of the key sustainability challenges and trade-offs in SSA, as discussed throughout these two volumes, the required information is typically multi- or transdisciplinary in nature (Chaps. 4, 8 Vol. 2). In most cases this would require data and analytical tools that are rather different from what a typical research student or a mono-disciplinary academic might be employing in their research.

G. P. von Maltitz (✉)
Council for Scientific and Industrial Research (CSIR), Pretoria, South Africa

© Springer Nature Singapore Pte Ltd. 2020
A. Gasparatos et al. (eds.), *Sustainability Challenges in Sub-Saharan Africa II*,
Science for Sustainable Societies, https://doi.org/10.1007/978-981-15-5358-5_9

Despite many university academics wanting to believe that their research has policy relevance, in reality much of this research (when undertaken in its original form) does not consider what is useful to policy-makers (Posner et al. 2016). The low policy relevance of much of this academic research, as well as the minimal knowledge of many academics on how to be strategic in research design, is commonly identified as the reason behind the poor linkages between research and policy (Stringer and Dougill 2013). Posner et al. (2016) identify a distinct disconnect between science and policy, which is in part due to the inability of scientists to identify relevant policy-makers and opportunities for timely engagement (Stringer and Dougill 2013) (Chap. 8 Vol. 2). Scientists and policy-makers are simply not speaking to each other on issues of mutual interest on a regular basis.

Clearly scientists need to understand how and why decision-makers use information and what types of information is most useful to them (Posner et al. 2016). However, there are also marked differences in the "language" and the "perspectives" between scientists and policy-makers, with scientific inquiry needing to challenge evidence on the one hand, whilst policy-making needing sound evidence (Stringer and Dougill 2013). Engels (2005: 8) takes this further by suggesting that many scientists find that "communicating with policymakers is a time-consuming and often frustrating activity for scientists; moreover, it usually goes unrewarded within academia." Clearly what drives much of academia currently (i.e. large number of high-impact research publications as the yardstick of academic success) may well discourage academics from engaging in what may appear to be a distraction. However, Scholes et al. (2017) suggest that this might be a misconception, as will be discussed below (Sect. 9.3.2.3).

A large proportion of the policy-relevant research is conducted outside traditional academic settings (e.g. universities), for example, in research institutes and councils funded through government contracts or international science-policy funding institutions.[1] In essence such organisations provide a bridge between the basic research that is undertaken at universities and the data/information needs of national and local government agencies (Chap. 8 Vol. 1; Chap. 8 Vol. 2). Historically institutes of this nature have had a distinctive advantage over universities in engaging and undertaking policy-relevant research. This was largely due to the mandate and structure of such organisations across themes and/or questions, rather than academic disciplines. This has allowed these research organisations to engage far more easily with multidisciplinary and interdisciplinary research to tackle sustainability challenges, compared to traditional universities that have more siloed and discipline-oriented structures. However, historically this research has been more difficult to get published and has been often presented in reports, policy briefs or institutional working paper series, which sometimes do not have the scientific rigour gained through peer review. Since these research institutes do not receive their base funding

---

[1]Although much of the policy-focused research in SSA is directed to meet national government needs, it is funded by international funds, including bi-lateral agreements with government departments, or alternately through direct funding to national and international research organisations.

through teaching, they are largely driven by funding opportunities and tend to focus on applied research which is sometimes in narrow areas where funding is available (Chap. 8 Vol. 1).

At the same time, policy-makers require a more diverse set of research outputs that go beyond the usual published, peer-reviewed journal papers that academics produce. Such outputs are not necessarily recognised for career advancement in academic institutions but are often the mainstay of national and international research institutes. However, recently many SSA universities have followed the global trends in developing interdisciplinary institutes, which has enabled them to move more easily into the integrated, multidisciplinary and policy-relevant research space (Chap. 8 Vol. 1; Chap. 8 Vol. 2). Furthermore, new scientific journals that encourage the publication of interdisciplinary and policy-relevant research are also emerging, making it easier for academics to publish such studies.

Notwithstanding the differences in perspective outlined above, forming strong and effective science-policy interfaces[2] to develop effective policy responses to the multiple sustainability challenges discussed throughout these two edited volumes is quite challenging in itself. On the one hand, solving the current multifaceted sustainability challenges facing SSA and achieving robust progress across the Sustainable Development Goals (SDGs) would require the development of policy responses that use both fundamental research and applied thinking. On the other hand, most research organisations, whether traditional universities or research institutes, are unable to achieve such outputs individually due to their inherent structures and mandates, as discussed above. Resource and capacity constraints further reduce the ability to generate appropriate knowledge and solutions to foster sustainability transitions at the local, national and regional scales (Chaps. 5, 8 Vol. 1). This is particularly evident in most SSA countries which are struggling with underfunded research systems, brain-drain and multiple other institutional and policy constraints (Chaps. 1, 8 Vol. 1). Even when substantial research data is available, scientists may well be insensitive to the political constraints hampering the uptake of their ideas. Thus translating data into a policy appropriate message and finding an appropriate forum for policy engagement remains rather challenging.

This chapter explores some of the aspects that can improve the effectiveness of science-policy interface processes to tackle sustainability challenges in SSA, especially related to scientific assessments. It uses the experience gained through the author's engagement in various recent science-policy interface processes related to sustainable development in SSA, including (a) the Africa Regional Assessment Report for the Intergovernmental Science-Policy Platform on Biodiversity and Ecosystem Services (IPBES) (Archer et al. 2018); (b) the IPBES Land Degradation and Restoration Assessment Report (LDR) (Scholes et al. 2018); (c) the

---

[2]For the purpose of this chapter, the term science-policy interface is defined as "social processes which encompass relations between scientists and other actors in the policy process, and which allow for exchanges, co-evolution, and joint construction of knowledge with the aim of enriching decision-making" (van den Hove 2007, 807).

South African National Coordinating Body for the United Nations Convention to Combat Desertification (UNCCD); (d) the World Atlas on Desertification (Cherlet et al. 2018); (e) the first National Action Program (NAP) of South Africa for the UNCCD (DEAT 2014); (f) the UNCCD Science-Policy Interface (SPI); and (g) various research projects considering the appropriateness of biofuels in SSA (e.g. see Chap. 5 Vol. 2). Further insights have also been obtained through interaction with colleagues intricately involved in the Intergovernmental Panel on Climate Change (IPCC), the Millennium Ecosystem Assessment (MA) and other national environmental assessments. Although participation in these processes does not necessarily make the author a specialist in the field of science-policy interfaces, it has allowed him to gain insights from numerous national and international initiatives. Thus much of the content of this chapter represents personal reflection on what remain complex and diverse processes.

The lens adopted in this chapter is that of sustainability as it relates to biodiversity conservation, ecosystems services and land management within the context of global environmental change. However, many of the insights provided should be relatively generally applicable to most science-policy issues related to sustainability and the SDGs more generally in SSA. Section 9.2 outlines some of the major considerations for science-policy interfaces. Section 9.3 discusses some of the key lessons learned during the author's engagement in science-policy interface processes in the past two decades. Section 9.4 identifies some of the main implications and proposals on how to improve the policy relevance of science-policy interfaces and harness their potential for catalysing progress for the SDGs.

9.2   Major Considerations for Science-Policy Interfaces

9.2.1   Ensure Scientific Relevance for Policy Options

The interface between science and policy differs between different spatial scales. For example, at the local scale, it is closer to a science-management or science decision-making interface, which informs local rules, policies or management options such as conservation areas or local municipalities (Chap. 8 Vol. 2). At the national level, the science-policy interface relates more to national legislation and policies, whilst at an international level, it may relate more to international conventions and the commitments of signature countries.

This suggests that at each spatial scale, there are changes in both the instruments available to policy-makers[3] and the scientific support for policy formulation. However, what remains common at all these levels is the need for evidence-based and

---

[3]Policy instruments are developed by the government as a means of implementing their policies and influencing the behaviour of citizens and the private sector (Howlett 1991; Bemelmans-Videc and Rist 1998).

objective information to assist policy-makers in developing appropriate policies and making sound decisions.

At the local level, for example, a standard response from many ecologists when asked why they engage in seemingly esoteric research (e.g. detailed studies for a specific species) is some variant of "if we do not understand the species, we will not know how to manage it". Conversely, conservation practitioners such as reserve managers tend to have very limited management options in a given local context.[4] Typically these detailed studies do not interrogate how the available management options should be changed. However, if the fundamental aim is to improve management, then a good understanding of the available management options should ideally underpin the study design. Conversely, when the scientific objectives have little to do with management goals (e.g. focus on evolutionary theory), it does not necessarily mean that the findings will not have management implications, even if they are opportunistic and only apparent far in the future. However, automatically assuming that research will assist management is naïve and shows a poor understanding of the management process. In the above example, achieving the sustainable management of the reserve may involve the inclusion of experts from domains that are far removed from the natural sciences such as law, law enforcement, international trade, development, sociology, tourism, economics and political science. All these disciplines are needed together with the good ecological understanding in order to find sustainable solutions for conservation areas in SSA, especially in contexts characterised by a lack of funding, land tenure disputes, poaching and high poverty incidence in surrounding communities.

At the national level, the policy options available to policy-makers are quite limited and are colloquially referred to as "sticks, carrots and sermons" (Bemelmans-Videc and Rist 1998). These instruments fall into some basic discrete categories such as legislative and regulatory, economic and fiscal, agreement-based and cooperative and information, communication and knowledge instruments (e.g. Bouwma et al. 2015) Legislation (and its related regulations) sets out what activities can and cannot take place as well as the possible types of sanctions for those that fail to adhere to these rules. This can be through simple rules and regulations such as dictating which plant species can be harvested without a permit and which must be protected. Alternatively, the rules can be more complex and, can for instance, demand that formal processes such as the implementation of environmental impact assessments (EIAs) or local development plans must be completed according to established procedures and standards before approving any substantial development to occur. Incentives and dis-incentives are typically linked to some form of financial assistance (e.g. subsidies) or taxes and other financial disincentives, respectively. National financial assistance can also be used to put in place "hard" and "soft" infrastructure to facilitate and assist on-the-ground activities. Providing research funding and ensuring that land managers have access to this research (e.g. through

---

[4]In conservation areas across SSA, reserve managers can functionally decide on fencing, water-point location, fire regimes and whether to cull either the studied species, its prey or its predators.

extension services) is another way that policy-makers can have an impact on promoting sustainability. Finally, policy options can also determine the organisational structure of the governance bodies themselves, for instance, central-izing decision-making and determining the amount of manpower dedicated to law enforcement, among other examples. The national-level policy space continually evolves and finds new and innovative mechanisms for using the available policy instruments in an attempt to achieve more effectively policy goals to enhance sustainability. This includes networking and working with the private sector using public-private partnerships or creating new market mechanisms such as emission trading schemes (Zito and Zito 2017).

At the international level, the policy options are even more limited than at the national and local levels, relying mainly on national governments to both buy into the respective policy processes and voluntarily ratify them. In principle the signatory countries to international environmental treaties are legally bound to these treaties, but in practice, for most of these treaties, there is little or no ability to sanction non-compliance. According to Chayes and Chayes (1995), conventions attempt to achieve compliance without enforcement. Although there have been substantive successes such as the wide adoption of the SDGs, the global greenhouse gas (GHG) emission targets set by the United Nations Framework Convention on Climate Change (UNFCCC), the land degradation neutrality targets of the UNCCD and the Aichi biodiversity targets set by the United Nations Convention on Biological Diversity (CBD), the actual progress achieving most of them has been far less successful (Maamoun 2018). For example, the ongoing land degradation, biodiversity loss and increasing GHG emissions illustrate the struggles to set and enforce globally binding targets (Scholes et al. 2018; Archer et al. 2018; Masson-Delmotte et al. 2018).[5] In the heart of this issue is the fact that it is voluntary for countries to be signatories to these international conventions and that specific resolutions are achieved through consensus. As a consequence, the specific goals are usually modest in that even if implemented, they would not reverse the trends that triggered the problem-solving effort (Susskind 2008).

Considering that the actual enforcement of sustainability targets and goals through international legislation is seldom available as a mechanism, these conven-tions tend to use "softer" approaches in achieving their goals (Paddock et al. 2011). One such approach is the provision of consolidated knowledge to underpin and inform the activities of the convention. This has been probably most visible within the UNFCCC and its use of the reports and science products of the Intergovernmen-tal Panel on Climate Change (IPCC) (e.g. Masson-Delmotte et al. 2018). The Millennium Ecosystem Assessment (MA) (Adeel et al. 2005) and more recently the IPBES assessments have provided similar support to the UN-CCD and CBD (e.g. Scholes et al. 2018; Archer et al. 2018). Another soft approach is to request

---

[5]The Montreal Protocol which effectively led to the banning of chlorofluorocarbons (CFCs) to prevent stratospheric ozone depletion is a rare exception. In this case there was actual enforcement to achieve a global environmental goal (i.e. ban of CFC emissions).

countries to establish national action programmes (NAP) and to undertake regular monitoring against these NAPs and other convention commitments. Such examples include the Intended Nationally Determined Contributions (INDCs) for the UNFCC and the National Biodiversity Strategies and Action Plans (NBSAPs) for the CBD. These are essentially used as mechanisms to assist countries to internalise their adherence to the convention as well as to help them better understand the global impact of the convention. The development of standardised indicators and the provision of guideline documents is another mechanism to assist signatory countries in implementing such conventions.

To be able to positively influence policy formulation, regardless of the spatial scale, scientific knowledge needs to link to the policy options available to policy-makers. However, it is not straightforward to transcend from good science to advice that will assist policy-makers in creating new policies or regulations. What local decision-makers (e.g. land managers) as well as national and international policy-makers require is data and information on the possible impacts of their decisions. This is typically provided through a monitoring and evaluation process, which in itself is often not scientific research but should be based on sound scientific under-standing, evidence and methods. Such an evidence-based and science-informed policy formulation process should be based on principles of adaptive management, where the monitoring and evaluation process, supplemented by additional contextual information, leads to policy adjustments to increase overall effectiveness (Fabricius and Cundill 2014). In such processes, scientific studies may be required to explain the behaviour of indicators during the monitoring process (especially if they respond in unexpected ways). Thus a key policy need is to design appropriate indicators and the underlying methodologies to objectively collect, analyse and communicate data for the different indicators (Chap. 8 Vol. 2). However, the scientific consensus on the best-suited indicators and methods is often elusive.

Further to appropriate indicators and methods, tracking the progress towards meeting policy objectives would also require baseline data against which to track the change. For example, tracking change against the baseline is needed at appro-priate intervals, with robust and rigorous methodologies to allow the detection of change. However, identifying the suitable criteria and indicators is often context-specific and has been found to be extremely difficult due to lack of agreement between scientists, lack of data, lack of reliable monitoring tools and methods, cost constraints and vested interests in specific methodologies (Orr 2011) (Chap. 1 Vol. 1). For example, the UNCCD and the CBD have taken alternative approaches, with the UNCCD opting for an absolute minimalistic set of indicators, whilst the CBD opting for a much larger set of indicators. In fact, for both biodiversity loss and land degradation, literally thousands of potential indicators and measures have been proposed over time (Orr 2011; Duelli and Obrist 2003). Both these approaches have advantages and drawbacks. However, almost always additional information is required before being able to develop a coherent "storyline" to aid policy-makers in policy formulation. One further complication is that indicators are seldom scal-able, and even if the same indicator can be used at all scales, the actual underlying data and methods may differ across scales.

Most of the above information echo the current situation on tracking progress for attaining the SDGs in SSA. As already discussed thoroughly in Chap. 1 Vol. 1, there is a lack of indicators and baseline data for many of the 169 individual SDG targets (Archer et al. 2018). Indeed, the sheer number of indicators, combined with current knowledge gaps and paucity of baseline data, render the process of tracking progress for the SDGs an extremely complicated and uncertain endeavour (Chap. 1 Vol. 1).

9.2.2 Adopt Appropriate Frameworks to Synthesise Policy-Relevant Data

The data requirements for policy formulation usually allude to a systems-based understanding of the specific policy domain (Van der Belt 2004). Probably the most important aspect is whether the data allows for a good understanding of the trend of change and if this constitutes "good" or "bad" change. This data is, however, ineffective for affecting policy formulation unless there is a good understanding of what factors drive the change. Furthermore, any policy change needs to consider the potential unintended consequences in other policy domains.

Such needs have catalysed the emergence and adoption of the pressure-state-response logic (Hammond et al. 1995) for synthesising and presenting policy-relevant data. In essence this process facilitates the development of a coherent "storyline" that can convey not only the change in a specific indicator (i.e. state) but also the reasons for the change (i.e. pressure) and the possible interventions (i.e. responses). This simple pressure-state-response conceptual framework has been refined into more complex frameworks such as the "drivers, pressures, state, impact and response" (DPSIR) framework (Carr et al. 2007). The MA (2005) developed the ecosystems services conceptual framework that shows the linkages between ecosystems and human well-being (Adeel et al. 2005) (Box 9.1). This link between the environment and humans has also been incorporated into modified DPSIR frameworks for the UNCCD (Orr 2011; Schwilch et al. 2011) and in a slightly different way forms the basis of the IPBES framework (Diaz et al. 2015a, b) (Box 9.1).

Such conceptual frameworks start from the basic premise that human actions cause change in the natural environment and that in turn this change affects humans. In many cases these impacts are considered to be negative, though some impacts can also be very positive. In reality, however, more complex phenomena emerge with any change, including multiple impacts, some of which are positive and some negative.[6] In any case such frameworks also assume that responses take place at many points along the arrows joining boxes (Box 9.1).

---

[6]For instance, landscape transformation for agricultural production can have many positive benefits related to food security and livelihoods (Chap. 3 Vol. 1). However, this often comes with negative impacts related to biodiversity loss, soil erosion and nutrient pollution (Chaps. 1, 6, 9 Vol. 1).

Box 9.1:   Linking the DPSIR and the MA Conceptual Frameworks
Linking the well-established DPSIR framework to the MA framework is seen as a way of better understanding the drivers of sustainable development and the linkages between human well-being and environmental goods and services. This framework was accepted in decision 22/UNCCD COP11 and has extensive similarities to the IPBES conceptual framework (Díaz et al. 2015a, b). Although the IPBES framework superficially looks very similar to the UNCCD framework, it has removed a specific "Response" block and replaced it with an "Anthropogenic Assets" block (Díaz et al. 2015a, b) (Fig. 9.1).

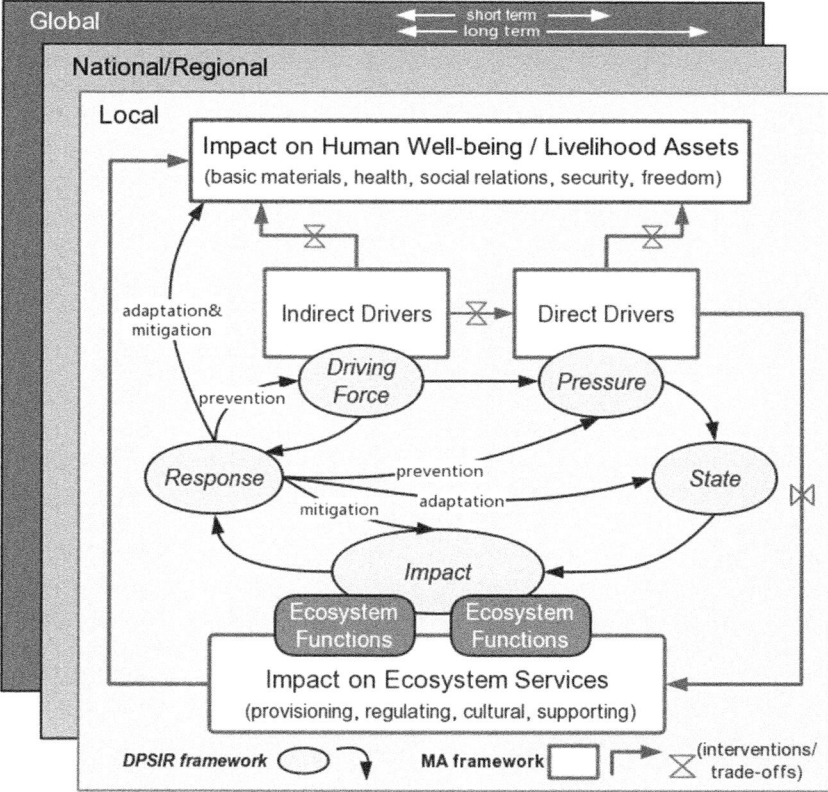

Fig. 9.1 Linkages between the DPSIR and the MA conceptual frameworks. Source: Adapted from MA (2005), FAO LADA (2009) and GEF KM:Land (2010)

Although the conceptual frameworks outlined above are quite useful for guiding the generation of policy-relevant data and storylines, they also have their fair share of problems. Firstly, even trained scientists struggle to define what issues should fit into the individual boxes of the DPSIR framework (or similar framework). For example, when asking scientists to allocate indicators into the modified UNCCD DPSIR framework, different scientists allocated the same indicators into different boxes, with some extreme instances of individual indicators being allocated to almost all boxes (Orr 2011). This is because any one of these issues may be a cause when viewed from one perspective but a consequence when viewed from a different perspective. For example, during the IPBES Land Degradation and Restoration assessment process, long cascading cause-effect chains for land degradation were found (Scholes et al. 2018).

Eventually the UNCCD undertook a concerted effort to review indicators against its enhanced DPSIR framework. What became apparent was that most global indicators tend to fall in the "State" component of the conceptual framework and that for any indicator it is necessary to derive relevant data for the other aspects of the DPSIR framework and develop a storyline around the indicator if it is to guide policy. Simply having an indicator for each of the boxes in a DPSIR framework does not guarantee that a full storyline can be developed, unless the indicators are specifically selected to "talk to each other". More importantly, understanding the storyline around any single indicator tends to be context-specific. For example, changes to a specific indicator are likely to have a different set of underlying drivers between regions. Furthermore, different indicators often require different skills and methods for their effective monitoring. "State" indicators might require methods from geospatial analysis, ecology and environmental sciences, whilst "Impacts" and "Drivers" methods from social, economic or political science. Thus it is important to adopt interdisciplinary or transdisciplinary approaches to populate such conceptual frameworks with data, interpret the different indicators and create the necessary storylines to provide sound policy advice.

Causal loop diagrams and systems modelling are an alternative to the conceptual frameworks following the DPSIR logic but also a more complex way of achieving system understanding (Sarriot et al. 2015). Such approaches, though seldom used due to their complexity, are powerful for sharing data between research and policy. In fact, the entire modelling process can be undertaken through joint teams that include social and natural scientists as well as policy-makers and other relevant stakeholders (Van der Belt 2004). Such processes can also capture the long-term impacts of slow variables[7] as well as help identify the possible unintended consequences of feedbacks that might not have been obvious through more linear and mono-disciplinary approaches.

---

[7]Variables can be considered as fast or slow depending on how rapidly they change in response to drivers of change. For instance, vegetation responds rapidly to rainfall (fast variable), but soil loss (slow variable) will have greater long-term impacts on productivity (Reynolds et al. 2007).

9.2.3 Consider and Articulate Trade-Offs

Complex trade-offs between societal and environmental outcomes are almost certain in coupled social-ecological systems (Chaps. 2, 9 Vol. 1). Depending on the context, these trade-offs can be both negative and positive (sometimes referred to as synergies). There is a clear need to both understand the nature and mechanisms of such trade-offs and identify ways to minimise any negative associated impact. Nexus-based approaches,[8] sometimes coupled with participatory policy formation processes using Decision Theatre approaches,[9] are some of the emerging relevant methodologies (White et al. 2017).

Currently, such methods are still in their infancy, both from a technical and a policy perspective. From a technical perspective, it is quite challenging to integrate model outcomes in a near real-time manner, whilst from a policy perspective, such processes require inter-departmental coordination and commitment to policy development (Simpson and Jewitt 2019). Most government departments and agencies still tend to operate in a very siloed manner, with limited incentives or procedures in place to enable interdepartmental decision-making (Chap. 8 Vol. 2). A further constraint is that although Nexus-based approaches are increasingly gaining recognition as important decision-support tools, there is limited funding for their development, partly due to their interdisciplinary nature.

The non-linear response of many environmental and socio-economic variables may result in critical thresholds, which if exceeded can result in catastrophic change (Holling 1973; Briske et al. 2017) (Chap. 7 Vol. 2). This potential non-linearity should be a critical component of trade-off analyses, but, unfortunately, we seldom know for most variables how they respond to increased pressures (i.e. nature of response curves). Furthermore, there is a need to consider the complex feedbacks from other components of the coupled socio-ecological system, making such modelling exercises both complex and highly uncertain (Van der Belt 2004).

Well-developed and operational Nexus-based tools should in theory be able to provide operational platforms that policy-makers can use to interrogate the likely outcomes from a range of plausible future scenarios. The underlying data and models should be unbiased and linked in models that can include, as best as possible, the complex interactions and feedback loops observed in these coupled social-ecological systems. Model outputs should be interpreted and ideally expressed in criteria that are important to allow policy-makers to make sound decisions over potential trade-offs. As such they should offer a range of socio-economic indicators (e.g. impacts on

---

[8]Nexus-based approaches signify research approaches that consider the interlinkages and feedbacks between sectors. The most common use of the term relates to research approaches, investigating the synergies and trade-offs of what is often phrased as the food-water-energy nexus, though it could be used for any other set of interacting issues (Howarth and Monasterolo 2016). Nexus modelling usually builds on ecological and sustainability science principles.

[9]Decision theater approaches entail the interaction of stakeholders from multiple sectors with real-time model simulations to find the best solutions to complex sustainability challenges (e.g. Boukherroub et al. 2016).

job creation, economic growth, equity) as well as environmental indicators (e.g. impacts on food production, water quality/quantity, biodiversity and ecosystem services).

Establishing and articulating the trade-offs between indicators is often problematic, as they tend to be complex, especially when it comes to trade-offs between short-term social needs and long-term environmental goals (Chapin et al. 2010) (Chap. 2 Vol. 1; Chap. 5 Vol. 2). For instance, in the IPBES Land Degradation and Restoration assessment, there was often an implicit need for trade-offs between biodiversity conservation and land conversion for intensive agricultural production to meet the food and raw material demand of the growing global population (Chaps. 1–3 Vol. 1).

An important political concern associated with trade-offs is the differentiated needs, requirements and responsibilities between developing and developed countries. For instance, although the IPBES Land Degradation and Restoration assessment acknowledged the local drivers of land degradation, it also highlighted the substantial effects of globalisation and changing consumption patterns in the developed world as indirect drivers of degradation in the developing world (Scholes et al. 2018). Similarly, many developing countries have argued that they suffer the consequences of climate change (Chaps. 1, 6 Vol. 1; Chaps. 1, 3 Vol. 2), despite not being primarily responsible for the problem, and thus unwilling to curb GHG emissions if it affects their economic growth (Ward and Mahowald 2014). Intergenerational trade-offs are also a major concern, as future generations will potentially reap the negative consequences of current decisions. However, these trade-offs are extremely difficult to be considered properly in science-policy interface processes, as future generations cannot be directly represented.

9.3   Scientific Assessments as Tools for Improving Science-Policy Interfaces

9.3.1   Characteristics of Scientific Assessments

"Scientific assessments" have gradually become a standard tool to guide international policy formulation related to environmental change. Even though the "science" of scientific assessments is still evolving, they have been identified as powerful tools to drive policy, especially global environmental policy. The origins of scientific assessments, especially for guiding international policy formulation, can be traced back to the 1985 Scientific Assessment of Stratospheric Ozone (WMO 1989). The climate change community picked up this approach, with the First Assessment Report of the IPCC completed in 1990, which is currently at its sixth iteration.

Based on personal experience from engagement in many assessment reports over the past two decades, as well as information from Scholes et al. (2017) and

IPBES (n.d.), it is possible to identify some key characteristics that separate assessment reports from academic knowledge syntheses. As discussed below these include unique aspects related to (a) commissioning, (b) scoping, (c) structure, (d) authoring and (e) reviewing.

Scientific assessments are usually formally commissioned in response to large and complex sustainability issues, which often entail complex trade-offs and highly emotive aspects that need a refined understanding. Although some early assessments including the first IPCC assessment report (Masson-Delmotte et al. 2018; IPCC n.d.) and the MA (MA 2005) were not commissioned by policy bodies, later IPCC and IPBES assessments were politically sanctioned and commissioned by politically representative bodies such as the UNFCCC, the CBD and other national actors.

The scope of the assessment is determined by a small appointed team, with the final assessment largely bound by the scoping document. This is often because the scoping document is sanctioned and agreed upon by the interested parties. Eventually, following the scoping document guidelines, assessment reports tend to break contentious issues down into relevant sub-sections and chapters. The final product is usually a long multi-chapter report, with thousands of references, and a short opening summary report usually termed the "Summary for Policy-makers" (SPM). If well-scoped, these individual chapters should capture all major issues identified by the relevant stakeholders.

Assessment reports are multi-authored by multidisciplinary teams, though the represented disciplines vary based on the focus of each chapter. Given the political nature of many assessments, there is often a requirement for authoring teams to be representative in terms of geographic regions, gender and academic disciplines, among others. Typically there are one to three coordinating lead authors and many more lead and contributing authors in each chapter.[10] Usually two to three chairs coordinate the assessment process and ensure the continuity between chapters. Authors are nominated by their national governments or mandated international organisations and tend to contribute on a voluntary basis. Funding is used under specific rules to convene meetings and to support the secretarial functions. Secretarial and administrative support staff is often the only full-time and paid personnel linked to large-scale scientific assessment.

A relatively rigid set of rules governs how scientific assessments are conducted. By nature the assessment reports are not new research per se but rather consolidate existing knowledge in novel ways that can inform policy formulation. There are also strict rules governing the content and approach of each chapter, the underlying data and conceptual frameworks, as well as how uncertainty is dealt with and

---

[10]Using the IPBES assessment reports as an example, each chapter typically consists of coordinating lead authors (CLA) who lead the team, coordinate writing activities between authors and do a substantial amount of the writing and editing. The lead authors (LA) are the main authors of each chapter and attend chapter workshops jointly with the CLAs. Sometimes there are contributing authors (CA) that provide specialist input in the form of text and/or analysis, but do not attend workshops, and are not always listed as chapter authors. In some cases interns and junior fellows may also be assigned to certain assessment reports and/or individual chapters.

communicated to readers. For recurring assessments such as the IPCC assessment reports, there is an effort to highlight the new literature and evidence since the previous assessment, with a strict final cut-off date for the included literature. However, there are substantial variations in what is considered relevant information. For example, the IPBES has allowed the inclusion of both published peer-reviewed and grey literature and indigenous knowledge, whilst the IPCC only allows the use of published literature. A very important consideration is the establishment and communication of the uncertainty around any given statement. These rules differ between assessments, with IPBES using a simple 4-box framework based on the quality of evidence and the level of scientific agreement, and the IPCC assessing the likelihood of an outcome, ranging from exceptionally unlikely (0–1% probability) to almost certain (99–100% probability).

The review of assessment reports is usually undertaken through an open review process, where any potential reviewer meeting certain requirements can contribute to the review. There are typically two iterations of the review, one after the first order draft and one after the second order draft. The responses to reviewers' comments are systematically documented and are usually made public. For the IPBES and IPCC assessment reports, there is a negotiation of the final text of the SPM with government representatives. This entails the full consideration and acceptance of the document sentence-by-sentence. Unless all policy-makers agree to a specific text, then it cannot be included. The final agreement over the text is typically reached through a consensus rather than voting and is a highly iterative process.

It is worth noting that scientific assessments can also be nationally focused. For instance, there have been various national environmental assessments in South Africa. Perhaps the two that have gained the highest attention were the assessment of African elephant population expansion in South Africa (Scholes and Smart 2008) and the assessment of proposed fracking of the Karoo shales for natural gas exploitation (Scholes et al. 2016) (see Sect. 9.3.3).

9.3.2 Lessons from International Scientific Assessments

9.3.2.1 Build an Effective Authoring Team

Scientific assessments are by no means the quickest or most time-efficient means of synthesizing scientific information to inform policy formulation. This is largely due to the need to achieve their main function of the scientific assessment (i.e. synthesis of scientific information) in a manner that is politically acceptable and takes into consideration multiple disciplines and viewpoints (Sect. 9.3.1). This is reflected in the effort of most scientific assessments to have a broad author representation in terms of region, gender and academic disciplines. This leads to complex social processes that, at times, can appear to be moving extremely slowly and ineffectively.

The initial author meetings tend to be relatively confusing, as large groups of experts (that often did not know each other beforehand) are brought together and divided into chapter teams. This is confounded by the different disciplinary and cultural backgrounds. It typically takes some time to establish the dynamics of working within these large teams, especially since many authors have never participated in such assessments and are therefore not accustomed to the tasks at hand, the functioning of the team and the broader assessment. Sound and experienced leadership is critical at this stage, especially as authors contribute voluntarily at their own cost (or that of their home institution) and cannot be coerced into providing timely input. Even though there are some potential benefits to being an author (see below), most authors typically have to fit the assessment activities into their already busy schedules. This means that any one author might not be available for the assessment activities at all times.

Most global assessments use English as the official working and document language. This gives native English speakers an automatic benefit and makes it a rather difficult endeavour for authors with poor competence in English.[11] There are also cultural differences that promote or inhibit effective engagement, especially through the different behaviour towards respect for authority, gender, age or simply norms on how to engage in group meetings. In many English-speaking cultures, there is a "no-holds-barred" approach to engagement, in that individuals feel free to engage on any issue, without any cultural inhibitions preventing them from expressing their opinion. However, it is not always easy for authors from some cultural backgrounds to openly express their opinions in group meetings, especially if there is a clear power differential.

Much of these initial author meetings is therefore spent around developing appropriate group dynamics. This obviously puts a huge pressure on the CLAs that coordinate the process, as it is critical to establish a good working relationship in just a few days before the authors return to their countries. These "bonding" efforts are also typically interrupted by multiple plenary sessions that seek to explain the overall rules of the assessment and ensure the effective coordination between chapters. In the case of the IPBES assessment reports, almost all authors were new to scientific assessment processes, which required a huge learning curve. In my experience as both a CLA and LA, by the end of the first author meeting, nobody was actually confident and ready to continue independently with the task at hand.

Author selection and allocation to chapters is a rather complicated process that requires substantial compromise to balance skills, experience and representation. On the one hand, the global experts in a specific theme do not necessarily choose to be involved in scientific assessments. On the other hand, this selection often entails criteria beyond having extensive expertise, as experts have to be nominated by their national government (or international organisation) and then be selected for the assessment. Although all authors tend to have some level of expertise in the theme

---

[11]This does not only refer to contributing to the written text but also to meaningfully interacting in group sessions. However, based on personal experience, some non-English speakers are able to interact far more effectively through written text rather than discussion.

of the specific chapter, it is highly probable that many of the leading international experts on the specific topic might not participate in a given chapter.

All of the above issues posed major challenges during the development of the IPBES Africa Regional Assessment Report. For example, despite the rich tradition in environmental research in some SSA countries such as South Africa, Kenya and Ghana, most of this expertise was on national (and usually rather narrow) issues rather than continent-wide issues.[12] At the same time, the assessment had to cover a rather wide set of environmental issues and areas ranging from tropical rainforests to deserts and from near wilderness areas to peri-urban and urban areas. Furthermore the geographical scope of the assessment was rather extensive, covering countries that are French-, English- and Portuguese-speaking, with different colonial legacies (e.g. British, German, Belgian, French, Portuguese). As a consequence author teams may not have had all the necessary skills to synthesise regional storylines in an adequate manner. This was further compounded by the fact that many of the experts having the widest experience on SSA environmental issues are not African and hence not part of the authoring teams.[13] Paradoxically some of the important debates within the chapter team often excluded authors that might have had a more extensive background in many parts of SSA or relevant experience from other SSA countries. On the positive side, the scientific assessment was first exposure to Africa-wide environmental issues for many of the participating African scientists, having thus a huge capacity-building effect.

The IPBES Land Degradation and Restoration Assessment Report, although global in nature (and thus potentially attractive to recruit globally-leading scholars), still suffered from some of the same constraints as the Africa Regional Assessment. Clearly not all of the key scholars in the field could be involved due to many of the issues outlined above. Furthermore, the relatively small chapter authoring teams did not always have the full needed set of regional and thematic expertise, making it necessary to include numerous contributing authors.

9.3.2.2 Navigate Team Dynamics

As discussed in Sect. 9.3.2.1, scientific assessments contain many large, diverse and dispersed groups. Team dynamics in such groups are complex and require both time and compromises to establish. Some chapter teams managed to achieve a good working relationship early in the assessment process (both within the IPBES Africa

---

[12]For example, each chapter contained a range of experts from across SSA. It was quite common to have radically different and fragmented expertise that was not encompassing the entire continent, e.g. an east African expert on the social aspects of environmental management in Kenya, a southern African expert on mammal ecology in Botswana, a north African expert on rangeland management in Morocco and a west African expert on mangrove ecology in the West African coast.

[13]This reflects a major element of post-colonial scientific assistance to SSA countries, in that scholars from donor countries of the global North are often engaged on global/continental-scale research, whilst their co-investigators from SSA focus only on national-level case studies.

Regional Assessment and the Land Degradation and Restoration Assessment), whilst others did not achieve this throughout the assessment process. As with many large and diverse teams, it was rather common for small sub-groups within the chapter to work well. Furthermore some individuals took on a disproportionately large amount of responsibility, whilst others had (willingly or unwillingly) limited engagement. In some cases the responsibilities between CLAs, LAs and CAs changed throughout the assessment, with many CAs making extensive contributions and in some instances being "promoted" to LAs or LAs promoted to CLAs.

Effective team dynamics and leadership are possibly two of the main factors behind chapter success. "Formal" leadership is exercised at two levels, (a) at the assessment level (by co-Chairs) and (b) at the chapter level (by CLAs). These individuals play a critical role throughout the assessment process, both by providing leadership and by ensuring continuity between (and within) chapters. They also draft the summary for policy-makers (SPM), engage with the IPBES secretariat and undertake the final negotiation and policy approval processes at the plenary meetings. Thus co-chairs and CLAs are implicitly responsible for mediating conflicts to ensure harmony within the chapter team and the overall assessment. Actually some conflicts are largely inevitable as assessments, by their nature, deal with complex issues. Often there are very divergent views on such complex issues, with an endless array of options as to how best to convey the main messages.

Fostering effective and frequent communication can help navigate team dynamics. The inclusion, in one way or another, of additional face-to-face meetings (sometimes just between CLAs) benefited greatly both the IPBES Africa Regional Assessment and the Land Degradation and Restoration Assessment. In hindsight, it would have been better to have more direct meetings within the authoring teams, as CLAs, due to their larger number of face-to-face meetings, had a far greater sense of community and engagement in the process compared to the LAs or CAs. The involvement of young scientists as "interns" within some chapters had mixed results, with some of them contributing more than many LAs and others remaining rather disengaged. This is, however, a valuable mechanism to build scientific capacity, as it was noted that many of the young scientists from the MA (in the early 2000s) were senior scientists and CLAs during the IPBES process (during the mid-2010s).

9.3.2.3 Ensure Team Commitment

The authors participating in most scientific assessments are not paid, apart from some direct costs related to meeting participation (e.g. travel, accommodation and subsistence for some participants). On the one hand, this reduces the monetary cost of the assessment to a fraction of what the true cost would have been. This creates huge value-addition and reduces the possibility for funding bodies to manipulate the process. On the other hand, it places a huge time burden on authors and especially the CLAs tasked to steer the respective chapters. For many participants the engagement on scientific assessments has to be accommodated on top of their existing work schedule, which is possibly a key reason why many skilled scientists and practitioners never get involved in scientific assessments. In a sense these activities are secondary to regular work, which naturally decreases commitment to the process for prolonged periods of time.

So the question arises, why would scientists get involved in scientific assessments? According to Scholes et al. (2017), the authors of scientific assessments are exposed to different transdisciplinary research approaches and can in theory impact policy and the society through their engagement. Furthermore, the individual chapters are often highly cited, thus advancing the careers of the authors. For younger scientists this experience is not necessarily a distraction from developing their publication portfolio, but on the contrary, the reasonable engagement in such activities can enhance their scientific profile. In fact evidence suggests that these young scientists may well outperform their peers, whilst the authors engaged in scientific assessments tend to have comparatively higher h-scores (Scholes et al. 2017). One more benefit of involvement in scientific assessments is that it allows the identification of critical gaps both in scientific research and in policy-relevant data. This can stimulate new publications that are often co-authored with other participants from the initial scientific assessment.

Furthermore, the extensive networking that takes place during the writing process offers substantial long-term career benefits to many authors. This is particularly true for CLAs who get to know each other extremely well throughout the course of the assessment, facilitating the possibility of longer-term collaboration between leading scientists. In many cases authors get to work with established experts, who they may only know by name from the academic literature. This is particularly valuable for young scientists or authors from developing countries, as it can improve their capacity to engage in collaborative research.

9.3.2.4 Ensure Quality and Timeliness of Output

As suggested above, scientific assessments are rather time-consuming and cumbersome as processes seeking to synthesise knowledge and develop a document. Working within imposed teams that are not necessarily created to offer the best mix of technical skills can be problematic and affect the quality and timely delivery of the final assessment report. Considering the size of the task at hand and despite (or possibly because of) the large number of authors, it is quite feasible that individual sub-sections are completely written by a single author that might sometimes not be well conversant in the specific topic. Most academics are generally very reluctant to tamper with the contribution of another author, as it can be perceived as an unethical practice. This usually increases the difficulty and workload of CLAs during the final editing, who again may not be experts on the parts of the text they have to edit. Even ensuring the traceability of individual text passages to their authors becomes increasingly difficult as the writing process unfolds, whilst it is often impossible for multiple authors to work on the same passages at the same time.

A rather challenging aspect is the need to sometimes discard text that has been worked on with great care, if it no longer fits the framework of the section and the overall goals of the chapter. Length limitations, duplication between chapters/sections or simply the gradually better understanding of the scope of each chapter mean that this happens more often than not. CLAs and LAs must be on the same page to identify such cases and mature enough to discard this information to fit the greater objective.

Furthermore it is not uncommon for text overlapping and trading between chapters, which requires between CLAs a good understanding of the overall assessment report structure. Such an example was the difficulty in differentiating between drivers and impacts in the IPBES Land Degradation and Restoration Assessment, which, although covered in separate chapters, are intricately linked. Other aspects such as land degradation processes tended to fall through the cracks between the drivers and impacts chapters. Even though it was recognised that their detailed coverage was not necessary as they were well documented elsewhere (and there was simply no space to maintain them in the chapters), many authors found it far easier to write about these processes rather than the state of the resource.

The technical coordinators and, where applicable, their support teams are key individuals in many scientific assessments. These individuals are primarily responsible for the success of the process and the quality and timely delivery of the output. The size of these coordinating bodies varies between assessments, with the IPBES Africa Regional Assessment having an entire technical support team, whilst in the Land Degradation and Restoration Assessment, this role was largely assumed by an individual. The technical coordination teams are not chapter authors, but they are essentially the "managers" that ensure that chapters actually progress. They often play a secretarial role and constitute the main interface between the author teams and the IPBES secretariat. Sometimes they contribute substantially to report editing (together with the co-Chairs) and ensure that the rules of the assessment are adhered to. All these functions and division of responsibilities with authors are essentially implemented to facilitate the assessment process and the delivery of the report.

9.3.2.5 Ensure Policy Relevance

The IPBES and IPCC processes are anchored on the notion that the scientific assessment outcomes should be policy-relevant rather than policy-prescriptive. In other words, the assessments should not push for specific policy options but rather highlight the advantages and disadvantages of the different options available to policy-makers in different contexts. Reminding authors of this differentiation is a recurring theme throughout the scientific assessment, especially during the development of the summaries for the individual chapters and the summary for policy-makers (SPM). Distinguishing between what is policy-prescriptive and policy-relevant is not easy for many participants, as most researchers do not consider this on a daily basis.

Furthermore, the assessment authors need to consider how best to convey the main synthesis points in each chapter (and the report as a whole) as a means of increasing the usefulness and impact of the policy messages. During the development of the SPMs, the CLAs usually consider various different ways to develop a narrative that can create the most policy-relevant storyline. Based on feedback from the co-chairs and the multidisciplinary expert panel (MEP), it is not uncommon to rehaul the first order draft of the SPM, by fully restructuring the overall layout of the key messages for the second-order draft. This restructuring sometimes means in this attempt to focus on the more policy-relevant findings, some of the issues that the authors deem important might end up receiving much less coverage in the final output.[14] Again this requires a certain degree of discipline and harmony within the authoring team in order to meet the greater goal, which is to increase the policy relevance of the final output.

The need for (and the aspects dictating) policy relevance becomes very obvious during the final approval of assessment reports. As mentioned above this approval is done by policy-makers representing their respective national governments. This is one of the unique features of many scientific assessment processes that most authors are not accustomed with. Despite being informed about the modalities of this defence process, for many CLAs and Las, the modalities of this process are rather unexpected. The reports of the four regional IPBES assessments and the Land Degradation and Restoration Assessment were all presented/defended during a 10-day plenary session in Medellin, Colombia (March 2018). Chapters were represented by the CLAs, who presented to the policy-makers each point contained within the SPMs. Even though most of the defence teams expected that country representatives would argue the science of the point, this was seldom the case. Rather, there was a lot of consideration around the exact wording of the statements. Effort was made to find the most acceptable formulations for the government delegations, to increase their ability to incorporate the underlying message into national processes. In a few instances, the original formulations of the messages were unacceptable to some delegations due to their national position on the issue or other government policies. In such cases the defence team and the national delegation(s) raising the concern would typically work jointly to find a more acceptable wording. In most cases there was not a discussion around the underlying science but simply on how the message could be best conveyed in a politically acceptable manner.

The degree to which some national delegations had interrogated the SPM was impressive. Some delegations traced some issues of concern back to the main assessment document and in some cases even back to the underlying references

---

[14]Such an example from the IPBES Land Degradation and Restoration Assessment was the need to highlight the drivers of land degradation and its impacts on humans. These impacts are essentially the main reasons why policy-makers should be concerned with land degradation in the first instance, whilst the drivers are the aspects that could be manipulated through policy change. One of the key messages that the SPM conveyed to policy-makers is the fact that land degradation is a global problem with global drivers. Many authors considered this point to be particularly important, as most chapters had already indicated that the prevailing perception of land degradation as a largely local issue is rather limiting, considering, for example, the extensive impact of globalisation and national/international policy on local land management practices.

used to derive the specific statements. However, crosschecking facts were rarely needed, as in most cases there was just a need to revise the wording in a way that was acceptable to the often conflicting requests from the national delegations. In fact, rather than an antagonistic relationship between authors and national delegations, there was a joint commitment to find mutually agreeable solutions. This process was slow and tedious, taking place during three working sessions per day and sometimes running through to the early morning hours. However, it is this policy approval and buy-in to the process that gives credibility to scientific assessment processes and their outcomes.

9.3.3   Lessons from National Scientific Assessments

South Africa has undertaken a few national assessments on contentious issues such as African elephant population expansion and shale gas fracking. These processes built on the experience acquired during international scientific assessments and especially those of the MA, IPBES and IPCC. In this sense many of the points discussed in Sect. 9.3.2 have been also relevant to these national assessments. For example, a key aspect of both aforementioned national scientific assessments was an extensive scoping to fully understand the range of the issues underpinning stake-holder concerns. Each legitimate concern was then allocated to an individual chapter, with a team of experts assembled to distil the best available scientific information relating to the issue. However, there were also some major differences.

In contrast to the international scientific assessments where there is usually a strong agreement between authors and policy-makers on the underlying issue (e.g. climate change, biodiversity loss), both of the cases mentioned above tackled issues that were exceptionally emotive in South Africa. Both elephant populations and shale gas fracking were hotly debated within the public space at that time, with different national interest groups taking strongly opposing viewpoints. For example, in the case of the elephant population, a major aspect of the debate was whether to cull the population or not. This was a strongly emotive issue with growing criticisms raised by the anti-hunting lobby but also with serious concerns raised by other stakeholder groups over the possible impacts of a growing elephant density on other ecological processes. For fracking, major concerns were raised over its environmental consequences (e.g. pollution of scarce water resources) as well as other less tangible impacts such as "sense of place". Opposing viewpoints focused on the potential development benefits of fracking, especially in an area such as the Karoo that is characterised by extreme poverty and few income opportunities.

The assessment processes had various positive impacts, such as removing the tension and emotion out of the respective debates. Opposing factions felt they had a voice in this process, which also ensured that the focus was on the best available evidence rather than the emotions.

It is worth mentioning that the outputs of national scientific assessments can be uptaken rather quickly on policy, considering the strong national interest about the topic. For example, the culling of elephants has ceased and populations have been allowed to expand naturally in South Africa, partly as a result of the assessment. Furthermore, artificial watering points have largely been discontinued. On the contrary, policy uptake can be slower and more difficult for the outcomes of international scientific assessments. For example, based on the outcomes of the IPBES Land Degradation and Restoration Assessment, the UNCCD science-policy interface recommended certain policy options to be taken up by the UNCCD. However, finding options and mechanisms for international policy interventions, even given the focused policy recommendations from the specific report, proved difficult. One reason for this is that achieving sustainable development in one domain, such as sustainable land management, may well need interventions in totally unrelated domains such as land tenure, peace and security, dietary choice, international trade, energy use policy, health or some other sector which cannot be directly addressed through one of the environmental conventions. This emphasises the need for holistic and interdisciplinary solutions for tackling the SDG objectives, as discussed in Sect. 9.4.

9.4   Policy Implications and Recommendations

As discussed throughout this chapter, a key element of science-policy interfaces is that scientists have to assist policy-makers to effectively tackle their policy objectives. As such it is important to establish clarity on what are the national or international policy needs and what policy outcomes should be achieved. This changes across scales and contexts, so it is important to reach a good understanding from both sides of the interface on what needs to be achieved.

For example, at the global scale, it was assumed that the linkages of biodiversity loss to the SDGs would be of primary importance to policy-makers during the IPBES process. This was in addition to specific policy targets of the different conventions, such as the Aichi targets for the CBD or the Land Degradation Neutrality (LDN) targets for the UNCCD. Unfortunately for the IPBES Land Degradation and Restoration Assessment process, the LDN targets and background documentation only became publicly available very late into the assessment process. This highlights the extreme importance of having a good understanding of the policy processes that the outcomes of the scientific assessments will feed into. In this sense scoping teams and policy-makers from the specific conventions and/or national bodies should work closely to tailor the assessments in a way to maximise policy uptake. The UNCCD science-policy interface and IPBES have forged stronger working relations since the release of the LDN report.

Furthermore, packaging the assessment outputs is important for achieving effective policy uptake. Considering the international uptake of the SDGs, it would be particularly useful to make strong links between the outputs of science-policy interfaces to the SDGs. For example, the links to SDGs received particular attention within the IPBES Land Degradation and Restoration Assessment and the Africa

Regional Assessment. In the former case, the authors created a table to map out the linkages between land degradation and each of the SDGs. Clearly SDGs 13 (Climate Action), 14 (Life Below Water) and 15 (Life on Land) were the principle goals that most environmental conventions supported. However, many different positive interactions were found with all the other SDGs (Scholes et al. 2017). The IPBES Africa Regional Assessment took a slightly different approach to map impacts with different policy targets, including the SDGs. In this case it was found that many of the SDGs would be positively or negatively impacted.

At the national level, there are specific issues that are more important to policy-makers. These are effectively a subset of the SDGs but with specific national concerns. For instance, in South Africa there are three underlying pillars to most policy options, namely, economic development, employment creation and equity (especially racially-based inequalities), which typically trump most other policy objectives. Such policy objectives are clearly well-aligned with the SDGs but have a slightly different language and focus. Policy options and suggestions that do not engage well (or worse conflict) with these national priorities are therefore unlikely to gain much traction with policy-makers, even if there is strong evidence of environmental consequences. Thus ensuring the strong linkage of the outputs of science-policy interfaces with such specific national priorities would be particularly important if the outputs of these processes are to be seriously uptaken.

Allied to the above is the need to ensure the credibility of science-policy interfaces and especially that scientific evidence is not manipulated to fuel vested interests (whether justified or not). In a sense, as policy often has a political undertone, it is important to ensure that this will not overrule the recommendations of the scientists and the overall objectivity of the science-policy interface. For example, in South Africa, many politicians view the redress of historical racially based practices as a primary short-term target. From a political perspective, they would look to the outputs of science-policy interface to enhance this objective. Understanding both the covert and overt objectives of policy-makers is therefore important. Similarly, in an era of false news and disinformation, scientists should put objective facts on the table even if they are not politically viable or acceptable. In South Africa this is again well illustrated in the current debates around land ownership and how to redress the historical land distribution problems. Yet, the underlying facts about the successes and failures of the land redistribution process are often at odds with political rhetoric. Long-term sustainability will likely be achieved when we find solutions that take both the political aspirations and the scientific understandings of the problem into account.

Finally, it is important to understand that by virtue of their structure science-policy interface, processes form partnerships and a common space to tackle sustainability challenges. As such they are ideal mechanisms to not only contribute to the development of solutions for the specific SDGs they are thematically linked to (see above) but directly contribute to progress for SDG 17 (Partnerships for the Goals).

9.5   Conclusions

Solving sustainability challenges in SSA requires the closer collaboration between scientists and policy-makers. This is required at all levels of policy formation, from the global to the local. A multitude of tools are maturing to facilitate such processes, alongside the evolution of traditional academic pursuits towards more integrated and multidisciplinary research that is vital to tackle sustainability challenges. Scientific assessments are gradually becoming the tool of choice to consolidate policy-relevant scientific knowledge that is linked to well-identified sustainability challenges. Scientific assessments are also becoming an effective means of identifying further research gaps in policy-relevant knowledge.

The science of the scientific assessment is evolving with new and novel approaches being piloted over time. Recently the IPBES process drew heavily on the experience accumulated through the IPCC processes but also introduced its own changes and improvements. Increasing dialogue and developing inclusive forums for science-policy engagement remains critical, and it is important that early career scientists as well as established scientists are involved. At the national level, forums for dialogue between scientists and policy-makers are probably one of the best mechanisms to facilitate science-policy engagement, but these seldom allow the space for true debate around contentious issues. Siloed approaches to tackling sustainable development challenges are common at the national level due to the structure of government. Finding approaches to work across government departments and scientific disciplines to national sustainability challenges remains difficult. Nexus-based tools look promising for helping policy-makers identify likely long-term impacts in complex social-ecological systems and facilitate inter-departmental coordinated policy development that can best deal with the complex trade-offs, achieving long-term sustainable development.

References

Adeel Z, Safriel U, Niemeijer D, White R (2005) Millennium Ecosystem Assessment. Ecosystems and human well-being: desertification synthesis. Island Press, Washington, DC

Archer E, Dziba L, Mulongoy K-J, Maoela MA, Walters MA, Biggs R, Cornier-Salem M-C, DeClerck F, Diaw C, Dunham AE, Failler P, Gordon C, Harhash K, Kasisi R, Kizito F, Nyingi W, Oguge N, Osman-Elasha B, Tito de Morais L, Assogbajo A, Egoh B, Halmy MW, Heubach K, Mensah A, Pereira L, Sitas N (2018) The regional assessment report on biodiversity and ecosystem services for Africa: summary for policymakers. Secretariat of the Intergovernmental Science-Policy Platform on Biodiversity and Ecosystem Services, Bonn. ISBN 978-3-947851-00-3

Bemelmans-Videc ML, Rist RC (1998) Carrots, sticks & sermons : policy instruments and their evaluation. Routledge, New Brunswick, 280p

Boukherroub T, Amours SD, Rönnqvist M (2016) Decision theaters: a creative approach for participatory planning in the forest sector. Information Systems Logistics and Supply Chain conference (ILS 2016). Bordeaux. http://ils2016conference.com/wp-content/uploads/2015/03/ILS2016_TD02_2.pdf. 16 January 2020

Bouwma IM, Gerritsen AL, Kamphorst DA, Kistenkas FH (2015) Policy instruments and modes of governance in environmental policies of the European Union; Past, present and future, WOt-technical report 60. Statutory Research Tasks Unit for Nature & the Environment (WOT Natuur & Milieu), Wageningen, 42 p

Briske DD, Illius AW, Anderies JM (2017) Nonequilibrium ecology and resilience theory. In: Briske D (ed) Rangeland systems. Springer series on environmental management. Springer, Cham

Carr ER, Wingard PM, Yorty SC, Thompson MC, Jensen NK, Roberson J (2007) Applying DPSIR to sustainable development. Int J Sustain Dev World Ecol 14:543–555

Chapin FS III, Carpenter SR, Kofinas GP, Folke C, Abel N, Clark WC, Olsson P, Stafford Smith DM, Walker B, Young OR, Berkes F, Biggs R, Grove JM, Naylor RL, Pinkerton E, Stephen W, Swanson FJ (2010) Ecosystem stewardship: sustainability strategies for a rapidly changing planet. Trends Ecol Evol 25:241–249

Chayes A, Chayes AH (1995) The new sovereignty: compliance with international regulatory agreements. Harvard University Press, Cambridge, MA/London, pp xii, 404

Cherlet M, Hutchinson C, Reynolds J, Hill J, Sommer S, von Maltitz G (eds) (2018) World atlas of desertification. Publication Office of the European Union, Luxembourg

DEAT (2014) National action programme combating land degradation to alleviate rural poverty. November 2004. Department of Environmental Affairs and Tourism. Republic of South Africa. Pretoria. South Africa

Díaz S et al (2015a) The IPBES conceptual framework – connecting nature and people, Current Opinion in Environmental Sustainability. Elsevier, pp 1–16. https://doi.org/10.1016/j.cosust.2014.11.002

Díaz S, Demissew S, Joly C, Lonsdale WM, Larigauderie A (2015b) A Rosetta Stone for nature's benefits to people. PLoS Biol 13:e1002040. https://doi.org/10.1371/journal.pbio.1002040

Duelli P, Obrist K (2003) Biodiversity indicators: the choice of values and measures. Agric Ecosyst Environ 98:87–98

Engels A (2005) The science-policy interface. Integr Assess J 5:7–26

Fabricius C, Cundill G (2014) Learning in adaptive management: insights from published practice, ecology and society. Resilience Alliance 19(1). https://doi.org/10.5751/ES-06263-190129

FAO-LADA (2009) Field manual for local level land degradation assessment in drylands. LADA-L part 1: methodological approach, planning and analysis. FAO, Rome, 76 pp

GEF KM:Land (2010) Project indicator profiles for the GEF Land Degradation Focal Area. Final report by the GEF MSP: ensuring impacts from SLM – Development of a Global Indicator System (KM:Land Initiative). Hamilton Ontario: UNU-INWEH, 67 pp. http://www.inweh.unu.edu/drylands/docs/KM-Land/KM-FAOLand_Indicator_Profiles_Final.pdf

Hammond A, Adriaanse A, Rodenburg E, Bryant D, Woodward R (1995) Environmental indicators: a systematic approach to measuring and reporting on environmental policy performance in the context of sustainable development. World Resources Institute, Washington, DC

Holling CS (1973) Resilience and stability of ecological systems. Annu Rev Ecol Syst 4:1–23

Howarth C, Monasterolo I (2016) Understanding barriers to decision making in the UK energy-food-water nexus: the added value of interdisciplinary approaches, Environmental Science and Policy. Elsevier Ltd 61:53–60. https://doi.org/10.1016/j.envsci.2016.03.014

Howlett M (1991) Policy instruments, policy styles, and policy implementation – national approaches to theories of instrument choice. Policy Stud J 19(2):1–21. https://doi.org/10.1111/j.1541-0072.1991.tb01878.x

IPBES (n.d.) Assessment guide. https://www.ipbes.net/document-library-categories/assessment-guide. Accessed 5 July 2019

IPCC (n.d.) History of the IPCC. https://www.ipcc.ch/about/history/. Accessed 5 July 2019

MA (2005) Ecosystems and human Well-being: scenarios, volume 2. Millennium ecosystem assessment. Island Press, Washington, DC

Maamoun N (2018) International environmental. Empirical Evidence of a Hidden Success, SSRN Electronic Journal. Elsevier BV, Agreements. https://doi.org/10.2139/ssrn.2990801

Masson-Delmotte V, Zhai P, Pörtner H.-O, Roberts D, Skea J, Shukla PR, Pirani A, Moufouma-Okia W, Péan C, Pidcock R, Connors S, Matthews JBR, Chen Y, Zhou X, Gomis MI, Lonnoy E, Maycock T, Tignor M, Waterfield T (eds.) (2018) Summary for policymakers. In: Global Warming of 1.5°C. An IPCC Special Report on the impacts of global warming of 1.5°C above pre-industrial levels and related global greenhouse gas emission pathways, in the context of strengthening the global response to the threat of climate change, sustainable development, and efforts to eradicate poverty. World Meteorological Organization, Geneva, 32 pp

Orr BJ (2011) Scientific review of the UNCCD provisionally accepted set of impact indicators to measure the implementation of strategic objectives 1, 2 and 3. White Paper – Version 1, 04 February 2011. Consultancy report for the CST of the UNCCD. 145 pp. (Same as ICCD/CST(S-2)/INF.1, but includes all Annexes). http://www.unccd.int/science/docs/Microsoft%20Word%20-%20White%20paper_Scientific%20review%20set%20of%20indicators_Ver1_31011%E2%80%A6.pdf

Paddock L, Qun D. Kotzé L, Markell D, Markowitz K, Zaelke D (eds) (2011) Compliance and enforcement in environmental law: toward a more effective implementation. IUCN Academy of Environmental Law. Edward Elgar, Cheltenham/Northampton

Posner SM, McKenzie E, Ricketts TH (2016) Policy impacts of ecosystem services knowledge. Proc Natl Acad Sci USA 113:1760–1765. https://doi.org/10.1073/pnas.1502452113

Reynolds JF, Smith DMS, Lambin EF, Turner BL, Mortimore M, Batterbury SPJ, Downing TE, Dowlatabadi H, Fernandez RJ, Herrick JE, Huber-Sannwald E, Jiang H, Leemans R, Lynam T, Maestre FT, Ayarza M, Walker B (2007) Global desertification: building a science for dryland development. Science 316(80):847–851. https://doi.org/10.1126/science.1131634

Sarriot E, Morrow M, Langston A, Weiss J, Landegger J, Tsuma L (2015) A causal loop analysis of the sustainability of integrated community case management in Rwanda. Soc Sci Med 131:147–155. https://doi.org/10.1016/J.SOCSCIMED.2015.03.014

Scholes RJ, Smart K (eds) (2008) Elephant management: a scientific assessment of South Africa. Wits University Press, Johannesburg. ISBN: 9781868144792

Scholes RJ, Lochner P, Schreiner G, Van der Walt S, de Jager M (eds) (2016) Shale gas development in the Central Karoo: a scientific assessment of the opportunities and risks. CSIR, Stellenbosch. ISBN: 978-0-7988-5631-7. https://doi.org/10.4102/abc.v47i2.2144

Scholes RJ, Schreiner GO, Snyman-Van der Walt L (2017) Scientific assessments: matching the process to the problem. Bothalia 47:9

Scholes R, Montanarella L, Brainich A, Barger N, Brink B, Cantele M, Erasmus B, Fisher J, Gardner T, Holland TG, Kohler F, Kotiaho JS, Von Maltitz G, Nangendo G, Pandit R, Parrotta J, Potts MD, Prince S, Sankaran M, Willemen L (2018) IPBES (2018): summary for policymakers of the assessment report on land degradation and restoration of the Intergovernmental SciencePolicy Platform on Biodiversity and Ecosystem Services. IPBES Secretariat, Bonn, 44 pages

Schwilch G, Bestelmeyer B, Bunning S, Critchley W, Herrick J, Kellner K, Liniger HP, Nachtergaele F, Ritsema CJ, Schuster B, Tabo R, Van Lynden G, Winslow M (2011) Experiences in monitoring and assessment of sustainable land management. Land Degrad Dev 22 (2):214–225. https://doi.org/10.1002/ldr.1040

Simpson GB, Jewitt GPW (2019) The development of the water-energy-food nexus as a framework for achieving resource security: a review. Front Environ Sci 7. https://doi.org/10.3389/fenvs.2019.00008

Stringer LC, Dougill AJ (2013) Channelling science into policy: enabling best practices from research on land degradation and sustainable land management in dryland Africa. J Environ Manag 114:328–335. https://doi.org/10.1016/j.jenvman.2012.10.025

Susskind L (2008) Strengthening the global environmental treaty systems. Issues Sci Technol 60:61–69

Van den Hove S (2007) A rationale for science–policy interfaces. Futures 39(7):807–826. https://doi.org/10.1016/j.futures.2006.12.004

Van der Belt M (2004) Mediated modelling. In: A systems dynamic approach to environmental consensus building. Island Press, Washington, DC. ISBN 9781559639613

Ward DS, Mahowald NM (2014) Contributions of developed and developing countries to global climate forcing and surface temperature change. Environ Res Lett 9:074008

White D, Jones J, Maciejewski R, Aggarwal R, Mascaro G, White DD, Jones JL, Maciejewski R, Aggarwal R, Mascaro G (2017) Stakeholder analysis for the food-energy-water nexus in Phoenix, Arizona: implications for nexus governance. Sustainability 9:2204. https://doi.org/10.3390/su9122204

WMO (1989) Scientific assessment of stratospheric ozone. Global ozone research and monitoring project – report no. 20. World Meteorological Organization

Zito AR, Zito AR (2017) New policy instruments. In: Oxford research Encyclopedia of politics. Oxford University Press, Oxford. https://doi.org/10.1093/acrefore/9780190228637.013.101

Chapter 10
Sustainability Challenges in Sub-Saharan Africa: Trade-Offs, Opportunities and Priority Areas for Sustainability Science

Alexandros Gasparatos, Abubakari Ahmed, Merle Naidoo, Alice Karanja, Osamu Saito, Kensuke Fukushi, and Kazuhiko Takeuchi

10.1 Linking Sustainability Challenges to the Sustainable Development Goals

The chapters contained in these two edited volumes have discussed some of the main sustainability challenges of sub-Saharan Africa (SSA). Collectively, these different sustainability challenges, and as an extent the content of the individual chapters, have touched on issues spanning the entire breadth of the Sustainable Development Goals (SDGs) (Juju et al. 2020).

Table 10.1 summarises the main sustainability challenges covered in each individual chapter and cross-maps these challenges and the underlying content across the

A. Gasparatos (✉) · A. Karanja
Institute for Future Initiatives (IFI), The University of Tokyo, Tokyo, Japan
e-mail: gasparatos@ifi.u-tokyo.ac.jp

A. Ahmed
Department of Planning, University for Development Studies, Wa, Ghana

M. Naidoo
Graduate Programme in Sustainability Science – Global Leadership Initiative (GPSS-GLI),
The University of Tokyo, Tokyo, Japan

O. Saito · K. Takeuchi
Institute for Global Environmental Studies (IGES), Hayama, Japan

Institute for Future Initiatives (IFI), The University of Tokyo, Tokyo, Japan
e-mail: o-saito@iges.or.jp; takeuchi@ifi.u-tokyo.ac.jp

K. Fukushi
Institute for Future Initiatives (IFI), The University of Tokyo, Tokyo, Japan

Institute for the Advanced Study of Sustainability (UNU-IAS), United Nations University,
Tokyo, Japan
e-mail: fukushi@ifi.u-tokyo.ac.jp

© Springer Nature Singapore Pte Ltd. 2020
A. Gasparatos et al. (eds.), *Sustainability Challenges in Sub-Saharan Africa II*,
Science for Sustainable Societies, https://doi.org/10.1007/978-981-15-5358-5_10

Table 10.1 Chapter themes and sustainability challenges across the Sustainable Development Goals (SDGs)

Chapter	Theme	SDG 1	SDG 2	SDG 3	SDG 4	SDG 5	SDG 6	SDG 7	SDG 8	SDG 9	SDG 10	SDG 11	SDG 12	SDG 13	SDG 14	SDG 15	SDG 16	SDG17
Vol. 1																		
1	Current status and major challenges for meeting the SDGs across sub-Saharan Africa (Juju et al. 2020)	✓	✓	✓	✓	✓	✓	✓	✓	✓	✓	✓	✓	✓	✓	✓	✓	✓
2	Current bioenergy pathways and priority areas for facilitating large-scale bioenergy transitions in sub-Saharan Africa (Johnson et al. 2020)	✓	✓	✓	-	-	-	✓	✓	-	-	✓	✓	✓	-	✓	-	-
3	History and drivers of industrial crop production in sub-Saharan Africa, and intersections with food security (Jarzebski et al. 2020)	✓	✓	-	-	✓	✓	✓	✓	✓	-	-	✓	-	-	✓	-	-
4	Patterns, drivers and outcomes of large-scale land acquisitions in sub-Saharan Africa, and impact mitigation potential of corporate social responsibility strategies (Antonelli et al. 2020)	✓	✓	-	-	-	-	✓	✓	-	-	-	✓	-	-	✓	-	-
5	Patterns of academic research and development assistance funding across SDGs, and key challenges/opportunities to attract and utilise funding effectively (Lopes et al. 2020)	-	-	✓	✓	✓	-	✓	✓	✓	-	-	-	-	-	-	✓	✓
6	Perceptions of community resilience to droughts and floods in semiarid areas of Northern Ghana (Boafo et al. 2020)	✓	✓	-	-	-	✓	-	-	✓	-	✓	-	✓	-	-	-	-
7	Influence of rural livelihoods on fuelwood procurement practices and mangrove degradation in coastal Guinea (Balde et al. 2020)	✓	✓	-	-	-	-	✓	-	✓	-	-	-	-	-	✓	-	-

	C1	C2	C3	C4	C5	C6	C7	C8	C9	C10	C11	C12	C13
8	Research collaboration between academic institutions and the private sector in Ghana, and barriers/solutions to improve partnerships (Mensah and Gordon 2020)	–	–	–	–	✓	–	–	–	✓	–	–	✓
9	Natural and anthropogenic drivers of past, present and future vegetation changes in forests and savannas of Central Africa (Aleman and Fayolle 2020)	✓	✓	–	✓	✓	–	✓	✓	–	✓	✓	–
10	Traditional knowledge in forest-agriculture landscapes of Cameroon, and their role in local livelihoods and biodiversity conservation (Mala et al. 2020)	✓	✓	–	✓	–	–	–	–	–	–	✓	–
Vol. 2													
1	Child malnutrition in the context of agricultural production, food security and nutrition in rural Rwanda (Sekiyama et al. 2020)	✓	✓	–	–	–	–	✓	–	–	–	–	–
2	Exposure of urban households to extreme weather events in Uganda, and its impact on household consumption (Akampumuza et al. 2020)	✓	✓	✓	–	–	–	–	✓	–	–	–	–
3	Traditional knowledge, practices and innovations for agricultural productivity in the context of climate change in rural Kenya (Ndalilo et al. 2020)	✓	✓	–	–	–	✓	–	–	–	✓	–	–
4	Cyclical failure of sanitation solutions in eastern Africa, and innovative approaches for sustaining sanitation service operation and maintenance (Gabrielsson et al. 2020)	–	–	✓	✓	✓	–	✓	–	–	–	–	–
5	Critical aspects of the production and adoption of ethanol as a clean	–	✓	–	–	–	✓	–	–	–	✓	✓	–

(continued)

Table 10.1 (continued)

Chapter	Theme	SDG 1	SDG 2	SDG 3	SDG 4	SDG 5	SDG 6	SDG 7	SDG 8	SDG 9	SDG 10	SDG 11	SDG 12	SDG 13	SDG 14	SDG 15	SDG 16	SDG 17
	cooking option in Malawi and Mozambique (Nyambane et al. 2020)		✓															
6	Landscape connectivity and eco-system services from an alien plant species in a semiarid area of southern Madagascar (Andriamparany et al. 2020)	✓		-	-	-	-	-	-	-	-	-	-	✓	-	✓	-	-
7	Societal impacts of mine decline and closure, and future options for mine-dependent communities in Zambia (Mfune et al. 2020)	✓	-	✓	-	-	-	-	✓	✓	-	-	✓	-	-	-	-	-
8	Knowledge co-production approaches for engaging with urban sustainability challenges in African cities (Patel et al. 2020)	-	-	-	-	-	-	-	-	-	-	✓	-	-	-	-	-	✓
9	Critical considerations for science-policy interfaces anchored on scientific assessments to tackle sustainability challenges in sub-Saharan Africa (von Maltiz 2020)	-	-	-	-	-	-	-	-	-	-	-	-	✓	-	✓	-	✓

Note: More detailed information about the links of each chapter to the specific SDGs is included in the introductory section and the dedicated policy implications/recommendations sub-section of each chapter relevant SDGs.[1] Table 10.1 aptly illustrates that individual sustainability challenges tend to span multiple SDGs. This suggests that sustainability challenges are rather multidimensional but also that by solving such challenges it is possible to achieve progress on multiple SDGs.

On the one hand, this renders most of the sustainability challenges covered in these two volumes as very difficult to be solved. Indeed, many scholars have suggested that such sustainability challenges are essentially "wicked" problems with a high degree of complexity, uncertainty and conflict and little consensus on the problem or the solution (Weber and Khademian 2008). This is especially true in the developing contexts of SSA characterised by low capacity and resource avail-ability to design and implement appropriate solutions for these challenges (Juju et al. 2020; Lopes et al. 2020). On the other hand, this also implies that by solving such multidimensional sustainability challenges, it is possible to harness the interlinkages between SDGs and achieve extensive progress in multiple sustainability domains. This creates important opportunities in the sense that well-designed solutions and interventions can have multiplier effects, therefore increasing their cost-effectiveness in contexts characterised by low resource availability (Lopes et al. 2020).

When looking more critically at the different chapters, it is possible to identify three common underpinning themes, namely, (a) the emergent trade-offs between energy, agriculture, environment and the economy (Sect. 10.2.1); (b) the low resilience and adaptive capacity to environmental and socioeconomic change (Sect. 10.2.2); and (c) the constraints and opportunities for designing and implementing solutions to multidimensional sustainability challenges (Sect. 10.2.3). Even though most chapters traverse through multiple of these underlying themes, to avoid confusion, we discuss below the main findings of each chapter through the lens of a single underpinning theme.

10.2 Underlying Chapter Themes

10.2.1 Emergent Trade-Offs Between Energy, Agriculture, Environment and the Economy

Sustainability challenges in SSA often entail trade-offs that cannot be easily delin-eated (Juju et al. 2020). Some of the most visible trade-offs discussed in these two volumes are between agricultural production, energy demand and use, environmen-tal conservation and human economic systems and livelihoods. For example, numer-ous chapters have pointed out that many of the current agricultural production and energy demand/use practices are inadvertently shaping landscapes and intersecting with environmental change throughout the continent. Indeed, some of these

---

[1]More detailed information about the links of each chapter to specific SDGs is included in the introductory and dedicated policy implications/recommendations sub-sections of each chapter.

prevailing agriculture and energy practices are major drivers of land use change, landscape transformation and ecosystem degradation (Aleman and Fayolle 2020; Balde et al. 2020; Nyambane et al. 2020), having multiple socioeconomic impacts (Jarzebski et al. 2020; Antonelli et al. 2020). However, such practices, though often unsustainable as they collide with environmental conservation and climate change adaptation/mitigation, cater for real policy concerns such as rural development, national economic growth and energy security (Jarzebski et al. 2020; Nyambane et al. 2020; Johnson et al. 2020; Juju et al. 2020).

Aleman and Fayolle (2020) suggest that human activity is one of the major drivers of the large-scale transformation and degradation of tropical forests and savannas in Central Africa, compounding the changes associated with climate change and other environmental factors. Tropical forests are degraded through logging, fuelwood harvesting and agricultural expansion, while savannas are specifically targeted for reforestation and the production of industrial crops, as they are perceived to have lower conservation value than forests. Some of these trends could become more pronounced under specific future climate scenarios, raising concerns about the long-term sustainability of these biomes. Thus, stronger efforts should seek to reverse such trends, for example, by targeting degraded areas for ecosystem restoration, expanding protected areas and promoting sustainable forest management.

Balde et al. (2020) identified how different agricultural production and energy procurement practices can cause mangrove degradation in coastal Guinea. In particular, rice agriculture in mangrove areas and fuelwood harvesting in upland and mangrove forests (for household use and livelihood activities such as salt-making) are two major drivers of landscape degradation. However, these activities are hugely important for local livelihoods and households' food and energy security. There is a real need to reduce such trade-offs, possibly through the adoption of better farming and energy utilisation practices and salt-making technologies.

Antonelli et al. (2020) highlight how biofuel demand in the European Union (and associated policies) has driven the surge in large-scale land acquisitions (LSLAs) in SSA for biofuel feedstock production (mainly *Jatropha*). However, they point out that, on many occasions, these LSLAs were unregulated, having multiple negative socioeconomic impacts to local communities that were often mediated through land-grabbing. This implies trade-offs related with energy security in the EU and national economic growth in SSA countries, with local food security, landscape transformation and loss of local livelihoods. Most of the EU investigated investors involved in such LSLAs did not adopt sustainability standards, and when they did the standards did not have provisions for land-related issues. This suggests both the need to expand the main certification standards for agroindustrial development to include provisions related to land and incentivise (or even require) investors to adopt them.

Jarzebski et al. (2020) describe the main characteristics of industrial crop production systems in SSA, the underlying drivers of their recent expansion and their main trade-offs with food security. They suggest that current industrial crop production practices can give rise to many trade-offs, ranging from trade-offs at the crop level (e.g. crops used for food vs. energy vs. other industrial uses) to trade-offs at the scale of production (e.g. large-scale vs. small-scale production systems), the policy goal (e.g. economic growth vs. energy security vs. rural development) or even the levels of the food security impact (e.g. multiple trade-offs and synergies between the different pillars of food security). They point to the need to generate a robust knowledge base on such trade-offs and synergies, as a means of harnessing the potential of these crops in SSA without compromising food security.

Nyambane et al. (2020) identify the different trade-offs inherent to the production and use of ethanol fuel in southern Africa. They identify how sugarcane has been a major driver of land use change at the production side (Dwangwa, Malawi), through the conversion of agricultural and forest land to a large sugarcane monoculture. However, while this large-scale land conversion reduced the available cropland and possibly affected the delivery of forest-related ecosystem services, it simultaneously provides biofuel feedstock that enhances national energy security and increases carbon storage capacity. At the demand side (Maputo, Mozambique), the study highlights the multiple considerations that consumers make when adopting ethanol fuel for cooking, such as costs, convenience, safety and market accessibility and stability, among others. These considerations represent essentially some of the trade-offs that consumers make when considering the characteristics of ethanol cooking options in relation to charcoal that is the main cooking option in the city. It is argued that trade-offs at the production and demand side need to be clearly evaluated in order to enhance the adoption and sustainability of ethanol for transport and cooking.

Johnson et al. (2020) outline many of the different trade-offs and challenges associated with modern bioenergy transitions in SSA, especially those related to clean cooking fuels and biofuels for transport. They identify that, more often than not, what seem like conflicting policy targets such as food security, rural development, energy security, national economic growth and climate change mitigation and adaption can be bridged through specific interventions and coordinated policy actions. By focusing on modern bioenergy, they acknowledge the possibility of trade-offs between individual policy concerns in some contexts but also the great potential to create synergies. They suggest that it is possible to facilitate modern bioenergy transitions in SSA and promote positive synergies through (a) identifying and strengthening positive SDG interlinkages in modern bioenergy transitions; (b) choosing the most appropriate scale, markets and production modes for modern bioenergy; (c) promoting integrated landscape approaches for feedstock production; and (d) fostering synergies between climate change mitigation and adaptation.

10.2.2 Low Resilience and Adaptive Capacity to Environmental and Socioeconomic Change

Many parts of SSA experience rapid environmental and/or socioeconomic change (Juju et al. 2020; Aleman and Fayolle 2020). Many chapters touched upon the fact that households and local communities are not always capable of coping success-fully with such long-term change or acute shocks. Indeed, chapters pointed to the low resilience and adaptive capacity of local communities to climatic hazards (e.g. Boafo et al. 2020; Akampumuza et al. 2020), landscape fragmentation (Andriamparany et al. 2020) and livelihood shocks (Mfune et al. 2020).

Andriamparany et al. (2020) indicate that landscape fragmentation is a major driver of biodiversity loss and ecosystem service degradation in southern Madagascar. They also point that severely fragmented and/or disturbed landscapes due to human activity may be vulnerable to alien invasive species that cause further biodiversity loss. They propose that hedges from introduced cacti (*Opuntia spp.*) could be one possible approach to enhance connectivity in agricultural landscapes and at the same time provide ecosystem services to local communities and other species, especially during the dry months. Such hedges could be a nature-based solution that can enhance the resilience of the landscapes and local communities to the ongoing climate change in the region.

Akampumuza et al. (2020) explore the exposure of urban household to droughts, floods, pests and diseases has an appreciable effect on household food security and consumption in eastern Uganda. The coping strategies available have differing potential to allow households to cope with these various shocks. In fact only some coping strategies are able to temporarily safeguard against household consumption declines. Furthermore, the effects of these shocks vary by gender of the household head, presumably due to their lower access to resources such as land and paid off-farm employment and thus their relative inability to adopt effective coping strategies. They suggest the need to develop and implement strategies that simultaneously support climate-smart food crop production and income diversification and strengthen the food supply and distribution system.

Boafo et al. (2020) elicit through a participatory approach the perceived resilience of local communities in Northern Ghana to floods and droughts. Despite some variation between resilience elements, communities and age groups, most respondents reported a rather low perceived resilience to these climatic hazards. This is particularly troubling when considering the increasing frequency and severity of such events in the semiarid areas of Western Africa. They suggest that bottom-up participatory approaches can be used as preplanning tools to identify priority areas and inform the development of context-specific interventions and solutions to enhance community resilience to floods and droughts.

Mfune et al. (2020) employ a resilience lens to unravel the history, evolution and impacts of copper mine closure in Kabwe (Zambia) on the local mine-dependent community. Mine closure was very unexpected with no measures in place to mitigate its negative environmental, social and economic outcomes. In fact, the unanticipated mine closure left a legacy of environmental degradation, unemployment, income decline, informal livelihoods and loss of infrastructure and social services that the region is still struggling to cope with. Despite efforts to reverse these negative outcomes, it has taken a rather long time. It is argued that there must be concerted effort to put in place appropriate mitigation strategies from the onset of mine development, as many of the observed negative effects were mediated by earlier failures to consider the eventuality of mine closure (and to plan against its possible impacts) at the levels of the national government, local government and mining company.

10.2.3   Constraints and Opportunities for Developing Solutions to Sustainability Challenges

Many chapters identified and discussed options to either directly solve or create preconditions for developing solutions to sustainability challenges in SSA. These options can range from individual interventions such as agricultural innovations (Ndalilo et al. 2020; Mala et al. 2020), sanitation solutions (Gabrielsson et al. 2020) and nutrition interventions (Sekiyama et al. 2020) to multi-level processes related to transdisciplinary knowledge generation and dissemination (von Maltitz 2020; Patel et al. 2020) and the development of broader enabling conditions to foster innovation (Mensah and Gordon 2020) and attract and effectively use funding (Lopes et al. 2020).

Mala et al. (2020) discuss how indigenous and local knowledge (ILK) is mobilised in the forest-agriculture systems of Centre-South Cameroon to meet multiple local objectives. They track such ILK practices to the strong connections of the local communities to the social-ecological system. They identify that the current supply of different technical, marketing and socio-organisational agricultural innovations does not always reflect the local needs, as agricultural innovations developed in other geographical contexts are usually promoted rather than local ones. They make the case that ILK can contribute to the development of locally appropriate policy/technology options and innovations for managing forest-agricultural systems, which can have positive outcomes to local livelihoods and biodiversity conservation, reconciling to some extent these often conflicting goals.

Ndalilo et al. (2020) discuss the uptake and potential of ILK practices and innovations geared towards enhancing agricultural productivity and food security in the face of climate change in coastal Kenya. They identity that local communities widely use ILK practices and innovations such as crop diversification, early planting, use of drought-tolerant and fast-growing local crop varieties, crop rotation, conservation tillage, domestication of wild food and medicinal plants and use of biopesticides. Despite some evidence of ILK erosion, the local communities mobilise their cultural values and customary resource management and governance systems to promote and preserve such ILK practices, as a means of enhancing their resilience to climate change.

Gabrielsson et al. (2020) focus on how to enhance the adoption of effective sanitation solutions in eastern Africa. First, they unravel how the sanitation problem (and its solutions) is commonly conceived in the region, and the current approaches and biases perpetuate a cyclical failure of sanitation interventions. A central of this failure is the uncritical adoption of imported sanitation solutions that do not always reflect local contexts, constraints, and communities' needs. Subsequently, they critically discuss the characteristics of certain practical solutions that are breaking out of this failure cycle by adopting new and innovative approaches to sanitation. Though different, all these successful sanitation models are characterised by adaptation to the local context, community participation, built-in mechanisms to ensure financial viability, use of culturally appropriate technologies and an emphasis on environmental sustainability.

Sekiyama et al. (2020) examine the relationship between household crop production, diet diversity and the nutritional status of children in areas that have received nutrition interventions in Rwanda. They find a high prevalence of stunting among children below 5 years old and that local diets are characterised by a limited variety and a high dependency on starchy foods. They argue that future interventions should have a broader focus, seeking to improve household agricultural production and intra-household resource allocation if they are to tackle effectively child malnutrition in the area. Central elements to achieve this would be to offer appropriate education to mothers regarding breastfeeding and weaning foods and leverage the potential of the plant varieties already produced in rural households.

Patel et al. (2020) outline how transdisciplinary modes of knowledge co-production can have ripple positive effects in defining and tackling urban sustainability challenges in South Africa. They outline how different transdisciplinary research projects were able to both generate new knowledge in urban contexts and create closer ties between academics, city officials and other stakeholders. They argue that such novel ways of co-producing knowledge can enhance the salience, credibility and legitimacy of the knowledge generated. Even though urban policy change is often slow, the outputs of such knowledge collaboration and co-production processes can increase the confidence and commitment of urban stakeholders in addressing urban sustainability challenges. In this sense, they can enhance stakeholder buy-in to ensure its usefulness and effective uptake.

Von Maltitz (2020) explores the interface between science and policy-making and how 'scientific assessments' can bridge this gap. By drawing on the experience gained through engagement in large scientific assessments on climate change, ecosystem services, desertification and land degradation, it is argued that such assessments offer many possibilities to bridge gaps between academia and policy and essentially contribute to the solution of major sustainability challenges in SSA. However, many aspects of scientific assessments must be managed properly, including (a) ensuring policy relevance, (b) ensuring the quality and usefulness of the final product and (c) organising effectively the internal processes of the assessment (e.g. build effective teams, navigate team dynamics, ensure author commitment).

Mensah and Gordon (2020) argue that partnerships between academic institutions, industry and government can play a major role in tackling sustainability challenges in SSA, especially through research co-design and co-development. They identify some major constraints and barriers in the productive engagement between universities, companies and the government and argue that the development of appropriate policies, institutional structures and processes (both internal and external) can strengthen partnerships, ensure their viability and promote their positive outcomes for tackling sustainability challenges. They argue that the systematic monitoring of funded research and development (R&D) activities could be a good start to identify success stories and best practices, as well as the institutional dynamics that hinder or support these partnerships. They also identify the need to enhance the capacities of individual researchers to become agents of change in national and subnational processes.

Lopes et al. (2020) explore the interface between funding, research and SDGs in SSA. They track the current research priorities related to the SDGs in the region, the SDGs most targeted by Overseas Development Assistance (ODA) and the factors that facilitate funding acquisition. They identify the clear mismatch between academic research priorities related to the SGDs and actual financial flows for SDGs. They argue that this mismatch poses potential risks for the effective resource allocation across the multiple sustainability challenges reflected by the SDGs. It is argued that various economic, institutional and political factors influence the acquisition and effective use of funding, with some of the domains that can be strengthened to overcome funding constraints being (a) capacity building, (b) liberalisation and deregulation, (c) regulation and incentives, (d) partnerships and (e) regional integration.

10.3 Mobilising Sustainability Science to Tackle Sustainability Challenges in Sub-Saharan Africa

The previous sections imply that due to the multidimensionality of sustainability challenges in SSA, there is a real need to mobilise new research approaches to both understand these challenges and design appropriate solutions. Many scholars have suggested that the emerging paradigm of sustainability science is ideal for such applications in developing and rapidly changing contexts such as SSA (Gasparatos et al. 2017; Burns and Weaver 2008). In fact, it has been argued that sustainability science should target some of the most pervasive grand challenges in SSA including poverty (Kates and Dasgupta 2007).

Sustainability science is characterised by (a) a problem-driven and solution-oriented approach, (b) an ability to link social and ecological systems and (c) an inter- and transdisciplinary perspective (Kates 2011; Komiyama and Takeuchi 2006). Sustainability science is well positioned to lead this research agenda, with an ever-increasing number of scholars mobilising it to both understand and offer solutions to sustainability challenges in SSA (Gasparatos et al. 2017; Aguirre-Bastos et al. 2019).

Most chapters in these two volumes have not embraced a comprehensive approach covering all of the main three elements mentioned above. However, practically all chapters have adopted at least one of these elements when analysing the respective sustainability challenges and/or offering relevant solutions. Table 10.2 indicates how each chapter has engaged with the three main elements of sustainability science.

Almost all chapters adopted a problem-driven and solutions-oriented approach. In terms of a problem-driven approach, all chapters clearly articulated the underlying sustainability challenges in their respective introductory sections and discussed

Table 10.2 Sustainability science elements reflected in each chapter

Chapter	Problem-driven and solution-oriented approach	Social-ecological systems approach	Inter- and/or transdisciplinary approach
Vol. 1			
Ch. 1 (Juju et al. 2020)	**NA**	**NA**	**NA**
Ch. 2 (Johnson et al. 2020)	√	-	√
Ch. 3 (Jarzebski et al. 2020)	√	√	√
Ch. 4 (Antonelli et al. 2020)	√	-	-
Ch. 5 (Lopes et al. 2020)	√	-	√
Ch. 6 (Boafo et al. 2020)	√	√	√
Ch. 7 (Balde et al. 2020)	√	√	-
Ch. 8 (Mensah and Gordon 2020)	√	-	√
Ch. 9 (Aleman and Fayolle 2020)	-	√	√
Ch. 10 (Mala et al. 2020)	√	√	√
Vol. 2			
Ch. 1 (Sekiyama et al. 2020)	√	-	√
Ch. 2 (Akampumuza et al. 2020)	√	√	-
Ch. 3 (Ndalilo et al. 2020)	√	√	√
Ch. 4 (Gabrielsson et al. 2020)	√	-	√
Ch. 5 (Nyambane et al. 2020)	√	√	√
Ch. 6 (Andriamparany et al. 2020)	-	√	√
Ch. 7 (Mfune et al. 2020)	√	-	√
Ch. 8 (Patel et al. 2020)	√	-	√
Ch. 9 (von Maltitz 2020)	-	-	√

policy implications and recommendations in dedicated sub-sections. Some chapters also actually adopted a clear solutions-oriented perspective, outlining concrete proposals for the design and/or implementation of specific technical or institutional solutions to sustainability challenges (e.g. Gabrielsson et al. 2020; Ndalilo et al. 2020; Mala et al. 2020; Patel et al. 2020; Sekiyama et al. 2020).

Many chapters took an integrated perspective linking social and ecological components in the specific study contexts. For example, several chapters outlined the close links between local communities and their supporting ecosystems (e.g. Balde et al. 2020; Mala et al. 2020; Ndalilo et al. 2020; Andriamparany et al. 2020) and/or clearly linked human activities with ecosystem change and degradation (e.g. Balde et al. 2020; Jarzebski et al. 2020; Nyambane et al. 2020; Aleman and Fayolle 2020; Andriamparany et al. 2020). Some chapters also discussed how environmental change can actually affect ecosystems and/or local communities (Aleman and Fayolle 2020; Boafo et al. 2020; Akampumuza et al. 2020).

Finally, almost all chapters adopted an interdisciplinary perspective, merging insights from the natural, social and engineering sciences. Indeed, between them the chapters used concepts and methods from very diverse academic fields such as economics, ecology, environmental sciences, nutrition, environmental engineering, sociology and geospatial analysis, to name a few. Some chapters adopted a more transdisciplinary approach by engaging deeply with different stakeholders, including ILK holders to understand the underlying systems and/or collect and analyse data (e.g. Mala et al. 2020; Ndalilo et al. 2020; Boafo et al. 2020; Balde et al. 2020). Other chapters argued strongly about the importance of partnerships to either generate/synthesise/disseminate knowledge (von Maltitz 2020; Patel et al. 2020), drive innovation (Mensah and Gordon 2020) or create the preconditions to attract and effectively manage funding (Lopes et al. 2020).

10.4 Afterword: Future Directions for Sustainability Research and Education in Sub-Saharan Africa

When reading critically each chapter and the underlying literature, we can identify three critical and interrelated needs for facilitating knowledge and solutions for sustainability challenges in SSA, namely:

- Increase the output and visibility of African scholars, and facilitate creative collaborations with external researchers.
- Invest in the development of state-of-the-art infrastructure for research and education.
- Create comprehensive educational curricula offering theoretical and practical tools to tackle sustainability challenges in Africa.
- Integrate more meaningfully African voices and perspectives in sustainability research and education.

First, despite the proliferation of sustainability science studies in SSA contexts, the actual number of studies produced solely by African scholars is still quite low (Elsevier 2015). In addition, scholars from SSA countries tend to be located mostly at the "margin" of the global sustainability science network, having low-intensity connections with the core of the network (Elsevier 2015). This is possibly because African scholars are either parts of larger multiauthor teams led by researchers from developed countries, or they publish in relatively lower-impact journals that are not captured by the main research search engines (Adebanwi 2016). This suggests the need to both improve the high-impact output produced and/or coordinated by African scholars and engage in more creative collaborations with non-African researchers (including from developing countries outside SSA). As discussed below, funding would definitely be a major hurdle for delivering high-impact and solutions-oriented research by African scholars, considering the meagre financial resources allocated for research in most SSA countries (Mensah and Gordon 2020; Ngongalah et al. 2018). Beyond funding, as discussed below, there is a simultaneous need to improve the capacity of young African scholars through stronger and more comprehensive educational curricula, mentorship and the inclusion of African voices in current academic paradigms (Kumwenda et al. 2017).

Despite the expanding sustainability literature in SSA, there have been only a few structured efforts to frame sustainability science purely with African voices and perspectives. For example, Burns and Weaver (2008) collate examples of sustainability science research from South Africa using diverse methodological approaches. Some scholars advocate that diverse worldviews should be merged to create a third space for research dialogue and educational curriculum development (Glasson et al. 2010). This is echoed in the processes of certain science-policy interfaces such as the Intergovernmental Science-Policy Platform on Biodiversity and Ecosystem Services (IPBES) (e.g. Pascual et al. 2017; Roué et al. 2016; von Maltitz 2020). Conversely, some scholars, both within and outside the field of sustainability science, have advocated for the decolonisation of science, calling for African scholars to raise their own voice and constructively transform sustainability science scholarship and education (Chilisa 2012, 2017; van Breda and Swilling 2019). Notwithstanding these two sides of the debate, the fact is that the interface between sustainability science, African research paradigms and ILK remains largely underdeveloped and needs to be strengthened appreciably to solve some of the most difficult sustainability challenges in the continent.

Second, most of the higher education institutions in SSA do not have adequate and state-of-the-art technical infrastructure. This applies to both hard (e.g. labs, computing, research facilities) and soft infrastructure (e.g. software for complex modelling). Indeed, with the exception of some countries such as South Africa, there is no access to such infrastructure without partnerships with international institutions or donations. Wide investment would be necessary to develop new (or upgrade existing) infrastructure to enable African researchers embark in cutting-edge research.

Third, there should be coordinated efforts to develop dedicated sustainable development and/or sustainability science educational curricula to foster a new generation of African researchers and practitioners proficient in sustainability. To achieve this, it has been argued that educational curricula across SSA should be reformed and realigned with sustainability science principles, concepts and themes at all levels of the educational system (Aguirre-Bastos et al. 2019). However, despite some successful efforts in certain countries such as South Africa and Ghana (Patel et al. 2020; Mensah and Gordon 2020), this is not the case in most other parts of SSA. The limited current progress could be attributed to the lack of relevant expertise locally to help in developing such curricula and the fact that the job market has not been tailored towards employing graduates with sustainability backgrounds (with the exception of some careers in academia and the civil society). Many prevailing challenges put further obstacles in developing such educational curricula, including the limited human, financial and infrastructural resources (Ighobor 2015; Aguirre-Bastos et al. 2019). Some ways forward would be to (a) catalyse shifts in education policy stressing the need to address sustainability challenges in SSA and meet the SDGs, (b) garner government endorsement/support and collaboration, (c) employ more people with sustainability backgrounds in academia and the broader education sector, (d) boost collaborative multi-stakeholder partnerships both nationally and internationally and (e) secure national and international funding opportunities for sustainability education.

Fourth, despite the proliferation of sustainability research in SSA, there have been rather few structured efforts to frame sustainability science purely with African voices and perspectives. For example, in one of the rare efforts, Burns and Weaver (2008) have collated examples of sustainability science research from South Africa using diverse methodological approaches. Some scholars have advocated that diverse worldviews should be merged to create a third space for research dialogue and educational curriculum development (Glasson et al. 2010). This has also been echoed in the modalities of certain science-policy interfaces such as the Intergovernmental Science-Policy Platform on Biodiversity and Ecosystem Services (IPBES) (e.g. Pascual et al. 2017; Roué et al. 2016; von Maltitz 2020). Conversely, some scholars, both within and outside the field of sustainability science, have advocated for the decolonisation of science, calling for African scholars to raise their own voice and constructively transform sustainability science scholarship and education (Chilisa 2012, 2017; van Breda and Swilling 2019). Notwithstanding these two sides of the debate, the fact is that the interface between sustainability science, African research paradigms and ILK remains largely underdeveloped and needs to be strengthened appreciably to solve some of the most difficult sustainability challenges in the continent.

However, none of the above might be achieved without first addressing the underlying global politics of knowledge production. Even though more and more international funding mechanisms facilitate the inclusion of African scholars in bilateral and multilateral sustainability research, these funding options are still rather limited, especially in view of the major research gaps and pressing sustainability challenges in the continent. More critically, African scholars cannot access and lead most of the currently available large international funding options without a partnership with a developed country. From this perspective, perhaps the most important way forward and pressing need is the development of new research funding mechanisms in SSA, by Africans and for Africans. This can go a long way in deconstructing the current politics of knowledge production and usher a new wave of research that can truly address the prevailing sustainability challenges in the continent.

Acknowledgements We acknowledge the support of the Japan Science and Technology Agency (JST) for project FICESSA, the Japan International Cooperation Agency (JICA) for project USiA and the Japan Society for the Promotion of Science (JSPS) for the Grant-in-Aid for Young Scientists A (Project 17H05037) and the Core-to-core project 'Establishment and Advancement of a Global Meta-Network on Sustainability Science' (Project 23001). Merle Naidoo, Alice Karanja and Abubakari Ahmed received Monbukagakusho PhD scholarships offered by the Japanese Ministry of Education, Culture, Sports, Science, and Technology (MEXT) through the Graduate Program in Sustainability Science – Global Leadership Initiative (GPSS-GLI).

References

Adebanwi W (2016) Rethinking knowledge production in Africa. Africa 86:350–353

Aguirre-Bastos C, Chaves-Chaparro J, Aricò S (2019) Co-designing science in Africa: first steps in assessing the sustainability science approach on the ground. UNESCO, Paris

Akampumuza P, Ggombe KM, Matsuda H (2020) Weather shocks, gender and household consumption: evidence from the urban households in Teso sub-region, Uganda. In: Gasparatos A, Ahmed A, Naidoo M, Karanja A, Fukushi K, Saito O, Takeuchi K (eds) Sustainability challenges in sub-Saharan Africa II: insights from eastern and southern Africa. Springer, Berlin

Aleman JC, Fayolle A (2020) Long-term vegetation change in Central Africa: the need for an integrated management framework for forests and savannas. In: Gasparatos A, Naidoo M, Ahmed A, Karanja A, Fukushi K, Saito O, Takeuchi K (eds) Sustainability challenges in Sub-Saharan Africa I: continental perspectives and insights from Western and Central Africa. Springer, Berlin

Andriamparany R, Lundberg J, Pyykönen M, Wurz S, Elmqvist T (2020) The effect of introduced *Opuntia* (Cactaceae) species on landscape connectivity and ecosystem service provision in southern Madagascar. In: Gasparatos A, Ahmed A, Naidoo M, Karanja A, Fukushi K, Saito O, Takeuchi K (eds) Sustainability challenges in Sub-Saharan Africa II: insights from eastern and southern Africa. Springer, Berlin

Antonelli M, Bracco S, Turvani ME, Vicario A (2020) Large-scale land acquisitions in Sub-Saharan Africa and corporate social responsibility (CSR): insights from Italian investments. In: Gasparatos A, Naidoo M, Ahmed A, Karanja A, Fukushi K, Saito O, Takeuchi K (eds) Sustainability challenges in Sub-Saharan Africa I: continental perspectives and insights from Western and Central Africa. Springer, Berlin

Balde BS, Karanja A, Kobayashi H, Gasparatos G (2020) Linking rural livelihoods and fuelwood demand from mangroves and upland forests in the coastal region of Guinea. In: Gasparatos A, Naidoo M, Ahmed A, Karanja A, Fukushi K, Saito O, Takeuchi K (eds) Sustainability challenges in Sub-Saharan Africa I: continental perspectives and insights from Western and Central Africa. Springer, Berlin

Boafo YA, Saito O, Jasaw GS, Yiran GAB, Lam RD, Mohan G, Kranjac-Berisavljevic G (2020) Perceived community resilience to floods and droughts induced by climate change in semi-arid Ghana. In: Gasparatos A, Naidoo M, Ahmed A, Karanja A, Fukushi K, Saito O, Takeuchi K (eds) Sustainability challenges in Sub-Saharan Africa I: continental perspectives and insights from Western and Central Africa. Springer, Berlin

Burns M, Weaver A (eds) (2008) Exploring sustainability science: a southern African perspective. African SUN MeDIA, Stellenbosch

Chilisa B (2012) Indigenous research methodologies. Sage, Thousand Oaks

Chilisa B (2017) Decolonising transdisciplinary research approaches: an African perspective for enhancing knowledge integration in sustainability science. Sustain Sci 12:813–827

Elsevier (2015) Sustainability science in a global landscape. Elsevier, Amsterdam

Gabrielsson S, Huston A, Gaskin S (2020) Reframing the challenges and opportunities for improved sanitation services in eastern Africa through sustainability science. In: Gasparatos A, Ahmed A, Naidoo M, Karanja A, Fukushi K, Saito O, Takeuchi K (eds) Sustainability challenges in Sub-Saharan Africa II: insights from eastern and southern Africa. Springer, Berlin

Gasparatos A, Takeuchi K, Elmqvist T, Fukushi K, Nagao M, Swanepoel F, Swilling M, Trotter D, von Blottnitz H (2017) Sustainability science for meeting Africa's challenges. Sustain Sci 12:631–848

Glasson GE, Mhango N, Phiri A, Lanier M (2010) Sustainability science education in Africa: negotiating indigenous ways of living with nature in the third space. Int J Sci Educ 32:125–141

Ighobor K (2015) Sustainable development goals in sync with Africa's priorities. Africa Renewal 29:3–5

Jarzebski MP, Ahmed A, Karanja A, Boafo YA, Balde BS, Chinangwa L, Degefa S, Dompreh EB, Saito O, Gasparatos A (2020) Linking industrial crop production and food security in sub-Saharan Africa: local, national and continental perspectives. In: Gasparatos A, Naidoo M, Ahmed A, Karanja A, Fukushi K, Saito O, Takeuchi K (eds) Sustainability challenges in Sub-Saharan Africa I: continental perspectives and insights from Western and Central Africa. Springer, Berlin

Johnson FX, Batidzirai B, Iiyama M, Ochieng CA, Olsson O, Gasparatos A (2020) Enabling sustainable bioenergy transitions in Sub-Saharan Africa: strategic issues for achieving climate-compatible development. In: Gasparatos A, Naidoo M, Ahmed A, Karanja A, Fukushi K, Saito O, Takeuchi K (eds) Sustainability challenges in Sub-Saharan Africa I: continental perspectives and insights from Western and Central Africa. Springer, Berlin

Juju D, Baffoe G, Dam Lam R, Karanja A, Naidoo M, Ahmed A, Jarzebski MP, Saito O, Fukushi K, Takeuchi K, Gasparatos A (2020) Sustainability challenges in sub-Saharan Africa in the context of the sustainable development goals (SDGs). In: Gasparatos A, Naidoo M, Ahmed A, Karanja A, Fukushi K, Saito O, Takeuchi K (eds) Sustainability challenges in Sub-Saharan Africa I: continental perspectives and insights from Western and Central Africa. Springer, Berlin

Kates RW (2011) What kind of a science is sustainability science? PNAS 108:19449–19450

Kates RW, Dasgupta P (2007) African poverty: a grand challenge for sustainability science. PNAS 104:16747–16750

Komiyama H, Takeuchi K (2006) Sustainability science: building a new discipline. Sustain Sci 1:1–6

Kumwenda S, El Hadji AN, Orondo PW, William P, Oyinlola L, Bongo GN, Chiwona B (2017) Challenges facing young African scientists in their research careers: a qualitative exploratory study. Malawi Med J 29(1):1–4

Lopes J, Somanje AN, Velez E, Dam Lam R, Saito O (2020) Determinants of foreign investment and international aid for meeting the sustainable development goals in Africa: a visual cognitive review of the literature. In: Gasparatos A, Naidoo M, Ahmed A, Karanja A, Fukushi K, Saito O, Takeuchi K (eds) Sustainability challenges in Sub-Saharan Africa I: continental perspectives and insights from Western and Central Africa. Springer, Berlin

Mala WA, Geldenhuys CJ, Prabhu R, Essouma FM (2020) Forest-agriculture in the Center-South region of Cameroon: how does traditional knowledge inform integrated management approaches? In: Gasparatos A, Naidoo M, Ahmed A, Karanja A, Fukushi K, Saito O, Takeuchi K (eds) Sustainability challenges in Sub-Saharan Africa I: continental perspectives and insights from Western and Central Africa. Springer, Berlin

Mensah AM, Gordon C (2020) Strategic partnerships between universities and non-academic institutions for sustainability and innovation: Insights from the University of Ghana

Mfune O, Kunda-Wamuwi CF, Chansa-Kabali T, Chisola MN, Manchisi J (2020) The legacy of mine closure in Kabwe, Zambia: what can resilience thinking offer to the mining sustainability discourse? In: Gasparatos A, Ahmed A, Naidoo M, Karanja A, Fukushi K, Saito O, Takeuchi K (eds) Sustainability challenges in Sub-Saharan Africa II: insights from Eastern and Southern Africa. Springer. Berlin

Ndalilo L, Wekesa C, Mbuvi MTE (2020) Indigenous and local knowledge practices and innovations for enhancing food security under climate change: examples from Mijikenda communities in coastal Kenya. In: Gasparatos A, Ahmed A, Naidoo M, Karanja A, Fukushi K, Saito O, Takeuchi K (eds) Sustainability challenges in Sub-Saharan Africa II: insights from Eastern and Southern Africa. Springer, Berlin

Ngongalah L, Niba RN, Wepngong EN, Musisi JM (2018). Research challenges in Africa–an exploratory study on the experiences and opinions of African researchers. bioRxiv 446328

Nyambane A, Johnson F, Romeu-Dalmau C, Ochieng C, Gasparatos A, Mudombi S, von Maltitz G (2020) Ethanol as a clean cooking alternative in Sub-Saharan Africa: Insights from sugarcane production and ethanol adoption sites in Malawi and Mozambique. In: Gasparatos A, Ahmed A, Naidoo M, Karanja A, Fukushi K, Saito O, Takeuchi K (eds) Sustainability Challenges in Sub-Saharan Africa II: insights from Eastern and Southern Africa. Springer, Berlin

Pascual U, Balvanera P, Díaz S, Pataki G, Roth E, Stenseke M, Watson RT et al (2017) Valuing nature's contributions to people: the IPBES approach. Curr Opin Environ Sustain 26:7–16

Patel Z, Marrengane N, Smit W, Anderson P (2020) Knowledge co-production in Sub-Saharan African cities: building capacity for the urban age. In: Gasparatos A, Ahmed A, Naidoo M, Karanja A, Fukushi K, Saito O, Takeuchi K (eds) Sustainability challenges in Sub-Saharan Africa II: insights from Eastern and Southern Africa. Springer, Berlin

Roué M, Césard N, Adou Yao YC, Oteng-Yeboah A (eds) (2016) Indigenous and local knowledge of biodiversity and ecosystem services in Africa. UNESCO, Paris

Sekiyama M, Matsuda H, Mohan G, Yanagisawa A, Sudo N, Amitani Y, Caballero Y, Matsuoka T, Imanishi H, Sasaki T (2020) Tackling child malnutrition by strengthening the linkage between agricultural production, food security and nutrition in rural Rwanda. In: Gasparatos A, Ahmed A, Naidoo M, Karanja A, Fukushi K, Saito O, Takeuchi K (eds) Sustainability challenges in Sub-Saharan Africa II: insights from Eastern and Southern Africa. Springer, Berlin

van Breda J, Swilling M (2019) The guiding logics and principles for designing emergent transdisciplinary research processes: learning experiences and reflections from a transdisciplinary urban case study in Enkanini informal settlement, South Africa. Sustain Sci 14:823–841

von Maltitz G (2020) Harnessing science-policy interface processes to tackle sustainability challenges in Sub-Saharan Africa. In: Gasparatos A, Ahmed A, Naidoo M, Karanja A, Fukushi K, Saito O, Takeuchi K (eds) Sustainability challenges in Sub-Saharan Africa II: insights from Eastern and Southern Africa. Springer, Berlin

Weber EP, Khademian AM (2008) Wicked problems, knowledge challenges, and collaborative capacity builders in network settings. Public Adm Rev 68:334–349

Index